Clashing Views on

Environmental Issues

THIRTEENTH EDITION

TAKING SIDES

Clashing Views on

Environmental Issues

THIRTEENTH EDITION

Selected, Edited, and with Introductions by

Thomas Easton
Thomas College

McGraw-Hill
Higher Education

Boston   Burr Ridge, IL   Dubuque, IA   New York   San Francisco   St. Louis
Bangkok   Bogotá   Caracas   Kuala Lumpur   Lisbon   London   Madrid   Mexico City
Milan   Montreal   New Delhi   Santiago   Seoul   Singapore   Sydney   Taipei   Toronto

McGraw-Hill
Higher Education

TAKING SIDES: ENVIRONMENTAL ISSUES,
THIRTEENTH EDITION

1 2 3 4 5 6 7 8 9 0 DOC/DOC 0 9 8

MHID: 0-07-351444-6
ISBN: 978-0-07-351444-4
ISSN: 1091-8825

Managing Editor: *Larry Loeppke*
Production Manager: *Faye Schilling*
Senior Developmental Editor: *Jill Peter*
Editorial Assistant: *Nancy Meissner*
Production Service Assistant: *Rita Hingtgen*
Permissions Coordinator: *Shirley Lanners*
Senior Marketing Manager: *Julie Keck*
Marketing Communications Specialist: *Mary Klein*
Marketing Coordinator: *Alice Link*
Project Manager: *Jane Mohr*
Design Specialist: *Tara McDermott*
Senior Administrative Assistant: *DeAnna Dausener*
Cover Graphics: *Kristine Jubeck*

Compositor: ICC Macmillan Inc.
Cover Image: © Getty Images/Photodisc Collection

Library of Congress Cataloging-in-Publication Data

Main entry under title:
Taking sides: clashing views on environmental issues/selected, edited, and with introductions by Thomas A. Easton.—13th ed.

Includes bibliographical references.
1. Environmental policy. 2. Environmental protection. I. Easton, Thomas A. *comp.*
363.7

Preface

Most fields of academic study evolve over time. Some evolve in turmoil, for they deal in issues of political, social, and economic concern. That is, they involve controversy.

It is the mission of the Taking Sides series to capture current, ongoing controversies and make the opposing sides available to students. This book focuses on environmental issues, from the philosophical to the practical. It does not pretend to cover all such issues, for not all provoke controversy or provoke it in suitable fashion. But there is never any shortage of issues that can be expressed as pairs of opposing essays that make their positions clearly and understandably.

The basic technique—presenting an issue as a pair of opposing essays—has risks. Students often display a tendency to remember best those essays that agree with the attitudes they bring to the discussion. They also want to know what the "right" answers are, and it can be difficult for teachers to refrain from taking a side, or from revealing their own attitudes. Should teachers so refrain? Some do not, but of course they must still cover the spectrum of opinion if they wish to do justice to the scientific method and the complexity of an issue. Some do, though rarely so successfully that students cannot see through the attempt.

For any Taking Sides volume, the issues are always phrased as yes/no questions. Which answer—yes or no—is the correct answer? Perhaps neither. Perhaps both. Perhaps we will not be able to tell for another hundred years. Students should read, think about, and discuss the readings and then come to their own conclusions without letting my or their instructor's opinions dictate theirs. The additional readings mentioned in the introductions and postscripts should prove helpful.

This edition of *Taking Sides: Clashing Views on Environmental Issues* contains 38 readings arranged in pro and con pairs to form 19 issues. For each issue, an *introduction* provides historical background and a brief description of the debate. The *postscript* after each pair of readings offers recent contributions to the debate, additional references, and sometimes a hint of future directions. Each part is preceded by an *Internet References* page that lists several links that are appropriate for further pursuing the issues in that part.

Changes to this edition   Half of this book consists of new material. Two issues, *Should the Endangered Species Act Be Strengthened?* (Issue 4) and *Is a Large-Scale Shift to Organic Farming the Best Way to Increase World Food Supply?* (Issue 15), were added for the 2008 partial revision of the twelfth edition. Our discussion of the precautionary principle (Issue 1) now focuses on risk analysis. The Superfund issue (Issue 18) is now *Is the Superfund Program Successfully Protecting Human Health from Hazardous Materials?* The global warming issue has been recast as *Is Global Warming Skepticism Just Smoke and Mirrors?* (Issue 8). The Environmental Racism issue (Issue 5) is now *Should the EPA Be Doing More to Fight Environmental Injustice?* Because of the new interest in energy issues, we have added *Is Wind Power Green?* (Issue 9), *Should Cars Be More Efficient?* (Issue 10), and *Do Biofuels Enhance Energy Security?* (Issue 11).

In addition, for five of the issues retained from the previous edition, one or both of the readings have been replaced. In all, 17 of the readings in this edition were not in the 12th edition.

A word to the instructor   An *Instructor's Resource Guide with Test Questions* (multiple-choice and essay) is available through the publisher for the instructor using Taking Sides in the classroom. Also available is a general guidebook, *Using Taking Sides in the Classroom*, which offers suggestions for adapting the pro-con approach in any classroom setting. An online version of *Using Taking Sides in the Classroom* and a correspondence service for Taking Sides adopters can be found at http://www.mhcls.com/usingts/.

Taking Sides: Clashing Views on Environmental Issues is only one title in the Taking Sides series. If you are interested in seeing the table of contents for any of the other titles, please visit the Taking Sides Web site at http://www.mhcls.com/takingsides/.

Thomas A. Easton
Thomas College

Contents In Brief

UNIT 1 Environmental Philosophy 1

Issue 1.   Is the Precautionary Principle a Sound Approach to Risk Analysis? 2

Issue 2.   Is Sustainable Development Compatible with Human Welfare? 22

Issue 3.   Should a Price Be Put on the Goods and Services Provided by the World's Ecosystems? 37

UNIT 2 Principles Versus Politics 63

Issue 4.   Should the Endangered Species Act Be Strengthened? 64

Issue 5.   Should the EPA Be Doing More to Fight Environmental Injustice? 81

Issue 6.   Can Pollution Rights Trading Effectively Control Environmental Problems? 96

UNIT 3 Energy Issues 117

Issue 7.   Should the Arctic National Wildlife Refuge Be Opened to Oil Drilling? 118

Issue 8.   Is Global Warming Skepticism Just Smoke and Mirrors? 136

Issue 9.   Is Wind Power Green? 166

Issue 10. Should Cars Be More Efficient? 190

Issue 11. Do Biofuels Enhance Energy Security? 205

Issue 12. Is It Time to Revive Nuclear Power? 220

UNIT 4 Food and Population 243

Issue 13. Do Falling Birthrates Pose a Threat to Human Welfare? 244

Issue 14. Is Genetic Engineering the Answer to Hunger? 258

Issue 15. Is a Large-Scale Shift to Organic Farming the Best Way to Increase World Food Supply? 270

UNIT 5 Toxic Chemicals 289

Issue 16. Should DDT Be Banned Worldwide? 290

Issue 17. Do Environmental Hormone Mimics Pose a Potentially Serious Health Threat? 311

Issue 18. Is the Superfund Program Successfully Protecting Human Health from Hazardous Materials? 328

Issue 19. Should the United States Reprocess Spent Nuclear Fuel? 343

Contents

Preface   v

Correlation Guide   xv

Introduction   xvii

UNIT 1   ENVIRONMENTAL PHILOSOPHY   1

Issue 1. Is the Precautionary Principle a Sound Approach to Risk Analysis?   2

YES: **Nancy Myers**, from "The Rise of the Precautionary Principle: A Social Movement Gathers Strength," *Multinational Monitor* (September 2004)   *4*

NO: **Bernard D. Goldstein**, from "The Precautionary Principle: Is It a Threat to Toxicological Science?" *International Journal of Toxicology* (January/February 2006)   *13*

Nancy Myers, communications director for the Science and Environmental Health Network, argues that because the precautionary principle "makes sense of uncertainty," it has gained broad international recognition as being crucial to environmental policy. Bernard D. Goldstein, Professor of Environmental and Occupational Health at the University of Pittsburgh, argues that although the precautionary principle is potentially valuable, it poses a risk that scientific (particularly toxicological) risk assessment will be displaced to the detriment of public health, social justice, and the field of toxicology itself.

Issue 2. Is Sustainable Development Compatible with Human Welfare?   22

YES: **Jeremy Rifkin**, from "The European Dream: Building Sustainable Development in a Globally Connected World," *E Magazine* (March/April 2005)   *24*

NO: **Ronald Bailey**, from "Wilting Greens," *Reason* (December 2002)   *32*

Jeremy Rifkin, president of the Foundation on Economic Trends, argues that Europeans pride themselves on their quality of life, and their emphasis on sustainable development promises to maintain that quality of life into the future. Environmental journalist Ronald Bailey states that sustainable development results in economic stagnation and threatens both the environment and the world's poor.

Issue 3. Should a Price Be Put on the Goods and Services Provided by the World's Ecosystems?   37

YES: **John E. Losey and Mace Vaughan**, from "The Economic Value of Ecological Services Provided by Insects," *BioScience* (April 2006)   *39*

NO: **Marino Gatto and Giulio A. De Leo**, from "Pricing Biodiversity and Ecosystem Services: The Never-Ending Story," *BioScience* (April 2000)   *51*

John E. Losey and Mace Vaughan argue that even conservative estimates of the value of the services provided by wild insects are enough to justify increased conservation efforts. They say that "everyone would benefit from the facilitation of the vital services these insects provide." Professors of applied ecology Marino Gatto and Giulio A. De Leo contend that the pricing approach to valuing nature's services is misleading because it falsely implies that only economic values matter.

UNIT 2   PRINCIPLES VERSUS POLITICS   63

Issue 4. Should the Endangered Species Act Be Strengthened?   64

YES: John Kostyack, from "Testimony before the Oversight Hearing on the Endangered Species Act," U.S. Senate Committee on Environment and Public Works (May 19, 2005)   *66*

NO: Monita Fontaine, from "Testimony before the Oversight Hearing on the Endangered Species Act," U.S. Senate Committee on Environment and Public Works (May 19, 2005)   *74*

Representing the National Wildlife Foundation, John Kostyack argues that the Endangered Species Act has been so successful that it should not be weakened but strengthened. Speaking for the National Endangered Species Act Reform Coalition, a group that represents those affected by the Endangered Species Act, Monita Fontaine argues that federal regulation under the ESA should be replaced by a system that relies more on voluntary and state species conservation efforts.

Issue 5. Should the EPA Be Doing More to Fight Environmental Injustice?   81

YES: Robert D. Bullard, from "Environmental Justice Programs," Statement before the Subcommittee on Superfund and Environmental Health, Senate Committee on Environment and Public Works (July 25, 2007)   *83*

NO: Granta Y. Nakayama, from "Environmental Justice Programs," Statement before the Subcommittee on Superfund and Environmental Health, Senate Committee on Environment and Public Works (July 25, 2007)   *89*

Professor Robert D. Bullard argues that despite a 1994 Executive Order directing the Environmental Protection Agency (EPA) to ensure that minority and poor communities not bear a disproportionate burden of pollution and other environmental ills, environmental justice still eludes many people. Granta Y. Nakayama, Assistant Administrator of the EPA's Office of Enforcement and Compliance, argues that the EPA is a "trailblazer" in the government's effort to achieve environmental justice and is planning to review its programs.

Issue 6. Can Pollution Rights Trading Effectively Control Environmental Problems?   96

YES: James Allen and Anthony White, from "Carbon Trading," *Electric Perspectives* (September/October 2005)   *98*

NO: Brian Tokar, from "Trading Away the Earth: Pollution Credits and the Perils of 'Free Market Environmentalism,'" *Dollars & Sense* (March/April 1996)   *106*

James Allen and Anthony White describe the European Union's Green-house Gas Emissions Trading Scheme and argue that it encourages investment in carbon-abatement technologies and depends on govern-mental commitments to reducing emissions despite possible adverse economic effects. Author, college teacher, and environmental activist Brian Tokar maintains that pollution credits and other market-oriented environ-mental protection policies do nothing to reduce pollution while transferring the power to protect the environment from the public to large corporate polluters.

UNIT 3   ENERGY ISSUES   117

Issue 7. Should the Arctic National Wildlife Refuge Be Opened to Oil Drilling?   118

YES: **Dwight R. Lee**, from "To Drill or Not to Drill," *The Independent Review* (Fall 2001)   *120*

NO: **Jeff Bingaman, et al.**, from Senate Energy Committee (October 24, 2005)   *129*

Professor of economics Dwight R. Lee argues that the economic and other benefits of Arctic National Wildlife Refuge (ANWR) oil are so great that even environmentalists should agree to permit drilling—and they probably would if they stood to benefit directly. The Minority Members of the Senate Energy Committee objected when the Committee approved a bill that would authorize oil and gas development in the Arctic National Wildlife Refuge. They argued that though the bill contained serious legal and environmental flaws, the greatest flaw lay in its choice of priorities: Wildernesss is to be preserved, not exploited.

Issue 8. Is Global Warming Skepticism Just Smoke and Mirrors?   136

YES: **Seth Schulman et al.**, from "Smoke, Mirrors & Hot Air: How ExxonMobil Uses Big Tobacco's Tactics to Manufacture Uncertainty on Climate Science," *Union of Concerned Scientists* (January 2007)   *138*

NO: **Ivan Osorio, Iain Murray, and Myron Ebell**, from "Liberal 'Scien-tists' Lead Jihad against Global-Warming Skeptics," *Human Events* (May 8, 2007)   *156*

The Union of Concerned Scientists argues that opposition to the idea that global warming is real, is due to human activities, and is a threat to human well-being has been orchestrated by ExxonMobil in a disinformation campaign very similar to the tobacco industry's efforts to convince the public that tobacco was not bad for health. Ivan Osorio, Iain Murray, and Myron Ebell, all of the Competitive Enterprise Institute, argue that the Union of Concerned Scientists is a liberal-funded partisan organization that distorts facts and attempts to discredit opponents with innuendo.

Issue 9. Is Wind Power Green?   166

YES: **Charles Komanoff**, from "Whither Wind?" *Orion* (September/October 2006)   *168*

NO: **Jon Boone**, from "The Wayward Wind?," speech given in the township of Perry, near Silver Lake, Wyoming County, New York (June 19, 2006)   *177*

Charles Komanoff argues that the energy needs of civilization can be met without adding to global warming if we both conserve energy and deploy large numbers of wind turbines. Jon Boone argues that wind power is better for corporate tax avoidance than for providing environmentally friendly energy. It is at best a placebo for our energy dilemma.

Issue 10. Should Cars Be More Efficient?   190

YES: **David Friedman**, from "CAFE Standards," Testimony before Committee on Senate Commerce, Science and Transportation (March 6, 2007)   *192*

NO: **Charli E. Coon**, from "Why the Government's CAFE Standards for Fuel Efficiency Should Be Repealed, Not Increased," The Heritage Foundation Backgrounder (July 11, 2001)   *200*

David Friedman, Research Director at the Union of Concerned Scientists, argues that the technology exists to improve the fuel efficiency standards for new cars and trucks and requiring improved efficiency can cut oil imports, save money, create jobs, and help with global warming. Charli E. Coon, Senior Policy Analyst with The Heritage Foundation, argues that the 1975 Corporate Average Fuel Economy (CAFE) program failed to meet its goals of reducing oil imports and gasoline consumption and has endangered human lives. It needs to be abolished and replaced with market-based solutions.

Issue 11. Do Biofuels Enhance Energy Security?   205

YES: **Bob Dinneen**, from "Testimony before Committee on Senate Energy and Natural Resources" (April 12, 2007)   *207*

NO: **Mark Anslow**, from "Biofuels—Facts and Fiction," *The Ecologist* (March 2007)   *213*

Bob Dinneen, president and CEO of the Renewable Fuels Association, the national trade association representing the U.S. ethanol industry, argues that government support of the renewable fuels industry has created jobs, saved consumers money, and reduced oil imports. The industry's potential is great, and continued support will contribute to ensuring America's future energy security. Journalist Mark Anslow argues that producing biofuels consumes more energy than it makes available for use, will take too long to make a contribution large enough to help fight global warming, generates dangerous quantities of waste, and because of government subsidies fails to make economic sense.

Issue 12. Is It Time to Revive Nuclear Power?   220

YES: **Michael J. Wallace**, from "Testimony before the United States Senate Committee on Energy & Natural Resources, Hearing on the Department of Energy's Nuclear Power 2010 Program" (April 26, 2005)   *222*

NO: **Karen Charman**, from "Brave Nuclear World, Part II" *World Watch* (July/August 2006)   *229*

Michael J. Wallace argues that because the benefits of nuclear power include energy supply and price stability, air pollution control, and greenhouse gas reduction, new nuclear power plant construction—with federal support—is essential. Karen Charman argues that nuclear power's drawbacks and the promise of clean, lower-cost, less dangerous alternatives greatly weaken the case for nuclear power.

UNIT 4   FOOD AND POPULATION   243

Issue 13. Do Falling Birthrates Pose a Threat to Human Welfare?   244

YES: **Michael Meyer**, from "Birth Dearth," *Newsweek* (September 27, 2004)   *246*

NO: **David Nicholson-Lord**, from "The Fewer the Better," *New Statesman* (November 8, 2004)   *252*

Michael Meyer argues that when world population begins to decline after about 2050, economies will no longer continue to grow, government benefits will decline, young people will have to support ever more elders, and despite some environmental benefits, quality of life will suffer. David Nicholson-Lord argues that the economic problems of population decline all have straightforward solutions. A less crowded world will not suffer from the environmental ills attendant on overcrowding and will, overall, be a roomier, gentler, less materialistic place to live, with cleaner air and water.

Issue 14. Is Genetic Engineering the Answer to Hunger?   258

YES: **Gerald D. Coleman**, from "Is Genetic Engineering the Answer to Hunger?" *America* (February 21, 2005)   *260*

NO: **Sean McDonagh**, from "Genetic Engineering Is Not the Answer," *America* (May 2, 2005)   *264*

Gerald D. Coleman argues that genetically engineered crops are useful, healthful, and nonharmful, and although caution may be justified, such crops can help satisfy the moral obligation to feed the hungry. Sean McDonagh argues that those who wish to feed the hungry would do better to address land reform, social inequality, lack of credit, and other social issues.

Issue 15. Is a Large-Scale Shift to Organic Farming the Best Way to Increase World Food Supply?   270

YES: **Brian Halweil**, from "Can Organic Farming Feed Us All?" *World Watch* (May/June 2006)   *272*

NO: **John J. Miller**, from "The Organic Myth: A Food Movement Makes a Pest of Itself," *National Review* (February 9, 2004)   *281*

Brian Halweil, senior researcher at the Worldwatch Institute, argues that organic agriculture is potentially so productive that it could sustainably increase world food supply, although the future may be more likely to see a mix of organic and nonorganic techniques. John J. Miller argues that organic farming is not productive enough to feed today's population, much less larger future populations, it is prone to dangerous biological contamination, and it is not sustainable.

UNIT 5   TOXIC CHEMICALS   289

Issue 16. Should DDT Be Banned Worldwide?   290

YES: **Anne Platt McGinn**, from "Malaria, Mosquitoes, and DDT," *World Watch* (May/June 2002)   *292*

NO: Donald R. Roberts, from "Statement before the U.S. Senate Committee on Environment & Public Works, Hearing on the Role of Science in Environmental Policy-Making" (September 28, 2005)   *300*

Anne Platt McGinn, a senior researcher at the Worldwatch Institute, argues that although DDT is still used to fight malaria, there are other, more effective and less environmentally harmful methods. She maintains that DDT should be banned or reserved for emergency use. Donald R. Roberts argues that the scientific evidence regarding the environmental hazards of DDT has been seriously misrepresented by anti-pesticide activists. The hazards of malaria are much greater and, properly used, DDT can prevent them and save lives.

Issue 17. Do Environmental Hormone Mimics Pose a Potentially Serious Health Threat?   311

YES: Michele L. Trankina, from "The Hazards of Environmental Estrogens," *The World & I* (October 2001)   *313*

NO: Michael Gough, from "Endocrine Disrupters, Politics, Pesticides, the Cost of Food and Health," *Daily Commentary* (December 15, 1997)   *319*

Professor of biological sciences Michele L. Trankina argues that a great many synthetic chemicals behave like estrogen, alter the reproductive functioning of wildlife, and may have serious health effects—including cancer—on humans. Michael Gough, a biologist and expert on risk assessment and environmental policy, argues that only "junk science" supports the hazards of environmental estrogens.

Issue 18. Is the Superfund Program Successfully Protecting Human Health from Hazardous Materials?   328

YES: Robert H. Harris, Jay Vandeven, and Mike Tilchin, from "Superfund Matures Gracefully," *Issues in Science & Technology* (Summer 2003)   *330*

NO: Randall Patterson, from "Not in Their Backyard," *Mother Jones* (May/June 2007)   *334*

Environmental consultants Robert H. Harris, Jay Vandeven, and Mike Tilchin argue that though the Superfund program still has room for improvement, it has made great progress in risk assessment and treatment technologies. Journalist Randall Patterson argues that the Superfund Program is not applied to some appropriate situations, largely because people resist its application.

Issue 19. Should the United States Reprocess Spent Nuclear Fuel?   343

YES: Phillip J. Finck, from "Statement before the House Committee on Science, Energy Subcommittee, Hearing on Nuclear Fuel Reprocessing," United States Senate (June 16, 2005)   *345*

NO: Matthew Bunn, from "The Case against a Near-Term Decision to Reprocess Spent Nuclear Fuel in the United States," United States Senate (June 16, 2005)   *352*

Phillip J. Finck argues that by reprocessing spent nuclear fuel, the United States can enable nuclear power to expand its contribution to the nation's energy needs while reducing carbon emissions, nuclear waste, and the need for waste repositories such as Yucca Mountain. Matthew Bunn argues that there is no near-term need to embrace nuclear spent fuel reprocessing, costs are highly uncertain, and there is a worrisome risk that the increased availability of bomb-grade nuclear materials will increase the risk of nuclear war and terrorism.

Contributors   362

Correlation Guide

The *Taking Sides* series presents current issues in a debate-style format designed to stimulate student interest and develop critical thinking skills. Each issue is thoughtfully framed with an issue summary, an issue introduction, and a postscript. The pro and con essays—selected for their liveliness and substance—represent the arguments of leading scholars and commentators in their fields.

Taking Sides: Clashing Views on Environmental Issues, 13/e is an easy-to-use reader that presents issues on important topics such as *environmental philosophy, energy issues, food and population* and *toxic chemicals*. For more information on *Taking Sides* and other *McGraw-Hill Contemporary Learning Series* titles, visit www.mhcls.com.

This convenient guide matches the issues in **Taking Sides: Environmental Issues, 13/e** with the corresponding chapters in two of our best-selling McGraw-Hill Environmental Science textbooks by Kaufmann/Cleveland and Cunningham/Cunningham.

Taking Sides: Environmental Issues, 13/e	Environmental Science, by Kaufmann/Cleveland	Environmental Science: A Global Concern, 10/e by Cunningham/Cunningham
Issue 1: Is the Precautionary Principle a Sound Approach to Risk Analysis?	**Chapter 24:** A Sustainable Future: Will Business as Usual Get Us There?	**Chapter 24:** Environmental Policy, Law, and Planning
Issue 2: Is Sustainable Development Compatible with Human Welfare?	**Chapter 1:** Environment and Society: A Sustainable Partnership? **Chapter 9:** Carrying Capacity: How Large a Population?	**Chapter 1:** Understanding Our Environment **Chapter 23:** Ecological Economics
Issue 3: Should a Price Be Put on the Goods and Services Provided by the World's Ecosystems?	**Chapter 10:** An Ecological View of the Economy **Chapter 16:** Agriculture: The Ecology of Growing Food	**Chapter 12:** Biodiversity: Preserving Landscapes
Issue 4: Should the Endangered Species Act Be Strengthened?	**Chapter 12:** Biodiversity: Species and So Much More	**Chapter 2:** Frameworks for Understanding: Science, Systems, and Ethics **Chapter 11:** Biodiversity **Chapter 12:** Biodiversity: Preserving Landscapes **Chapter 18:** Water Pollution **Chapter 24:** Environmental Policy, Law, and Planning
Issue 5: Should the EPA Be Doing More to Fight Environmental Injustice?	**Chapter 11:** The Driving Forces of Environmental Change	**Chapter 2:** Frameworks for Understanding: Science, Systems, and Ethics
Issue 6: Can Pollution Rights Trading Effectively Control Environmental Problems?	**Chapter 24:** A Sustainable Future: Will Business as Usual Get Us There?	**Chapter 23:** Ecological Economics
Issue 7: Should the Arctic National Wildlife Refuge Be Opened to Oil Drilling?	**Chapter 11:** The Driving Forces of Environmental Change **Chapter 20:** Fossil Fuels: The Lifeblood of the Global Economy **Chapter 24:** A Sustainable Future: Will Business as Usual Get Us There?	**Chapter 19:** Conventional Energy **Chapter 20:** Sustainable Energy

continued

Taking Sides: Environmental Issues, 13/e	Environmental Science, by Kaufmann/Cleveland	Environmental Science: A Global Concern, 10/e by Cunningham/Cunningham
Issue 8: Is Global Warming Skepticism Just Smoke and Mirrors?	**Chapter 1:** Environment and Society: A Sustainable Partnership? **Chapter 13:** Global Climate Change: A Warming Planet	**Chapter 1:** Understanding Our Environment **Chapter 2:** Frameworks for Understanding: Science, Systems, and Ethics **Chapter 3:** Matter, Energy, and Life **Chapter 15:** Air, Weather, and Climate **Chapter 17:** Water Use and Management
Issue 9: Is Wind Power Green?	**Chapter 10:** An Ecological View of the Economy **Chapter 21:** Nuclear Power **Chapter 22:** Renewable Energy and Energy Efficiency	**Chapter 20:** Sustainable Energy
Issue 10: Should Cars Be More Efficient?	**Chapter 11:** The Driving Forces of Environmental Change **Chapter 19:** Air Pollution: Costs and Benefits of Clean Air **Chapter 20:** Fossil Fuels: The Lifeblood of the Global Economy **Chapter 24:** A Sustainable Future: Will Business as Usual Get Us There?	**Chapter 16:** Air Pollution **Chapter 20:** Sustainable Energy
Issue 11: Do Biofuels Enhance Energy Security?	**Chapter 22:** Renewable Energy and Energy Efficiency	**Chapter 20:** Sustainable Energy
Issue 12: Is It Time to Revive Nuclear Power?	**Chapter 9:** Carrying Capacity: How Large a Population? **Chapter 20:** Fossil Fuels: The Lifeblood of the Global Economy **Chapter 21:** Nuclear Power **Chapter 22:** Renewable Energy and Energy Efficiency	**Chapter 19:** Conventional Energy
Issue 13: Do Falling Birthrates Pose a Threat to Human Welfare?	**Chapter 9:** Carrying Capacity: How Large a Population?	**Chapter 1:** Understanding Our Environment **Chapter 7:** Human Populations
Issue 14: Is Genetic Engineering the Answer to Hunger?	**Chapter 16:** Agriculture: The Ecology of Growing Food	**Chapter 9:** Food and Agriculture **Chapter 10:** Pest Control
Issue 15: Is a Large-Scale Shift to Organic Farming the Best Way to Increase World Food Supply?	**Chapter 16:** Agriculture: The Ecology of Growing Food	**Chapter 9:** Food and Agriculture **Chapter 10:** Pest Control
Issue 16: Should DDT Be Banned Worldwide?	**Chapter 5:** The Flow of Energy in Biological Systems: Why Does It Matter? **Chapter 16:** Agriculture: The Ecology of Growing Food	**Chapter 10:** Pest Control **Chapter 18:** Water Pollution
Issue 17: Do Environmental Hormone Mimics Pose a Potentially Serious Health Threat?		
Issue 18: Is the Superfund Program Successfully Protecting Human Health from Hazardous Materials?		**Chapter 8:** Environmental Health and Toxicology **Chapter 10:** Pest Control **Chapter 13:** Restoration Ecology **Chapter 18:** Water Pollution **Chapter 21:** Solid, Toxic, and Hazardous Waste
Issue 19: Should the United States Reprocess Spent Nuclear Fuel?	**Chapter 21:** Nuclear Power	**Chapter 14:** Geology and Earth Resources **Chapter 19:** Conventional Energy

Introduction

Environmental Issues: The Never-Ending Debate

One of the courses I teach is "Environmentalism: Philosophy, Ethics, and History." I begin by explaining the roots of the word "ecology," from the Greek *oikos* (house or household), and assigning the students to write a brief paper about their own household. How much, I ask them, do you need to know about the place where you live? And why?

The answers vary. Some of the resulting papers focus on people—roommates if the "household" is a dorm room, spouses and children if the students are older, parents and siblings if they live at home—and the need to cooperate and get along, and perhaps the need not to overcrowd. Some pay attention to houseplants and pets, and occasionally even bugs and mice. Some focus on economics—possessions, services, and their costs, where the checkbook is kept, where the bills accumulate, the importance of paying those bills, and of course the importance of earning money to pay those bills. Some focus on maintenance—cleaning, cleaning supplies, repairs, whom to call if something major breaks. For some the emphasis is operation—garbage disposal, grocery shopping, how to work the lights, stove, fridge, and so on. A very few recognize the presence of toxic chemicals under the sink and in the medicine cabinet and the need for precautions in their handling. Sadly, a few seem to be oblivious to anything that does not have something to do with entertainment.

Not surprisingly, some students object initially that the exercise seems trivial. "What does this have to do with environmentalism?" they ask. Yet the course is rarely very old before most are saying, "Ah! I get it!" That nice, homey microcosm has a great many of the features of the macrocosmic environment, and the multiple ways people can look at the microcosm mirror the ways people look at the macrocosm. It's all there, as is the question of priorities: What is important? People or fellow creatures or economics or maintenance or operation or waste disposal or food supply or toxics control or entertainment? Or all of the above?

And how do you decide? I try to illuminate this question by describing a parent trying to teach a teenager not to sit on a woodstove. In July, the kid answers, "Why?" and continues to perch. In August, likewise. And still in September. But in October or November, the kid yells "Ouch!" and jumps off in a hurry.

That is, people seem to learn best when they get burned.

This is surely true in our homely *oikos*, where we may not realize our fellow creatures deserve attention until houseplants die of neglect or cockroaches invade the cupboards. Economics comes to the fore when the phone gets cut off, repairs when a pipe ruptures, air quality when the air conditioner breaks or strange fumes rise from the basement, waste disposal when the bags pile up and begin to stink or the toilet backs up. Toxics control suddenly matters when a child or pet gets into the rat poison.

In the larger *oikos* of environmentalism, such events are paralleled by the loss of a species, or an infestation by another, by floods and droughts, by lakes turned into cesspits by raw sewage, by air turned foul by industrial smokestacks, by groundwater contaminated by toxic chemicals, by the death of industries and the loss of jobs, by famine and plague and even war.

If nothing is going wrong, we are not very likely to realize there is something we should be paying attention to. And this too has its parallel in the larger world. Indeed, the history of environmentalism is in important part a history of people carrying on with business as usual until something goes obviously awry. Then, if they can agree on the nature of the problem (Did the floor cave in because the joists were rotten or because there were too many people at the party?), they may learn something about how to prevent recurrences.

The Question of Priorities

There is of course a crucial "if" in that last sentence: *If people can agree.* . . . It is a truism to say that agreement is difficult. In environmental matters, people argue endlessly over whether anything is actually wrong, what its eventual impact will be, what if anything can or should be done to repair the damage, and how to prevent recurrence. Not to mention who's to blame and who should take responsibility for fixing the problem! Part of the reason is simple: Different things matter most to different people. Individual citizens may want clean air or water or cheap food or a convenient commute. Politicians may favor sovereignty over international cooperation. Economists and industrialists may think a few coughs (or cases of lung cancer, or shortened life spans) a cheap price to pay for wealth or jobs.

No one now seems to think that protecting the environment is not important. But different groups—even different environmentalists—have different ideas of what "environmental responsibility" means. To a paper company cutting trees for pulp, it may mean leaving a screen of trees (a "beauty strip") beside the road and minimizing erosion. To hikers following trails through or within view of the same tract of land, that is not enough; they want the trees left alone. The hikers may also object to seeing the users of trail bikes and all-terrain-vehicles on the trails. They may even object to hunters and anglers, whose activities they see as diminishing the wilderness experience. They may therefore push for protecting the land as limited-access wilderness. The hunters and anglers object to that, of course, for they want to be able to use their vehicles to bring their game home, or to bring their boats to their favorite rivers and lakes. They also argue, with some justification, that their license fees support a great deal of environmental protection work.

To a corporation, dumping industrial waste into a river may make perfect sense, for alternative ways of disposing of waste are likely to cost more and diminish profits. Of course, the waste renders the water less useful to wildlife or downstream humans, who may well object. Yet telling the corporation it cannot dump may be seen as depriving it of property. A similar problem arises when regulations prevent people and corporations from using land—and making money—as they had planned. Conservatives have claimed that environmental regulations thus violate the Fifth Amendment to the U.S. Constitution, which says "No person shall . . . be deprived of . . . property, without due process of law; nor shall private property be taken for public use, without just compensation."

One might think the dangers of such things as dumping industrial waste in rivers are obvious. But scientists can and do disagree, even given the same evidence. For instance, a chemical in waste may clearly cause cancer in laboratory animals. Is it therefore a danger to humans? A scientist working for the company dumping that chemical in a river may insist that no such danger has been proven. Yet a scientist working for an environmental group such as Greenpeace may insist that the danger is obvious since carcinogens do generally affect more than one species.

Scientists are human. They have not only employers but also values, rooted in political ideology and religion. They may feel that the individual matters more than corporations or society, or vice versa. They may favor short-term benefits over long-term benefits, or vice versa.

And scientists, citizens, corporations, and government all reflect prevailing social attitudes. When America was expanding westward, the focus was on building industries, farms, and towns. If problems arose, there was vacant land waiting to be moved to. But when the expansion was done, problems became more visible and less avoidable. People could see that there were "trade-offs" involved in human activity: more industry meant more jobs and more wealth, but there was a price in air and water pollution and human health (among other things).

Nowhere, perhaps, are these trade-offs more obvious than in Eastern Europe. The former Soviet Union was infamous for refusing to admit that industrial activity was anything but desirable. Anyone who spoke up about environmental problems risked jail. The result, which became visible to Western nations after the fall of the Iron Curtain in 1990, was industrial zones where rivers had no fish, children were sickly, and life expectancies were reduced. The fate of the Aral Sea, a vast inland body of water once home to a thriving fishery and a major regional transportation route, is emblematic: Because the Soviet Union wanted to increase its cotton production, it diverted for irrigation the rivers which delivered most of the Aral Sea's fresh water supply. The Sea then began to lose more water to evaporation than it gained, and it rapidly shrank, exposing sea-bottom so contaminated by industrial wastes and pesticides that wind-borne dust is now responsible for a great deal of human illness. The fisheries are dead, and freighters lie rusting on bare ground where once waves lapped.

The Environmental Movement

The twentieth century saw immense changes in the conditions of human life and in the environment that surrounds and supports human life. According to historian J. R. McNeill, in *Something New Under the Sun: An Environmental History of the Twentieth-Century World* (W. W. Norton, 2000), the environmental impacts that resulted from the interactions of burgeoning population, technological development, shifts in energy use, politics, and economics in that period are unprecedented in both degree and kind. Yet a worse impact may be that we have come to accept as "normal" a very temporary situation that "is an extreme deviation from any of the durable, more 'normal,' states of the world over the span of human history, indeed over the span of earth history." We are thus not prepared for the inevitable and perhaps drastic changes ahead.

Environmental factors cannot be denied their role in human affairs. Nor can human affairs be denied their place in any effort to understand environmental change. As McNeill says, "Both history and ecology are, as fields of knowledge go, supremely integrative. They merely need to integrate with each other."

The environmental movement, which grew during the twentieth century in response to increasing awareness of human impacts, is a step in that direction. Yet environmental awareness reaches back long before the modern environmental movement. When John James Audubon (1785–1851), famous for his bird paintings, was young, he was an enthusiastic slaughterer of birds (a few of which he used as models for the paintings). Later in life, he came to appreciate that birds were diminishing in numbers, as were the American bison, and he called for conservation measures. His was a minority voice, however. It was not till later in the century that George Perkins Marsh warned in *Man and Nature* (1864) that "We are, even now, breaking up the floor and wainscoting and doors and window frames of our dwelling, for fuel to warm our bodies and seethe our pottage, and the world cannot afford to wait till the slow and sure progress of exact science has taught it a better economy." The Earth, he said, was given to man for "usufruct" (to use the fruit of), not for consumption or waste. Resources should remain to benefit future generations. Stewardship was the point, and damage to soil and forest should be prevented and repaired. He was not concerned with wilderness as such; John Muir (1838–1914; founder of the Sierra Club) was the first to call for the preservation of natural wilderness, untouched by human activities. Marsh's ideas influenced others more strongly. In 1890, Gifford Pinchot (1865–1946) found "the nation . . . obsessed by a fury of development. The American Colossus was fiercely intent on appropriating and exploiting the riches of the richest of all continents." Under President Theodore Roosevelt, he became the first head of the U.S. Forest Service and a strong voice for conservation (not to be confused with preservation; Gifford's conservation meant using nature but in such a way that it was not destroyed; his aim was "the greatest good of the greatest number in the long run"). By the 1930s, Aldo Leopold (1887–1948), best known for his concept of the "land ethic" and his book, *A Sand County Almanac*, could argue that we had a responsibility not only to maintain the environment but also to repair damage done in the past.

The modern environmental movement was kick-started by Rachel Carson's *Silent Spring* (Houghton Mifflin, 1962). In the 1950s, Carson realized that the use of pesticides was having unintended consequences—the death of non-pest insects, food-chain accumulation of poisons and the consequent loss of birds, and even human illness—and meticulously documented the case. When her book was published, she and it were immediately vilified by pesticide proponents in government, academia, and industry (most notably, the pesticides industry). There was no problem, the critics said; the negative effects if any were worth it, and she—a *woman* and a nonscientist—could not possibly know what she was talking about. But the facts won out. A decade later, DDT was banned and other pesticides were regulated in ways unheard of before Carson spoke out.

Other issues have followed or are following a similar course.

The situation before Rachel Carson and *Silent Spring* is nicely captured by Judge Richard Cudahy, who in "Coming of Age in the Environment," *Environmental Law,* Winter 2000, writes, "It doesn't seem possible that before 1960 there was no 'environment'—or at least no environmentalism. I can even remember the Thirties, when we all heedlessly threw our trash out of car windows, burned coal in the home furnace (if we could afford to buy any), and used a lot of lead for everything from fishing sinkers and paint to no-knock gasoline. Those were the days when belching black smoke meant a welcome end to the Depression and little else."

Historically, humans have felt that their own well-being mattered more than anything else. The environment existed to be used. Unused, it was only wilderness or wasteland, awaiting the human hand to "improve" it and make it valuable. This is not surprising at all, for the natural tendency of the human mind is to appraise all things in relation to the self, the family, and the tribe. An important aspect of human progress has lain in enlarging our sense of "tribe" to encompass nations and groups of nations. Some now take it as far as the human species. Some include other animals. Some embrace plants as well, and bacteria, and even landscapes.

The more limited standard of value remains common. Add to that a sense that wealth is not just desirable but a sign of virtue (the Puritans brought an explicit version of this with them when they colonized North America; see Lynn White, Jr., "The Historical Roots of Our Ecological Crisis," *Science,* 1967), and it is hardly surprising that humans have used and still use the environment intensely. People also tend to resist any suggestion that they should restrain their use out of regard for other living things. Human needs, many insist, must come first.

The unfortunate consequences include the loss of other species. Lions vanished from Europe about 2,000 years ago. The dodo of Mauritius was extinguished in the 1600s (see the American Museum of Natural History's account at http://www.amnh.org/exhibitions/expeditions/treasure_fossil/Treasures/Dodo/dodo. html?dinos). The last of North America's passenger pigeons died in a Cincinnati zoo in 1914 (see http:// www.amnh.org/exhibitions/expeditions/treasure_fossil/Treasures/ Passenger_Pigeons/pigeons.html?dinos). Concern for such species was at first limited to those of obvious value to humans. In 1871, the U.S. Commission on Fish and Fisheries was created and charged with finding solutions to the decline in food fishes and promoting aquaculture. The first federal legislation designed to protect game animals was the Lacey Act of 1900. It was not until 1973 that the U.S. Endangered Species Act was adopted to shield all species from the worst human impacts.

Other unfortunate consequences of human activities include dramatic erosion, air and water pollution, oil spills, accumulations of hazardous (including nuclear) waste, famine, and disease. Among the many "hot stove" incidents that have caught public attention are the following:

- The Dust Bowl—in 1934 wind blew soil from drought-stricken farms in Oklahoma all the way to Washington, D.C.;
- Cleveland's Cuyahoga River caught fire in the 1960s;
- The Donora, Pennsylvania, smog crisis—in one week of October 1948, 20 died and over 7,000, were sickened;
- The London smog crisis in December 1952—4,000 dead;
- The Torrey Canyon and Exxon Valdez oil spills, which fouled shores and killed seabirds, seals, and fish;
- Love Canal, where industrial wastes seeped from their burial site into homes and contaminated groundwater;
- Union Carbide's toxics release at Bhopal, India—3,800 dead and up to 100,000 ill, according to Union Carbide; others claim a higher toll;
- The Three Mile Island and Chernobyl nuclear accidents;
- The decimation of elephants and rhinoceroses to satisfy a market for tusks and horns;
- The loss of forests—in 1997, fires set to clear Southeast Asian forest lands produced so much smoke that regional airports had to close;
- Ebola, a virus which kills nine tenths of those it infects, apparently first struck humans because growing populations reached into its native habitat;
- West Nile Fever, a mosquito-borne virus with a much less deadly record, was brought to North America by travelers or immigrants from Egypt;
- Acid rain, global climate change, and ozone depletion, all caused by substances released into the air by human activities.

The alarms have been raised by many people in addition to Rachel Carson. For instance, in 1968 (when world population was only a little over half of what it is today), Paul Ehrlich's *The Population Bomb* (Ballantine Books) described the ecological threats of a rapidly growing population and Garrett Hardin's influential essay, "The Tragedy of the Commons," *Science* (December 13,1968) described the consequences of using self-interest alone to guide the exploitation of publicly owned resources (such as air and water). (In 1974, Hardin introduced the unpleasant concept of "lifeboat ethics," which says that if there are not enough resources to go around, some people must do without.) In 1972, a group of economists, scientists, and business leaders calling themselves "The Club of Rome" published *The Limits to Growth* (Universe Books), an analysis of population, resource use, and pollution trends that predicted difficult times within a century; the study was redone as *Beyond the Limits to Growth: Confronting Global Collapse, Envisioning a Sustainable Future* (Chelsea Green, 1992) and again as *Limits to Growth: The 30-Year Update* (Chelsea Green, 2004), using more powerful computer models, and came to very similar conclusions. Among the most recent books is Jared Diamond's *Collapse: How Societies Choose to Fail or Succeed* (Viking, 2005), which uses historical cases to illuminate the roles of human biases and choices in dealing with environmental problems. Among Diamond's important points is the idea that in order to cope successfully with such problems, a society may have to surrender cherished traditions.

The following list of selected U.S. and UN laws, treaties, conferences, and reports illustrates the national and international responses to the various cries of alarm:

1967   The U.S. Air Quality Act set standards for air pollution.

1968   The UN Biosphere Conference discussed global environmental problems.

1969   The U.S. Congress passed the National Environmental Policy Act, which (among other things) required federal agencies to prepare environmental impact statements for their projects.

1970   The first Earth Day demonstrated so much public concern that the Environmental Protection Agency (EPA) was created; the Endangered Species Act, Clean Air Act, and Safe Drinking Water Act soon followed.

1971   The U.S. Environmental Pesticide Control Act gave the EPA authority to regulate pesticides.

1972   The UN Conference on the Human Environment, held in Stockholm, Sweden, recommended government action and led to the UN Environment Programme.

1973   The Convention on International Trade in Endangered Species of Wild Fauna and Flora (CITES) restricted trade in threatened species; because enforcement was weak, however, a black market flourished.

1976   The U.S. Resource Conservation and Recovery Act and the Toxic Substances Control Act established control over hazardous wastes and other toxic substances.

1979   The Convention on Long-Range Transboundary Air Pollution addressed problems such as acid rain (recognized as crossing national borders in 1972).

1982   The Law of the Sea addressed marine pollution and conservation.

1982   The second UN Conference on the Human Environment (the Stockholm +10 Conference) renewed concerns and set up a commission to prepare a "global agenda for change," leading to the 1987 Brundtland Report (*Our Common Future*).

1983   The U.S. Environmental Protection Agency and the U.S. National Academy of Science issued reports calling attention to the prospect of global warming as a consequence of the release of greenhouse gases such as carbon dioxide.

1987   The Montreal Protocol (strengthened in 1992) required nations to phase out use of chlorofluorocarbons (CFCs), the chemicals responsible for stratospheric ozone depletion (the "ozone hole").

1987   The Basel Convention controlled cross-border movement of hazardous wastes.

1988   The UN assembled the Intergovernmental Panel on Climate Change, which would report in 1995, 1998, and 2001 that the dangers of global warming were real, large, and increasingly ominous.

1992   The UN Convention on Biological Diversity required nations to act to protect species diversity.

1992   The UN Conference on Environment and Development (also known as the Earth Summit), held in Rio de Janeiro, Brazil, issued a broad call for environmental protections.

1992   The UN Convention on Climate Change urged restrictions on carbon dioxide release to avoid climate change.

1994   The UN Conference on Population and Development, held in Cairo, Egypt, called for stabilization and reduction of global population growth, largely by improving women's access to education and health care.

1997   The Kyoto Protocol attempted to strengthen the 1992 Convention on Climate Change by requiring reductions in carbon dioxide emissions, but U.S. resistance limited success.

2001   The UN Stockholm Convention on Persistent Organic Pollutants required nations to phase out use of many pesticides and other chemicals. It took effect May 17, 2004, after ratification by over fifty nations (not including the United States and the European Union).

2002   The UN World Summit on Sustainable Development, held in Johannesburg, South Africa, brought together representatives of governments, nongovernmental organizations, businesses, and other groups to examine "difficult challenges, including improving people's lives and conserving our natural resources in a world that is growing in population, with ever-increasing demands for food, water, shelter, sanitation, energy, health services and economic security."

2003   The World Climate Change Conference held in Moscow, Russia, concluded that global climate is changing, very possibly because of human activities, and the overall issue must be viewed as one of intergenerational justice. "Mitigating global climate change will be possible only with the coordinated actions of all sectors of society."

2005   The UN Millennium Project Task Force on Environmental Sustainability released its report, *Environment and Human Well-Being: A Practical Strategy*.

2005   The UN Millennium Ecosystem Assessment released its report, *Ecosystems and Human Well-Being: Synthesis* (http://www.millenniumassessment. org/ en/index.aspx) (Island Press).

2005   The UN Climate Change Conference held in Montreal, Canada, marked the taking effect of the Kyoto Protocol, ratified in 2004 by 141 nations (not including the United States and Australia).

2007   The Intergovernmental Panel on Climate Change (IPCC) released its Fourth Assessment Report, asserting that global warming is definitely due to human releases of carbon dioxide, the effects on both nature and humanity will be profound, and mitigation, though possible, will be expensive.

Rachel Carson would surely have been pleased by these responses, for they suggest both concern over the problems identified and determination to solve those problems. But she would just as surely have been frustrated, for a simple listing of laws, treaties, and reports does nothing to reveal the endless wrangling and the way political and business forces try to block progress whenever it is seen as interfering with their interests. Agreement on banning chlorofluorocarbons was relatively easy to achieve because CFCs were not seen as essential to civilization and there were substitutes available. Restraining greenhouse gas emissions is harder because we see fossil fuels as essential and though substitutes may exist, they are so far more expensive.

The Globalization of the Environment

Years ago, it was possible to see environmental problems as local. A smoke-stack belched smoke and made the air foul. A city sulked beneath a layer of smog. Bison or passenger pigeons declined in numbers and even vanished. Rats flourished in a dump where burning garbage produced clouds of smoke and runoff contaminated streams and groundwater and made wells unusable. Sewage, chemical wastes, and oil killed the fish in streams, lakes, rivers, and harbors. Toxic chemicals such as lead and mercury entered the food chain and affected the health of both wildlife and people.

By the 1960s, it was becoming clear that environmental problems did not respect borders. Smoke blows with the wind, carrying one locality's contamination to others. Water flows to the sea, carrying sewage and other wastes with it. Birds migrate, carrying with them whatever toxins they have absorbed with their food. In 1972, researchers were able to report that most of the acid rain falling on Sweden came from other countries. Other researchers have shown that the rise and fall of the Roman Empire can be tracked in Greenland, where glaciers preserve lead-containing dust deposited over the millennia—the amount rises as Rome flourished, falls with the Dark Ages, and rises again with the Renaissance and Industrial Revolution. Today we know that pesticides and other chemicals can show up in places (such as the Arctic) where they have never been used, even years after their use has been discontinued. The 1979 Convention on Long-Range Trans-boundary Air Pollution has been strengthened several times with amendments to address persistent organic pollutants, heavy metals, and other pollutants.

We are also aware of new environmental problems that exist only in a global sense. Ozone depletion, first identified in the stratosphere over Antarctica, threatens to increase the amount of ultraviolet light reaching the ground, and thereby increase the incidence of skin cancer and cataracts, among other things. The cause is the use by the industrialized world of chlorofluorocar-bons (CFCs) in refrigeration, air conditioning, aerosol cans, and electronics (for cleaning grease off circuit boards). The effect is global. Worse yet, the cause is rooted in northern lands such as the United States and Europe, but the worst effects may be felt where the sun shines brightest—in the tropics, which are dominated by developing nations. A serious issue of justice or equity is therefore involved.

A similar problem arises with global warming, which is also rooted in the industrialized world and its use of fossil fuels. The expected climate effects will hurt worst the poorer nations of the tropics, and perhaps worst of all those on low-lying South Pacific islands, which are expecting to be wholly inundated by rising seas. People who depend on the summertime melting of winter snows and mountain glaciers (including the citizens of California) will also suffer, for already the snows are less and the glaciers are vanishing.

Both the developed and the developing world are aware of the difficulties posed by environmental issues. In Europe, "green" political parties play a major and growing part in government. In Japan, some environmental regulations are more demanding than those of the United States. Developing nations understandably place dealing with their growing populations high on their list of priorities, but they play an important role in UN conferences on environmental issues, often demanding more responsible behavior from developed nations such as the United States (which often resists these demands; it has refused to ratify international agreements such as the Kyoto Protocol, for example).

Western scholars have been known to suggest that developing nations should forgo industrial development because if their huge populations ever attain the same per-capita environmental impact as the populations of wealthier lands, the world will be laid waste. It is not hard to understand why the developing nations object to such suggestions; they too want a better standard of living. Nor do they think it fair that they suffer for the environmental sins of others.

Are global environmental problems so threatening that nations must surrender their sovereignty to international bodies? Should the United States or Europe have to change energy supplies to protect South Pacific nations from inundation by rising seas? Should developing nations be obliged to reduce birthrates or forgo development because their population growth is seen as exacerbating pollution or threatening biodiversity?

Questions such as these play an important part in global debates today. They are not easy to answer, but their very existence says something important about the general field of environmental studies. This field is based in the science of ecology. Ecology focuses on living things and their interactions with each other and their surroundings. It deals with resources and limits and coexistence. It can see problems, their causes, and even potential solutions. And it can turn its attention to human beings as easily as it can to deer mice.

Yet human beings are not mice. We have economies and political systems, vested interests, and conflicting priorities and values. Ecology is only one part of environmental studies. Other sciences—chemistry, physics, climatology, epidemiology, geology, and more—are involved. So are economics, history, law, and politics. Even religion can play a part.

Unfortunately, no one field sees enough of the whole to predict problems (the chemists who developed CFCs could hardly have been expected to realize what would happen if these chemicals reached the stratosphere). Environmental studies is a field for teams. That is, it is a holistic, multidisciplinary field.

This gives us an important basic principle to use when evaluating arguments on either side of any environmental issue: Arguments that fail to recognize the complexity of the issue are necessarily suspect. On the other hand, arguments that endeavor to convey the full complexity of an issue may be impossible to understand; a middle ground is essential for clarity, but any reader or student must realize that something important may be being left out.

Current Environmental Issues

In 2001, the National Research Council's Committee on Grand Challenges in Environmental Sciences published *Grand Challenges in Environmental Sciences* (National Academy Press, 2001) in an effort to reach "a judgment regarding the most important environmental research challenges of the next generation—the areas most likely to yield results of major scientific and practical importance if pursued vigorously now." These areas include the following:

- Biogeochemical cycles (the cycling of plant nutrients, the ways human activities affect them, and the consequences for ecosystem functioning, atmospheric chemistry, and human activities)
- Biological diversity
- Climate variability
- Hydrologic forecasting (groundwater, droughts, floods, etc.)
- Infectious diseases
- Resource use
- Land use
- Reinventing the use of materials (e.g., recycling)

Similar themes appeared when *Issues in Science and Technology* celebrated its twentieth anniversary with its Summer 2003 issue. The editors noted that over the life of the magazine to date, some problems have hardly changed, nor has our sense of what must be done to solve them. Others have been affected, sometimes drastically, by changes in scientific knowledge, technological capability, and political trends. In the environmental area, the magazine paid special attention to:

- Biodiversity
- Overfishing
- Climate change
- The Superfund program
- The potential revival of nuclear power
- Sustainability

Many of the same basic themes were reiterated when *Science* magazine (published weekly by the American Association for the Advancement of Science) published in November and December 2003 a four-week series on the "State of the Planet," followed by a special issue on "The Tragedy of the Commons." In the introduction to the series, H. Jesse Smith began with these words: "Once in a while, in our headlong rush toward greater prosperity, it is wise to ask ourselves whether or not we can get there from here. As global population increases, and the demands we make on our natural resources grow even faster, it becomes ever more clear that the well-being we seek is imperiled by what we do."

Among the topics covered in the series were:

- Human population
- Biodiversity
- Tropical soils and food security
- The future of fisheries
- Freshwater resources
- Energy resources
- Air quality and pollution
- Climate change
- Sustainability
- The burden of chronic disease

Many of the topics on these lists are covered in this book. There are of course a great many other environmental issues—many more than can be covered in any one book such as this one. I have not tried to deal here with invasive species, the removal of dams to restore populations of anadromous fishes such as salmon, the depletion of aquifers, floodplain development, urban planning, the debate over "Pleistocene rewilding" (see, e.g., C. Josh Donlan, "Restoring America's Big, Wild Animals," *Scientific American*, June 2007), or many others. My sample of the variety available begins with the more philosophical issues. For instance, there is considerable debate over the "precautionary principle," which says in essence that even if we are not sure that our actions will have unfortunate consequences, we should take precautions just in case (see Issue 1). This principle plays an important part in many environmental debates, from those over the value of protecting endangered species (Issue 4) or the wisdom of opening the Arctic National Wildlife Refuge to oil drilling (Issue 7) to the folly (or wisdom) of reprocessing nuclear waste (Issue 19).

I said above that many people believed (and still believe) that nature has value only when turned to human benefit. One consequence of this belief is that it may be easier to convince people that nature is worth protecting if one can somehow calculate a cash value for nature "in the raw." Some environmentalists object to even trying to do this, on the grounds that economic value is not the only value, or even the value that should matter. (See Issue 3.)

What other values might we consider? Past editions of this book have considered whether nature has a value all its own, or a right to exist unmolested, and whether human property rights should take precedence. Here we discuss whether we should strive for social justice (Issue 5) and whether market forces can solve environmental problems such as pollution (see Issue 6).

Should we be concerned about the environmental impacts of specific human actions or products? Here too we can consider opening the Arctic National Wildlife Refuge to oil drilling (Issue 7), as well as the conflict between the value of DDT for preventing malaria and its impact on ecosystems (Issue 16), the hormone-like effects of some pesticides and other chemicals on both wildlife and humans (Issue 17), and the hazards of global warming (Issue 8).

World hunger is a major problem, with people arguing that it may be eased by genetic engineering (Issue 14) or organic farming (Issue 15).

Waste disposal is a problem area all its own. It encompasses both nuclear waste (Issue 19) and hazardous waste (Issue 18). A new angle on hazardous waste comes from the popularity of the personal computer—or more specifically, from the huge numbers of PCs that are discarded each year.

What solutions are available? Some are specific to particular issues, as wind power (Issue 9), automobile efficiency (Issue 10), biofuels (Issue 11), or a revival of nuclear power (Issue 12) may be to the problems associated with fossil fuels. Some are more general, as we might expect as soon as we hear someone speak of population growth as a primary cause of environmental problems (Issue 13) (there is some truth to this, for if the human population were small enough, its environmental impact—no matter how sloppy people were—would also be small).

Some analysts argue that whatever solutions we need, government need not impose them all. Private industry may be able to do the job if government can find a way to motivate industry, as with the idea of tradable pollution rights (Issue 6).

The overall aim, of course, is to avoid disaster and enable human life and civilization to continue prosperously into the future. The term for this is "sustainable development" (see Issue 2), and it was the chief concern of the UN World Summit on Sustainable Development, held in Johannesburg, South Africa, in August 2002. Exactly how to avoid disaster and continue prosperously into the future are the themes of the UN Millennium Ecosystem Assessment report, *Ecosystems and Human Well-Being: Synthesis* (http://www.millenniumassessment.org/en/index.aspx) (Island Press, 2005). The main findings of this report are that over the past half century meeting human needs for food, fresh water, fuel, and other resources has had major negative effects on the world's ecosystems; those effects are likely to grow worse over the next half century and will pose serious obstacles to reducing global hunger, poverty, and disease; and although "significant changes in policies, institutions, and practices can mitigate many of the negative consequences of growing pressures on ecosystems, . . . the changes required are large and not currently under way." Also essential will be improvements in knowledge about the environment, the ways humans affect it, and the ways humans depend upon it, as well as improvements in technology, both for assessing environmental damage and for repairing and preventing damage, as emphasized by Bruce Sterling in "Can Technology Save the Planet?" *Sierra* (July/August 2005). Sterling concludes, perhaps optimistically, that "When we see our historical predicament in its full, majestic scope, we will stir ourselves to great and direly necessary actions. It's not beyond us to think and act in a better way. Yesterday's short-sighted habits are leaving us, the way gloom lifts with the dawn." George Musser, introducing the special September 2005 "Crossroads for Planet Earth" issue of *Scientific American* in "The Climax of Humanity," notes that the next few decades will determine our future. Sterling's optimism may be fulfilled, but if we do not make the right choices, the future may be very bleak indeed.

Internet References . . .

The Earth Day Network

The Earth Day Network (EDN) promotes environmental citizenship and helps activists in their efforts to change local, national, and global environmental policies. It also makes the Ecological Footprint quiz (http://www.earthday.net/Footprint/index. asp) available.

http://www.earthday.net/

The Natural Resources Defense Council

The Natural Resources Defense Council is one of the most active environmental research and advocacy organizations. Its home page lists its concerns as clean air and water, energy, global warming, toxic chemicals, ocean health, and much more.

http://www.nrdc.org/

The United Nations Environment Programme

The United Nations Environment Programme "works to encourage sustainable development through sound environmental practices everywhere. Its activities cover a wide range of issues, from atmosphere and terrestrial ecosystems, the promotion of environmental science and information, to an early warning and emergency response capacity to deal with environmental disasters and emergencies."

http://www.unep.org/

The International Institute for Sustainable Development

The International Institute for Sustainable Development advances sustainable development policy and research by providing information and engaging in partnerships worldwide. It says it "promotes the transition toward a sustainable future. We seek to demonstrate how human ingenuity can be applied to improve the well-being of the environment, economy and society."

http://www.iisd.org/default.aspx

The Ecosystems Services Project

The Ecosystems Services Project studies the services people obtain from their environments, the economic and social values inherent in these services, and the benefits of considering these services more fully in shaping land management policies.

http://www.ecosystemservicesproject.org/

Environmental Philosophy

*E*nvironmental debates are rooted in questions of values—what is right? what is just?—and inevitably political in nature. It is worth stressing that people who consider themselves to be environmentalists can be found on both sides of most of the issues in this book. They differ in what they see as their own self-interest and even in what they see as humanity's long-term interest.

Understanding the general issues raised in this section is useful preparation for examining the more specific controversies that follow in later sections.

* Is the Precautionary Principle a Sound Approach to Risk Analysis?

* Is Sustainable Development Compatible with Human Welfare?

* Should a Price Be Put on the Goods and Services Provided by the World's Ecosystems?

ISSUE 1

Is the Precautionary Principle a Sound Approach to Risk Analysis?

YES: Nancy Myers, from "The Rise of the Precautionary Principle: A Social Movement Gathers Strength," *Multinational Monitor* (September 2004)

NO: Bernard D. Goldstein, from "The Precautionary Principle: Is It a Threat to Toxicological Science?" *International Journal of Toxicology* (January/February 2006)

ISSUE SUMMARY

YES: Nancy Myers, communications director for the Science and Environmental Health Network, argues that because the precautionary principle "makes sense of uncertainty," it has gained broad international recognition as being crucial to environmental policy.

NO: Bernard D. Goldstein, professor of Environmental and Occupational Health at the University of Pittsburgh, argues that although the precautionary principle is potentially valuable, it poses a risk that scientific (particularly toxicological) risk assessment will be displaced to the detriment of public health, social justice, and the field of toxicology itself.

The traditional approach to environmental problems has been reactive. That is, first the problem becomes apparent—wildlife or people sicken and die, or drinking water or air tastes foul. Then researchers seek the cause of the problem, and regulators seek to eliminate or reduce that cause. The burden is on society to demonstrate that harm is being done and that a particular cause is to blame.

An alternative approach is to presume that *all* human activities—construction projects, new chemicals, new technologies, etc.—have the potential to cause environmental harm. Therefore, those responsible for these activities should prove in advance that they will not do harm and should take suitable steps to prevent any harm from happening. A middle ground is occupied by the "precautionary principle," which has played an increasingly important part in environmental law ever since it first appeared in Germany in the mid-1960s. On the international scene, it has been applied to climate change, hazardous waste management, ozone depletion, biodiversity, and fisheries management. In 1992 the

Rio Declaration on Environment and Development, listing it as Principle 15, codified it thus:

> In order to protect the environment, the precautionary approach shall be widely applied by States according to their capabilities. When there are threats of serious or irreversible damage, lack of full scientific certainty shall not be used as a reason for postponing cost-effective measures to prevent environmental degradation.

Other versions of the principle also exist, but all agree that when there is reason to think—not absolute proof—that some human activity is or might be harming the environment, precautions should be taken. Furthermore, the burden of proof should be on those responsible for the activity, not on those who may be harmed. This has come to be broadly accepted as a basic tenet of ecologically or environmentally sustainable development. See Paul L. Stein, "Are Decision-Makers Too Cautious With the Precautionary Principle?" *Environmental and Planning Law Journal* (February 2000), and Marco Martuzzi and Roberto Bertollini, "The Precautionary Principle, Science and Human Health Protection," *International Journal of Occupational Medicine and Environmental Health* (January 2004).

The precautionary principle also contributes to thinking in the areas of risk assessment and risk management in general. Human activities—the manufacture of chemicals and other products; the use of pesticides, drugs, and fossil fuels; the construction of airports and shopping malls; and even agriculture—can damage health and the environment. Some people insist that action need not be taken against any particular activity until and unless there is solid, scientific proof that it is doing harm, and even then risks must be weighed against each other. Others insist that mere suspicion should be grounds enough for action.

Since solid, scientific proof can be very difficult to obtain, the question of just how much proof is needed to justify action is vital. Not surprisingly, if action threatens an industry, that industry's advocates will argue against taking precautions, generally saying that more proof is needed. A good example can be found in Stuart Pape, "Watch Out for the Precautionary Principle," *Prepared Foods* (October 1999): "In recent months, U.S. food manufacturers have experienced a rude introduction to the 'Precautionary Principle.' . . . European regulators have begun to adopt extreme definitions of the Principle in order to protect domestic industries and place severe restrictions on the use of both old and new materials without justifying their action upon sound science."

In the following selections, Nancy Myers, communications director for the Science and Environmental Health Network, argues that the precautionary principle has gained broad international recognition as crucial to environmental policy for good reason. "The essence of the Precautionary Principle is that when lives and the future of the planet are at stake, people must act on . . . clues and prevent as much harm as possible, despite imperfect knowledge and even ignorance." Professor Bernard D. Goldstein argues that although the precautionary principle is potentially valuable, it poses a risk that scientific (particularly toxicological) risk assessment will be displaced to the detriment of public health, social justice, and the field of toxicology itself.

3

YES

Nancy Myers

The Rise of the Precautionary Principle: A Social Movement Gathers Strength

Ed Soph is a jazz musician and professor at the University of North Texas in Denton, a growing town of about 100,000 just outside Dallas, Texas. In 1997, Ed and his wife Carol founded Citizens for Healthy Growth, a Denton group concerned about the environment and future of their town. The Sophs and their colleagues—the group now numbers about 400—are among the innovative pioneers who are implementing the Precautionary Principle in the United States.

The Sophs first came across the Precautionary Principle in 1998, in the early days of the group's campaign to prevent a local copper wire manufacturer, United Copper Industries, from obtaining an air permit that would have allowed lead emissions. Ed remembers the discovery of the Wingspread Statement on the Precautionary Principle—a 1998 environmental health declaration holding that "When an activity raises threats of harm to human health or the environment, precautionary measures should be taken even if some cause and effect relationships are not fully established scientifically"—as "truly a life-changing experience." Using the Precautionary Principle as a guide, the citizens refused to be drawn into debates on what levels of lead, a known toxicant, might constitute a danger to people's health. Instead, they pointed out that a safer process was available and insisted that the wise course was not to issue the permit. The citizens prevailed.

The principle helped again in 2001, when a citizen learned that the pesticides 2,4-D, simazine, Dicamba and MCPP were being sprayed in the city parks. "The question was, given the 'suspected' dangers of these chemicals, should the city regard those suspicions as a reassurance of the chemicals' safety or as a warning of their potential dangers?" Ed recalls. "Should the city act out of ignorance or out of common sense and precaution?"

Soph learned that the Greater Los Angeles School District had written the Precautionary Principle into its policy on pesticide use and had turned to Integrated Pest Management (IPM), a system aimed at controlling pests without the use of toxic chemicals. The Denton group decided to advocate for a similar policy. They persuaded the city's park district to form a focus group of park users

From *Multinational Monitor*, September 2004, pp. 9–15. Copyright © 2004 by Essential Information. Reprinted by permission.

and organic gardening experts. The city stopped spraying the four problem chemicals and initiated a pilot IPM program.

The campaign brought an unexpected economic bonus to the city. In the course of their research, parks department staff discovered that corn gluten was a good turf builder and natural broadleaf herbicide. But the nearest supplier of corn gluten was in the Midwest, and that meant high shipping costs for the city. Meanwhile, a corn processing facility in Denton was throwing away the corn gluten it produced as a byproduct. The parks department made the link, and everyone was pleased. The local corn company was happy to add a new product line; the city was happy about the expanded local business and the lower price for a local product; and the environmental group chalked up another success.

The citizens of Denton, Texas, did not stop there. They began an effort to improve the community's air pollution standards. They got arsenic-treated wood products removed from school playgrounds and parks and replaced with nontoxic facilities. "The Precautionary Principle helped us define the problems and find the solutions," Ed says.

But, as he wrote in an editorial for the local paper, "The piecemeal approach is slow, costly and often more concerned with mitigation than prevention." Taking a cue from Precautionary Principle pioneers in San Francisco, they also began lobbying for a comprehensive new environmental code for the community, based on the Precautionary Principle.

In June 2003, San Francisco's board of supervisors had become the first government in the United States to embrace the Precautionary Principle. A new environmental code drafted by the city's environment commission put the Precautionary Principle at the top, as Article One. Step one in implementing the code was a new set of guidelines for city purchasing, pointing the way toward "environmentally preferable" purchases by careful analysis and choice of the best alternatives. The White Paper accompanying the ordinance pointed out that most of the city's progressive environmental policies were already in line with the Precautionary Principle, and that the new code provided unity and focus to the policies rather than a radically new direction.

That focus is important; too often, environmental matters seem like a long, miscellaneous and confusing list of problems and solutions.

Likewise in Denton, the Precautionary Principle has not been a magic wand for transforming policy, but it has put backbone into efforts to enact truly protective and far-sighted environmental policies. Ed Soph points out that, in his community as in others, growth had often been dictated by special interests in the name of economic development, and the environment got short shrift.

"Environmental protection and pollution prevention in our city have been a matter, not of proactive policy, but of reaction to federal and state mandates, to the threat of citizens' lawsuits, and to civic embarrassment. Little thought is given to future environmental impacts," he told the city council when he argued for a new environmental code.

He added, "The toxic chemical pollution emitted by area industries has been ignored or accepted for all the ill-informed or selfish reasons that we are too familiar with. The Precautionary Principle dispels that ignorance and empowers concerned citizens with the means to ensure a healthier future."

The Precautionary Principle has leavened the discussion of environmental and human health policy on many fronts—in international treaty negotiations and global trade forums, in city resolutions and national policies, among conservationists and toxicologists, and even in corporate decision making.

Two treaties negotiated in 2000 incorporated the principle for the first time as an enforceable measure. The Cartagena Protocol on Biosafety allows countries to invoke the Precautionary Principle in decisions on admitting imports of genetically modified organisms. It became operative in June 2003. The Stockholm Convention on Persistent Organic Pollutants prescribes the Precautionary Principle as a standard for adding chemicals to the original list of 12 that are banned by the treaty. This treaty went into force in February 2004.

Making Sense of Uncertainty

Understanding the need for the Precautionary Principle requires some scientific sophistication. Ecologists say that changes in ecological systems may be incremental and gradual, or surprisingly large and sudden. When change is large enough to cause a system to cross a threshold, it creates a new dynamic equilibrium that has its own stability and does not change back easily. These new interactions become the norm and create new realities.

Something of this new reality is evident in recently observed changes in patterns of human disease:

- Chronic diseases and conditions affect more than 100 million men, women, and children in the United States—more than a third of the population. Cancer, asthma, Alzheimer's disease, autism, birth defects, developmental disabilities, diabetes, endometriosis, infertility, multiple sclerosis and Parkinson's disease are becoming increasingly common.
- Nearly 12 million children in the United States (17 percent) suffer from one or more developmental disabilities. Learning disabilities alone affect at least 5 to 10 percent of children in public schools, and these numbers are increasing. Attention deficit hyperactivity disorder conservatively affects 3 to 6 percent of all school children. The incidence of autism appears to be increasing.
- Asthma prevalence has doubled in the last 20 years.
- Incidence of certain types of cancer has increased. The age-adjusted incidence of melanoma, non-Hodgkins lymphoma, and cancers of the prostate, liver, testis, thyroid, kidney, breast, brain, esophagus and bladder has risen over the past 25 years. Breast cancer, for example, now strikes more women worldwide than any other type of cancer, with rates increasing 50 percent during the past half century. In the 1940s, the lifetime risk of breast cancer was one in 22. Today's risk is one in eight and rising.
- In the United States, the incidence of some birth defects, including male genital disorders, some forms of congenital heart disease and obstructive disorders of the urinary tract, is increasing. Sperm density is declining in some parts of the United States and elsewhere in the world.

These changes in human health are well documented. But proving direct links with environmental causative factors is more complicated.

Here is how the scientific reasoning might go: Smoking and diet explain few of the health trends listed above. Genetic factors explain up to half the population variance for several of these conditions—but far less for the majority of them—and in any case do not explain the changes in disease incidence rates. This suggests that other environmental factors play a role. Emerging science suggests this as well. In laboratory animals, wildlife and humans, considerable evidence documents a link between environmental contamination and malignancies, birth defects, reproductive disorders, impaired behavior and immune system dysfunction. Scientists' growing understanding of how biological systems develop and function leads to similar conclusions.

But serious, evident effects such as these can seldom be linked decisively to a single cause. Scientific standards of certainty (or "proof") about cause and effect are high. These standards may never be satisfied when many different factors are working together, producing many different results. Sometimes the period of time between particular causes and particular results is so long, with so many intervening factors, that it is impossible to make a definitive link. Sometimes the timing of exposure is crucial—a trace of the wrong chemical at the wrong time in pregnancy, for example, may trigger problems in the child's brain or endocrine system, but the child's mother might never know she was exposed.

In the real world, there is no way of knowing for sure how much healthier people might be if they did not live in the modern chemical stew, because the chemicals are everywhere—in babies' first bowel movement, in the blood of U.S. teenagers and in the breastmilk of Inuit mothers. No unexposed "control" population exists. But clearly, significant numbers of birth defects, cancers and learning disabilities are preventable.

Scientific uncertainty is a fact of life even when it comes to the most obvious environmental problems, such as the disappearance of species, and the most potentially devastating trends, such as climate change. Scientists seldom know for sure what will happen until it happens, and seldom have all the answers about causes until well after the fact, if ever. Nevertheless, scientific knowledge, as incomplete as it may be, provides important clues to all of these conditions and what to do about them.

The essence of the Precautionary Principle is that when lives and the future of the planet are at stake, people must act on these clues and prevent as much harm as possible, despite imperfect knowledge and even ignorance.

Environmental Failures

A premise of Precautionary Principle advocates is that environmental policies to date have largely not met this challenge. Part of the explanation for why they have not is that the dimensions of the emerging problems are only now becoming apparent. The limits of the earth's assimilative capacity are much clearer now than they were when the first modern environmental legislation was enacted 30 years ago.

Another part of the explanation is that, although some environmental policies are preventive, most have focused on cleaning up messes after the fact—what environmentalists call "end of pipe" solutions. Scrubbers on power plant stacks, catalytic converters on tailpipes, recycling and super-sized funds dedicated to detoxifying the worst dumps have not been enough. The Precautionary Principle holds that earlier, more comprehensive and preventive approaches are necessary. Nor is it enough to address problems only after they have become so obvious that they cannot be ignored—often, literally waiting for the dead bodies to appear or for coastlines to disappear under rising tides.

The third factor in the failure of environmental policies is political, say Precautionary Principle proponents. After responding to the initial burst of concern for the environment, the U.S. regulatory system and others like it were subverted by commercial interests, with the encouragement of political leaders and, increasingly, the complicity of the court system. Environmental laws have been subjected to an onslaught of challenges since the 1980s; many have been modified or gutted, and all are enforced by regulators who have been chastened by increasing challenges to their authority by industry and the courts.

The courts, and now increasingly international trade organizations and agreements like the World Trade Organization (WTO) and the North American Free Trade Agreement (NAFTA), have institutionalized an anti-precautionary approach to environmental controls. They have demanded the kinds of proof and certainty of harms and efficacy of regulation that science often cannot provide.

False Certainties

Ironically, one tool that has proved highly effective in the battle against environmental regulations was one that was meant to strengthen the enforcement of such laws: quantitative risk assessment. Risk assessment was developed in the 1970s and 1980s as a systematic way to evaluate the degree and likelihood of harmful side effects from products and technologies. With precise, quantitative risk assessments in hand, regulators could more convincingly demonstrate the need for action. Risk assessments would stand up in court. Risk assessments could "prove" that a product was dangerous, would cause a certain number of deaths per million, and should be taken off the market.

Or not. Quantitative risk assessment, which became standard practice in the United States in the mid-1980s and was institutionalized in the global trade agreements of the 1990s, turned out to be most useful in "proving" that a product or technology was not inordinately dangerous. More precisely, risk assessments presented sets of numbers that purported to state definitively how much harm might occur. The next question for policymakers then became: How much harm is acceptable? Quantitative risk assessment not only provided the answers; it dictated the questions.

As quantitative risk assessment became the norm, commercial and industrial interests were increasingly able to insist that harm must be proven "scientifically"—in the form of a quantitative risk assessment demonstrating harm in excess of acceptable limits—before action was taken to stop a process or product. These exercises were often linked with cost-benefit assessments that heavily weighted the immediate monetary costs of regulations and gave little, if any, weight to costs to the environment or future generations.

Although risk assessments tried to account for uncertainties, those projections were necessarily subject to assumptions and simplifications. Quantitative risk assessments usually addressed a limited number of potential harms, often missing social, cultural or broader environmental factors. These risk assessments have consumed enormous resources in strapped regulatory agencies and have slowed the regulatory process. They have diverted attention from questions that could be answered: Do better alternatives exist? Can harm be prevented?

The slow pace of regulation, the insistence on "scientific certainty," and the weighting toward immediate monetary costs often give the benefit of doubt to products and technologies, even when harmful side effects are suspected. One result is that neither international environmental agreements nor national regulatory systems have kept up with the increasing pace and cumulative effects of environmental damage.

A report by the European Environment Agency in 2001 tallied the great costs to society of some of the most egregious failures to heed early warnings of harm. Radiation, ozone depletion, asbestos, Mad Cow disease and other case studies show a familiar pattern: "Misplaced 'certainty' about the absence of harm played a key role in delaying preventive actions," the authors conclude.

They add, "The costs of preventive actions are usually tangible, clearly allocated and often short term, whereas the costs of failing to act are less tangible, less clearly distributed and usually longer term, posing particular problems of governance. Weighing up the overall pros and cons of action, or inaction, is therefore very difficult, involving ethical as well as economic considerations."

The Precautionary Approach

As environmentalists looked at looming problems such as global warming, they were appalled at the inadequacy of policies based on quantitative risk assessment. Although evidence was piling up rapidly that human activities were having an unprecedented effect on global climate, for example, it was difficult to say when the threshold of scientific certainty would be crossed. Good science demanded caution about drawing hard and fast conclusions. Yet, the longer humanity waited to take action, the harder it would be to reverse any effect. Perhaps it was already too late. Moreover, action would have to take the form of widespread changes not only in human behavior but also in technological development. The massive shift away from fossil fuels that might yet mitigate the effects of global warming would require rethinking the way humans produce and use energy. Nothing in the risk-assessment-based approach to policy prepared society to do that.

The global meetings called to address the coming calamity were not helping much. Politicians fiddled with blame and with protecting national economic interests while the globe heated up. Hard-won and heavily compromised agreements such as the 1997 Kyoto agreement on climate change were quickly mired in national politics, especially in the United States, the heaviest fossil-fuel user of all.

In the United States and around the globe, a different kind of struggle had been going on for decades: the fight for attention to industrial pollution in communities. From childhood lead poisoning in the 1930s to Love Canal in the 1970s, communities had always faced an uphill battle in proving that pollution and toxic products were making them sick. Risk assessments often made the case that particular hazardous waste dumps were safe, or that a single polluting industry could not possibly have caused the rash of illnesses a community claimed. But these risk assessments missed the obvious fact that many communities suffered multiple environmental assaults, compounded by other effects of poverty. A landmark 1987 report by the United Church of Christ coined the term "environmental racism" and confirmed that the worst environmental abuses were visited on communities of color. This growing awareness generated the international environmental justice movement.

In early 1998, a small conference at Wingspread, the Johnson Foundation's conference center in Racine, Wisconsin, addressed these dilemmas head-on. Participants groped for a better approach to protecting the environment and human health. At that time, the Precautionary Principle, which had been named in Germany in the 1970s, was an emerging precept of international law. It had begun to appear in international environmental agreements, gaining reference in a series of protocols, starting in 1984, to reduce pollution in the North Sea; the 1987 Ozone Layer Protocol; and the Second World Climate Conference in 1990.

At the Rio Earth Summit in 1992, precaution was enshrined as Principle 15 in the Rio Declaration on Environment and Development: "In order to protect the environment, the precautionary approach shall be widely applied by states according to their capabilities. Where there are threats of serious or irreversible damage, lack of full scientific certainty shall not be used as a reason for postponing cost-effective measures to prevent environmental degradation."

In the decade after Rio, the Precautionary Principle began to appear in national constitutions and environmental policies worldwide and was occasionally invoked in legal battles. For example:

- The Maastricht Treaty of 1994, establishing the European Union, named the Precautionary Principle as a guide to EU environment and health policy.

- The Precautionary Principle was the basis for arguments in a 1995 International Court of Justice case on French nuclear testing. Judges cited the "consensus flowing from Rio" and the fact that the Precautionary Principle was "gaining increasing support as part of the international law of the environment."

- At the World Trade Organization in the mid-1990s, the European Union invoked the Precautionary Principle in a case involving a ban on imports of hormone-fed beef.

THE U.S. CHAMBER OF COMMERCE:
POLICY BRIEF AGAINST THE PRECAUTIONARY PRINCIPLE

Objective

To ensure that regulatory decisions are based on scientifically sound and technically rigorous risk assessments, and to oppose the adoption of the Precautionary Principle as the basis for regulation.

Summary of the Issue

The U.S. Chamber supports a science-based approach to risk management; where risk is assessed based on scientifically sound and technically rigorous analysis. Under this approach, regulatory actions are justified where there are legitimate, scientifically ascertainable risks to human health, safety, or the environment. That is, the greater the risk, the greater the degree of regulatory scrutiny. This standard has served the nation well, and has led to astounding breakthroughs in the fields of science, health care, medicine, biotechnology, agriculture, and many other fields. There is, however, a relatively new theory, known as the Precautionary Principle that is gaining popularity among environmentalists and other groups. The Precautionary Principle says that when the risks of a particular activity are unclear or unknown, assume the worst and avoid the activity. It is essentially a policy of risk avoidance.

The regulatory implications of the Precautionary Principle are huge. For instance, the Precautionary Principle holds that since the existence and extent of global warming and climate change are not known one should assume the worst and immediately restrict the use of carbon-based fuels. However the nature and extent of key environmental, health, and safety concerns requires careful scientific and technical analysis. That is why the U.S. Chamber has long supported the use of sound science, cost-benefit analysis, risk assessment and a full understanding of uncertainty when assessing a particular regulatory issue.

The Precautionary Principle has been explicitly incorporated into various laws and regulations in the European Union and various international bodies. In the United States, radical environmentalists are pushing for its adoption as a basis for regulating biotechnology, food and drug safety, environmental protection, and pesticide use.

U.S. Chamber Strategy

- Support a science-based approach to risk management, where risk is assessed based on scientifically sound and technically rigorous standards.
- Oppose the domestic and international adoption of the Precautionary Principle as a basis for regulatory decision making.
- Educate consumers, businesses, and federal policymakers about the implications of the Precautionary Principle.

The Wingspread participants believed the Precautionary Principle was not just another weak and limited fix for environmental problems. They believed it could bring far-reaching changes to the way those policies were formed and implemented. But action to prevent harm in the face of scientific uncertainty alone did not translate into sound policies protective of the environment and human health. Other norms would have to be honored simultaneously and as an integral part of a precautionary decision-making process. Several other principles had often been linked with the Precautionary Principle in various statements of the principle or in connection with precautionary policies operating in Northern European countries. The statement released at the end of the meeting, the Wingspread Statement on the Precautionary Principle, was the first to put four of these primary elements on the same page—acting upon early evidence of harm, shifting the burden of proof, exercising democracy and transparency, and assessing alternatives. These standards form the basis of what has come to be known as the overarching or comprehensive Precautionary Principle or approach:

> When an activity raises threats of harm to human health or the environment, precautionary measures should be taken even if some cause and effect relationships are not fully established scientifically.
>
> In this context the proponent of an activity, rather than the public, should bear the burden of proof.
>
> The process of applying the Precautionary Principle must be open, informed and democratic and must include potentially affected parties. It must also involve an examination of the full range of alternatives, including no action.

The conference generated widespread enthusiasm for the principle among U.S. environmentalists and academics as well as among some policymakers. That was complemented by continuing and growing support for the principle among Europeans as well as ready adoption of the concept in much of the developing world. And in the years following Wingspread, the Precautionary Principle has gained new international status.

Bernard D. Goldstein  **NO**

The Precautionary Principle: Is It a Threat to Toxicological Science?

The central figure in developing the *1970 US Clean Air Act,* Senator Edmund Muskie, rather plaintively asked that the scientific information underlying this act be obtained from a one-handed scientist. He was tired of hearing scientists saying "on the one hand . . . , yet on the other hand. . . ." The Senator was expressing the desire of lawmakers for indisputable scientific facts on which to base regulatory actions. Unfortunately, this is not the way science works, particularly in situations where there is uncertainty concerning the questions being asked, or the terms being used.

Lack of clarity in definition or consistency in use is a particular problem when considering the Precautionary Principle. My discussion of the impact of the Precautionary Principle will inevitably fall into the "on the one hand . . . , yet on the other hand . . ." category that made Senator Muskie so unhappy. I will argue that the precautionary principle is both good public health and that it is causing public health problems; that it will both increase and decrease the importance of toxicological science in decision making; and that the European Community (EC) is both using the Precautionary Principle in some innovative ways to protect public health and the environment, and that the EC is intentionally abusing the precautionary principle by using it as a means to erect artificial trade barriers not warranted by toxicological science and risk analysis.

Definitions of the Precautionary Principle

The precautionary principle was developed primarily in Europe, beginning in the 1980s in Germany. One of the first times it received major prominence was in 1992 at the United Nations Conference on Environment and Development in Rio de Janeiro. The primary goal of this meeting was to establish an environmental agenda that would guide nations toward sustainable development. At this meeting the precautionary principle was adopted in the following form:

From *International Journal of Toxicology,* vol. 25, no. 1, January/February 2006, pp. 3–7. Copyright © 2006 by American College of Toxicology. Reprinted by permission of Taylor & Francis Journals. Via Rightslink.

"Nations shall use the precautionary approach to protect the environment. Where there are threats of serious or irreversible damage, scientific uncertainty shall not be used to postpone cost-effective measures to prevent environmental degradation."

It is informative to compare this definition with the more recent Wingspread Statement:

"When an activity raises threat [of harm] to human health or the environment, precautionary measures should be taken even if some cause and effect relationships are not fully established scientifically."

Wingspread is notable for a number of differences from the earlier Rio declaration. The Rio declaration includes a statement that the threat of harm should be serious and irreversible, but no threat of harm is specified in Wingspread; the Rio declaration states that the action should be cost-effective, but cost-effectiveness is not part of theWingspread statement; the whole thrust of the Wingspread statement is positive in that measures should be taken, whereas in Rio it is negative in that the lack of uncertainty should not postpone taking measures; and finally the Rio statement focuses solely on the environment whereas the Wingspread statement goes beyond the environment to the broad field of public health.

The problem of appropriately defining the precautionary principle is not simply academic. Like sustainable development, as a concept the precautionary principle is supportable by everyone. But the precautionary principle also has been adopted in various national and international treaties and by governmental entities in a way that provides legal teeth to what is otherwise a nebulous policy concept. The Wingspread approach is beginning to take hold in the United States where the Precautionary Principle has been approved by the San Francisco and Berkeley City Councils and is under consideration by other local authorities. Where laws and actions are at stake, definitions become very important. The European approach, which often infuriates its trading partners including the United States, is to state that the precautionary principle is enshrined within European policy—but to never define it. Rather, a policy or action is judged as to whether it will result in a precautionary approach. This has led to what Carruth and I call the Cynical American Definition:

"The Precautionary Principle is a nebulous doctrine invented by Europeans as a means to erect a trade barrier against any item produced more efficiently elsewhere."

There are many different actions that can be taken under the heading of the precautionary principle, or the precautionary approach as it is often described. It is useful to consider such actions under the public health and prevention nomenclature of primary and secondary prevention. Simply put, primary prevention is an approach that focuses on there never being a problem in the first place; e.g., an individual does not start smoking or, even better, there are no cigarettes being produced. Secondary prevention depends upon the early detection of a problem, such as detecting high blood pressure before someone has a stroke or is otherwise symptomatic. Much of what is done as a result of environmental risk assessment and risk management is secondary prevention, particularly when it is related to the standard approaches to dealing with known toxic chemicals. Many risk assessors would argue that the precautionary principle is already part of risk assessment. Among the potentially precautionary standard risk assessment practices are 10-fold "safety" factors; the 95% upper confidence limit; and the use of exposure models containing prudent default assumptions. The primary preventive approaches of the precautionary principle go well beyond more stringent risk assessment. Its components include taking preventive action in the face of uncertainty; shifting the burden of proof to the proponents of an activity such as industry of government; exploring a wide range of alternatives; and increasing public involvement in decision making.

The Precautionary Principle and World Trade

The implications of how to define the precautionary principle have nowhere been more evident than for some of the trade issues to which it is being applied. In order to keep a level playing field, various world trade agreements depend upon harmonizing risk assessments so that a nation cannot just arbitrarily decide that another nation's product is unhealthy. Unfortunately, there are a variety of examples in which it appears that the European community has used the precautionary principle as a means to exclude items from elsewhere in the world. Before describing these as inappropriate uses of toxicological science, let me first emphasize that we in the United States have much to learn from the EC as it wrestles with core issues of democracy that impact on the control of environmental hazards. The current debates about the governance relationships among citizens, states, and the multistate European confederation are unparalleled in the US since the Federalist Papers; and are of profound global importance.

As I do not want current broader EU-US issues to cloud the discussion, I will start with an issue that does not involve the United States. An example of action based upon the precautionary principle that fits under the heading of secondary prevention is the adoption by the EC of a very stringent standard for aflatoxin. This was done in essence by adding an additional safety factor for the genotoxicity of aflatoxins. For all nondairy products designated for human consumption, the EC standard is one-fifth that of the United States and for milk it is one-tenth of that of the United States. This has resulted in the exclusion of an estimated $700 million yearly of trade from Sub-Saharan Africa to the advantage of European growers. These Sub-Saharan nations are among the poorest in the world, with relatively fixed trade routes following old colonial patterns. The difference in risk is less than one cancer per year in Europe. The Joint FAO/WHO Expert Committee on Food Additives (JECFA), the expert body of the Food and Agricultural Organization and the World Health Organization, at the request of the Codex Alimentarius Commission, reviewed the difference between the US and European standard and found that it was of no health significance. There is also concern that the EC standard is halting the usual harvesting of Brazil nuts in the Amazon area with the result that the trees are being cut down rather than providing resources for indigenous groups.

An example of a trade barrier related to primary prevention aspects of the precautionary principle is that of the EU's ban on beef from cattle that have previously been treated with estrogenic growth agents. Canada and the US brought this issue to the World Trade Organization (WTO) who decided in their favor. The EU argued that despite the lack of recognition of risk by JECFA, or even its own scientific bodies, the precautionary principle permitted them to ban beef from hormone-treated animals as a risk could not be ruled out, and that a formal risk assessment was not needed. Among the arguments advanced by Canada and the US were that the EC was inconsistent in not having limits on natural estrogens, and was inconsistent in permitting the use in swine of antibiotics whose residues in pork were carcinogenic in laboratory animals. The WTO ruling basically said that the precautionary principle was not yet generally accepted as a legal doctrine capable of supplanting risk assessment. Much of the subsequent EU activity in justifying the precautionary principle can be seen as a means of reversing the WTO opinion. The use of the precautionary principle as a trade barrier will likely be central to the forthcoming WTO hearings on the EU's ban of genetically modified foods.

The use of a primary preventive precautionary approach also frequently occurs in US law. An interesting example demonstrating a switch to primary precaution is the hazardous air pollutant (HAP) provisions of the *1990 Clean Air Act* amendments. Prior to 1990, to regulate a compound as a HAP, the Environmental Protection Agency (EPA) had to go through an involved and time-consuming process. It began by listing a compound as one reasonably anticipated to cause adverse health effects (and is not regulated by setting an ambient standard). After suitable hearings, the EPA would then use risk-based and other considerations to decide which sources of the listed compound to regulate. The burden of proof was on the government to show the potential for adverse consequences. Relatively few compounds had been regulated under this approach, primarily those believed to be human carcinogens. In 1990, faced with impatience at this slow approach, and the recognition that common air contaminants responsible for many tons of emissions were uncontrolled, Congress dramatically changed the HAP regulatory process. Congress listed over 180 compounds to be regulated for which the only rule-making procedure was the onerous one of attempting to remove them from the list by demonstrating safety, i.e., the burden of proof was switched to the industry. Further, Congress specified that emission sources should be required to use the maximum available control technology (MACT), reasoning that all sources should do as well as the best (defined operationally as the upper 12th percentile). Risk considerations are taken into account only secondarily. If, after MACT, there is still a one in a million risk to the maximally exposed individual (MEI), additional steps are to be taken. These steps are currently under consideration and likely will go through court challenges in the next few years. The use of the MEI as the target in a public health bill is problematic—it does not matter whether the source is in the middle of a populous major metropolitan area or upwind of the Mojave Desert.

If a similar switch in the burden of proof, and away from risk based approaches, was under consideration today, the debate would be phrased in terms of the precautionary principle. As it is now 15 years from the passage of the *1990 Clean Air Act* amendments, the change in the HAP provisions allows an opportunity to consider the effectiveness of the precautionary principle in action. Carruth and I have pointed out potential public health shortfalls to the precautionary approach taken in the HAP amendments. From the vantage point of a toxicologist interested in understanding the mechanism of chemical effects, the most important is that once a compound is listed, why would there be any interest in studying it further? In fact, there does appear to be a decline in EPA's research interest in HAPs in recent years. Another issue is that the compounds on the HAP list have varying degrees of toxicity, e.g., benzene and toluene. But if they are both on the list and subject to the same controls, what is the incentive to use the least toxic agent? Also, nonlisted compounds are likely to be in the large universe of compounds for which there is less toxicological information and even more likelihood for unwanted surprise. Yet the HAP amendments may well push toward use of nonlisted compounds as it allows the industry source to avoid regulation. There is also a concern that the MACT standards, which have taken years and much debate to establish, will inhibit investment in even better technology—for example, air pollution control technology that might result from an advance in materials science. All of these concerns can be counteracted by provisions enacted into the 1990 HAP amendments—but whether they will be is uncertain. Accordingly, despite the *1990 Clean Air Act* HAP amendments likely leading to a decrease in tonnage of air pollutants emitted, it is still questionable whether there will be any significant lessening of toxicity directly caused by HAPs.

Registration Evaluation and Authorization of Chemicals (Reach)

Currently under consideration by the EU is a new far-reaching act that is firmly based upon the precautionary principle (Registration, Evaluation, and Authorisation of Chemicals 2005). It is a work in progress, with amendments being acted upon that may change the details. REACH does have provisions for prioritization of chemicals based upon exposure, or based upon hazard—but not based upon risk. In many ways, REACH is the antithesis of the *US Toxic Substances Control Act* through its requirement of a substantial amount of safety data for all chemicals, new or existing in commerce. Through this requirement it presumably will counteract the major problems we have had in the US with existing chemicals, such as the counterproductive move toward using MTBE and other oxygenated fuels without adequate toxicological testing. Also, through its insistence on the testing of all chemicals, REACH presumably will avoid the potential problem posed by unlisted chemicals in the HAP provisions. Like the *1990 US Clean Air Act* HAP amendments, REACH is justified by its proponents in part on the basis of frustration with the slowness of existing regulatory approaches. This expectation of a speedier regulatory response has not been borne out for the US HAP amendments—15 years after the passage of the act much remains unfinished, including the inevitable court cases.

Alan Boobis has pointed out that there are far fewer toxicologists in Europe than in the US, and has questioned whether there are sufficient toxicologists to accomplish the task. Accordingly, REACH may greatly increase the demand for toxicologists; although in my view, much of these funds would be better spent on developing the scientific basis for newer equivalents of the Ames test that would be effective in primary prevention. A similar criticism can be made of the current approach to testing high production volume chemicals.

Toxicology as Primary Prevention

Toxicology should be given more recognition as a primary precautionary approach. An excellent example is the Ames test, which was developed based upon scientific advances in understanding basic bacterial genetics; the role of mutagenesis in carcinogenesis; and the importance of liver metabolism in activating indirect carcinogens. To develop new consumer and industrial products, the chemical industry routinely uses the Ames test, and similar tests, to weed out potentially harmful chemicals from their development processes. The focusing of new chemical development and marketing on less harmful products is an excellent example of a primary preventive precautionary approach achieved through advances in toxicological sciences.

Unfortunately, the contributions of toxicology to primary precautionary approaches through safety assessment are not well understood by many who advocate the precautionary principle. Some are distrustful of scientists and are opposed to what they see as a technocracy that is controlling their lives. This is not surprising, in view of the frequency with which industry or government has used scientific uncertainty as a basis to stall needed actions. The justifiable concern is that without a precautionary approach, industry will operate under the CATNIP principle—Cheapest Available Technology Not Involving Prosecution.

Communication with Those Who Advocate the Precautionary Principle

Potentially problematic is the lack of communication between those active in the field of toxicology or environmental risk assessment and those most strongly advocating the precautionary principle. Of note is that of the 34 signers of the Wingspread Statement on the Precautionary Principle described above, none appear to have been members of the Society of Toxicology or of the Society for Risk Analysis. This lack of communication among different disciplines concerned with environmental protection may well limit understanding of the role of toxicological science in preventing disease. As this could be of grave consequence to our field, our reaching out to those who are advocates of the precautionary principle is particularly important. Let me emphasize that many, and perhaps most, active supporters of the precautionary principle are well aware of the role of science in preventing disease, and that there is a growing literature on the subject of science and the precautionary principle.

Conclusion

Among the major external threats to the field of toxicology are animal rights activists who would rather put humans and animal pets at risk rather than subject any laboratory animal to study; so called ethicists who seem opposed to any controlled experimental exposures of human subjects to chemical agents, even at levels at which we are exposed in our homes and in gasoline stations; the whole field of toxic torts; and, I would add, the precautionary principle. Each of these threats has their potentially valuable side in promoting the use of toxicology or developing new approaches—such as in vitro techniques that can replace laboratory animals. Similarly, although the precautionary principle can be a threat to the field of toxicology, it also challenges us to think carefully about the goals of toxicology and of the importance of broadly communicating about what we do and why we do it.

POSTSCRIPT

Is the Precautionary Principle a Sound Approach to Risk Analysis?

Ronald Bailey, in "Precautionary Tale," *Reason* (April 1999), defines the precautionary principle as "precaution in the face of any actions that may affect people or the environment, no matter what science is able—or unable—to say about that action." "No matter what science says" is not quite the same thing as "lack of full scientific certainty." Indeed, Bailey turns the precautionary principle into a straw man and thereby endangers whatever points he makes that are worth considering. One of those points is that widespread use of the precautionary principle would hamstring the development of the Third World. Roger Scruton, in "The Cult of Precaution," *National Interest* (Summer 2004), calls the Precautionary Principle "a meaningless nostrum" that is used to avoid risk and says it "clearly presents an obstacle to innovation and experiment," which are essential. Bernard D. Goldstein and Russellyn S. Carruth remind us in "Implications of the Precautionary Principle: Is It a Threat to Science?" *International Journal of Occupational Medicine and Environmental Health* (January 2004), that there is no substitute for proper assessment of risk. Jonathan Adler, in "The Precautionary Principle's Challenge to Progress," in Ronald Bailey, ed., *Global Warming and Other Eco-Myths* (Prima, 2002), argues that because the precautionary principle does not adequately balance risks and benefits, "the world would be safer without it." Peter M. Wiedemann and Holger Schutz, "The Precautionary Principle and Risk Perception: Experimental Studies in the EMF Area," *Environmental Health Perspectives* (April 2005), report that "precautionary measures may trigger concerns, amplify . . . risk perceptions, and lower trust in public health protection." Cass R. Sunstein, *Laws of Fear: Beyond the Precautionary Principle* (Cambridge, 2005), criticizes the precautionary principle in part because, he says, people overreact to tiny risks. John D. Graham, dean of the Frederick S. Pardee RAND Graduate School, argues in "The Perils of the Precautionary Principle: Lessons from the American and European Experience," Heritage Lecture #818 (delivered January 15, 2004, at the Heritage Foundation, Washington, D.C.), that the precautionary principle is so subjective that it permits "precaution without principle" and threatens innovation and public and environmental health. It must therefore be used cautiously.

The 1992 Rio Declaration emphasized that the precautionary principle should be "applied by States according to their capabilities" and that it should be applied in a cost-effective way. These provisions would seem to preclude the draconian interpretations that most alarm the critics. Yet, say David Kriebel et al., in "The Precautionary Principle in Environmental Science," *Environmental Health Perspectives* (September 2001), "environmental scientists should be aware of the policy uses of their work and of their social responsibility to do science that protects human health and the environment." Businesses are also con-flicted, writes Arnold Brown in "Suitable Precautions," *Across the Board* (January/February 2002), because the precautionary principle tends to slow decision making, but he maintains that "we will all have to learn and practice anticipation." Professor Goldstein mentions the European Union's REACH (Registration Evaluation and Authorization of Chemicals) program. It took effect on June 1, 2007. See http://ec.europa.eu/environment/chemicals/reach/reach_intro.htm for details.

ISSUE 2

Is Sustainable Development Compatible with Human Welfare?

YES: Jeremy Rifkin, from "The European Dream: Building Sustainable Development in a Globally Connected World," *E Magazine* (March/April 2005)

NO: Ronald Bailey, from "Wilting Greens," *Reason* (December 2002)

ISSUE SUMMARY

YES: Jeremy Rifkin, president of the Foundation on Economic Trends, argues that Europeans pride themselves on their quality of life, and their emphasis on sustainable development promises to maintain that quality of life into the future.

NO: Environmental journalist Ronald Bailey states that sustainable development results in economic stagnation and threatens both the environment and the world's poor.

Over the last 30 years, many people have expressed concerns that humanity cannot continue indefinitely to increase population, industrial development, and consumption. The trends and their impacts on the environment are amply described in numerous books, including historian J. R. McNeill's *Something New Under the Sun: An Environmental History of the Twentieth-Century World* (W. W. Norton, 2000).

"Can we keep it up?" is the basic question behind the issue of sustainability. In the 1960s and 1970s, this was expressed as the "Spaceship Earth" metaphor, which said that we have limited supplies of energy, resources, and room and that we must limit population growth and industrial activity, conserve, and recycle in order to avoid crucial shortages. "Sustainability" entered the global debate in the early 1980s, when the United Nations secretary general asked Gro Harlem Brundtland, a former prime minister and minister of environment in Norway, to organize and chair the World Commission on Environment and Development and produce a "global agenda for change." The resulting report, *Our Common Future* (Oxford University Press, 1987), defined *sustainable development* as "development that meets the needs of the present without compromising the ability of future generations to meet their own needs." It recognized that limits on population size and resource use cannot be known precisely; that problems may arise not suddenly but rather gradually, marked by rising costs; and that limits may be redefined by changes in technology. The report also recognized that limits exist and must be taken into account when governments, corporations, and individuals plan for the future.

The Brundtland report led to the UN Conference on Environment and Development held in Rio de Janeiro in 1992. The Rio conference set sustainability firmly on the global agenda and made it an essential part of efforts to deal with global environmental issues and promote equitable economic development. In brief, sustainability means such things as cutting forests no faster than they can grow back, using groundwater no faster than it is recharged by precipitation, stressing renewable energy sources rather than exhaustible fossil fuels, and farming in such a way that soil fertility does not decline. In addition, economics must be revamped to take into account environmental costs as well as capital, labor, raw materials, and energy costs. Many add that the distribution of the Earth's wealth must be made more equitable as well.

Given continuing growth in population and demand for resources, sustainable development is clearly a difficult proposition. Some think it can be done, but others think that for sustainability to work, either population or resource demand must be reduced. Not surprisingly, many people see sustainable development as in conflict with business and industrial activities, private property rights, and such human freedoms as the freedoms to have many children, to accumulate wealth, and to use the environment as one wishes. Economics professor Jacqueline R. Kasun, in "Doomsday Every Day: Sustainable Economics, Sustainable Tyranny," *The Independent Review* (Summer 1999), goes so far as to argue that sustainable development will require sacrificing human freedom, dignity, and material welfare on a road to tyranny. Lester R. Brown, "Picking up the Tab," *USA Today Magazine* (May 2007), suggests that because governments currently spend some $700 billion a year subsidizing environmentally destructive activities such as automobile driving, one essential step toward sustainability is to end the subsidies.

In the following selections, Jeremy Rifkin, president of the Foundation on Economic Trends, argues that Europeans pride themselves on a quality of life that in some ways exceeds that of Americans, and their emphasis on inclusivity, diversity, sustainable development, social rights, and individual human rights promises to maintain that quality of life into the future. Ronald Bailey argues that preserving the environment, eradicating poverty, and limiting economic growth are incompatible goals. Indeed, vigorous economic growth provides wealth for all and leads to environmental protection.

YES

Jeremy Rifkin

The European Dream: Building Sustainable Development in a Globally Connected World

A growing number of Americans are beginning to wonder why Europe has leaped ahead of the U.S. to become the most environmentally advanced political space in the world today. To understand why Europe has left America behind in the race to create a sustainable society, we need to look at the very different dreams that characterize the American and European frame of mind.

Ask Americans what they most admire about the U.S.A. and they will likely cite the individual opportunity to get ahead—at least until recently. The American Dream is based on a simple but compelling covenant: Anyone, regardless of the station to which they are born, can leverage a good public education, determination and hard work to become a success in life. We can go from "rags to riches."

Ask Europeans what they most admire about Europe and they will invariably say "the quality of life." Eight out of 10 Europeans say they are happy with their lives and when asked what they believe to be the most important legacy of the 20th century, 58 percent of Europeans picked their quality of life, putting it second only to freedom in a list of 11 legacies.

While the American Dream emphasizes individual success, the European Dream emphasizes collective well-being. The reason for this lies in the divergent histories of the two continents. America's founders came over from Europe 200 years ago in the waning days of the Protestant Reformation and the early days of the European Enlightenment. They took these two streams of European thought, froze them in time, and kept them alive in their purest form until today. Americans are the most devoutly Christian and Protestant people in the industrial world and the fiercest champions of the capitalist marketplace and the nation-state.

Both the Protestant Reformation and the Enlightenment emphasized the central role of the individual in history. John Calvin exhorted the faithful that every person stands alone with his or her God. Adam Smith, in turn, argued that all individuals pursue their own self interest in the marketplace. This individualist strain fit the American context far better than it did the European setting. In a wide-open frontier, every new immigrant did indeed stand alone and had to secure his or her survival with little or no social supports.

Americans, even today, are taught by his or her parents that to be free they must learn to be self-sufficient and independent, and that they cannot depend on others.

Europeans, however, never fully bought the idea of the individual alone in the universe. Europe was already densely populated and without a frontier by the late 18th century. Walled cities and tightly packed human settlement demanded a more communal way of life. While Americans defined freedom in terms of individual autonomy and mobility, Europeans defined freedom by their communal relationships.

In America there was enough cheap and free land and resources so that newcomers could become rich. In Europe, well-defined class boundaries—a remnant of the feudal aristocracy—made it far more difficult for an individual born in a lesser station of life to rise to the top and become wealthy. So while Americans preferred to pursue happiness individually, Europeans pursued happiness collectively by emphasizing the quality of life of the community. Today, Americans devote less than 11 percent of their Gross Domestic Product (GDP) to social benefits, compared to 26 percent in Europe.

Doing It Better

So, what does Europe do better than America? It works hard to create a remarkably high quality of life for all of its people. The European Dream focuses on inclusivity, diversity, sustainable development, social rights and universal human rights. And it works. While Americans are 28 percent wealthier per capita than Europeans, in many ways, Europeans experience a higher quality of life, clear evidence that, in the long run, cooperation rather than competition is sometimes a surer path to happiness.

Europe and the U.S. have nearly opposite approaches to the question of environmental stewardship. At the heart of the difference is the way Americans and Europeans perceive risk. We Americans take pride in being a risk-taking people. We come from immigrant stock, people who risked their lives to journey to the new world and start over, often with only a few coins in their pockets and a dream of a better life. When Europeans and others are asked what they most admire about Americans, our risk-taking, "can-do" attitude generally tops the list. Where others see difficulties and obstacles, Americans see opportunities.

Our optimism is deeply entwined with our faith in science and technology. It has been said that Americans are a nation of tinkerers. When I was growing up, the engineer was held in as high esteem as the cowboy, admired for his efforts to improve the lot of society and contribute to the progress and welfare of civilization.

On the other side of the water, the sensibilities are different. It's not that Europeans aren't inventive. One could make the case that over the course of history Europe has produced most of the great scientific insights and not a few of the major inventions. But with their longer histories, Europeans are far more mindful of the dark side of science and technology.

Saying No to GE Foods

In recent years, the European Union (EU) has turned upside down the standard operating procedure for introducing new technologies and products into the marketplace and society, much to the consternation of the United States. The turnaround started with the controversy over genetically engineered (GE) foods and the introduction of genetically modified organisms (GMOs). The U.S. government gave the green light to the widespread introduction of GE foods in the mid 1990s, and by the end of the decade more than half of America's agricultural land was given over to GE crops. No new laws were enacted to govern the potential harmful environmental and health impacts. Instead, existing statutes were invoked, and no special handling or labeling of the products was required.

In Europe, massive opposition to GMOs erupted across the continent. Farmers, environmentalists and consumer organizations staged protests and political parties and governments voiced concern. A defacto moratorium on the planting of GE crops and sale of GE food products was put into effect. Meanwhile, the major food processors, distributors and retailers pledged not to sell any products containing GE traits.

The EU embarked on a lengthy review process to assess the environmental and health risks of introducing GE food products. In the end, it established tough new protections designed to mitigate the potential harm. The measures included procedures to segregate and track GE grain and food products from the fields to the retail stores to ensure against contamination; labeling of GMOs at every stage of the food process to ensure transparency; and independent testing as well as more rigorous testing requirements by the companies producing GE seeds and other GMOs.

The EU is forging ahead on a wide regulatory front, changing the very conditions and terms by which new scientific and technological pursuits and products are introduced into the marketplace and the environment. Its bold initiatives put the EU far ahead of the rest of the world. Behind all of its newfound regulatory zeal is the looming question of how best to model global risks and create a sustainable and transparent approach to economic development.

Ensuring Safety

In May of 2003, the EU proposed sweeping new regulatory controls on chemicals to mitigate toxic impacts on the environment and human and animal health. The proposed new law would require new companies to register and test for the safety of more than 30,000 chemicals at an estimated cost to the producers of nearly eight billion Euros. Under existing rules, 99 percent of the total volume of chemicals sold in Europe have not passed through any environmental and health testing and review process. In the past, there was no way to even know what kind of chemicals were being used by industry, making it nearly impossible to track potential health risks. The new regulations will change all of that. The "REACH" system—which stands for Registration, Evaluation and Authorization of Chemicals—requires the companies to conduct safety and environmental tests to prove that the products they are producing are safe. If they can't, the products will be banned from the market.

The new procedures represent an about face to the way the chemical industry is regulated in the U.S. In America, new chemicals are generally assessed to be safe and the burden is primarily put on the consumer, the public or the government to prove that they cause harm. The EU has reversed the burden of proof. Former EU Environmental Commissioner Margot Wallstrom makes the point: "No longer do public authorities need to prove they [the products] are dangerous. The onus is now on industry to prove that the products are safe."

Making companies prove that their chemical products are safe before they are sold is a revolutionary change. It's very difficult to conceive of the U.S. entertaining the kind of risk prevention regulatory regime that the EU has rolled out. In a country where corporate lobbyists spend millions of dollars influencing congressional legislation, the chances of ever having a similar regulatory regime to the one being implemented in Europe would be nigh on impossible.

BUILDING THE HYDROGEN ECONOMY

At the very top of the list of environmental priorities for the EU is the plan to become a fully integrated renewable-based hydrogen economy by mid-century. The EU has led the world in championing the Kyoto Protocol on Climate Change, and to ensure compliance it has made a commitment to produce 22 percent of its electricity and 12 percent of all of its energy using renewable sources by 2010. Although a number of member states are lagging behind on meeting their renewable energy targets, the very fact that the EU has set benchmarks puts it far ahead of the U.S. in making the shift from fossil fuels to renewable energy sources. The Bush administration has consistently fought Congressional attempts to establish similar benchmarks for ushering in a U.S.-based renewable energy regime.

In June of 2003, EU President Romano Prodi said, "It is our declared goal of achieving a step-by-step shift toward a fully integrated hydrogen economy, based on renewable energy sources, by the middle of the century." He added that creating this economy would be the next critical step in integrating Europe after the introduction of the Euro.

The European hydrogen game plan is being implemented with a sense of history in mind. Great Britain became the world's leading power in the 19th century because it was the first country to harness its vast coal reserves with steam power. The U.S., in turn, became the world's preeminent power in the 20th century because it was the first country to harness its vast oil reserves with the internal-combustion engine. The multiplier effects of both energy revolutions were extraordinary. The EU is determined to lead the world into the third great energy revolution of the modern era.—*J.R.*

GMOs and chemical products represent just part of the new "risk prevention" agenda taking shape in Brussels. In early 2003, the EU adopted a new rule prohibiting electronics manufacturers from selling products in the EU that contain mercury, lead and other heavy metals. Another new regulation requires the manufacturers of all consumer electronics and household appliances to cover the costs for recycling their products. American companies complain that compliance with the new regulations will cost them hundreds of millions of dollars a year.

All of these strict new rules governing risk prevention would come as a shock to Americans who believe that the U.S. has the most vigilant regulatory oversight regime in the world for governing risks to the environment and public health. Although that was the case 30 years ago, it no longer is today.

The EU is the first governing institution in history to emphasize human responsibilities to the global environment as a centerpiece of its political vision. Europe's new sensitivity to global risks has led it to champion the Kyoto Protocol on climate change, the Biodiversity Treaty, the Chemical Weapons Convention and many others. The U.S. government has refused, to date, to ratify any of the above agreements.

A New Era

In Europe, intellectuals are increasingly debating the question of the great shift from a risk-taking age to a risk-prevention era. That debate is virtually non-existent in the U.S., where risk-taking is seen as a virtue. The new European intellectuals argue that vulnerability is the underbelly of risks. A sense of vulnerability can motivate people to band together in common cause. The EU stands as a testimonial to collective political engagement arising from a sense of risk and shared vulnerability.

What's changed qualitatively in the last half century since the dropping of the atomic bombs on Hiroshima and Nagasaki is that risks are now global in scale, open ended in duration, incalculable in their consequences and not compensational. Their impact is universal, which means that no one can escape their potential effects. Risks have now become truly democratized, making everyone vulnerable. When everyone is vulnerable, then traditional notions of calculating and pooling risks become virtually meaningless. This is what European academics call a risk society.

Americans aren't there yet. While some academics speak to global risks and vulnerabilities and a significant minority of Americans express their concerns about global risks, from climate change to loss of biodiversity, the sense of utter vulnerability just isn't as strong on this side of the Atlantic. Europeans say we have blinders on. In reality, it's more nuanced than that. Call it delusional, but the sense of personal empowerment is so firmly embedded in the American mind, that even when pitted against growing evidence of potentially overwhelming global threats, most Americans shrug such notions off as overly pessimistic and defeatist. "Individuals can move mountains." Most Americans believe that. Fewer Europeans do.

The EU has already institutionalized a litmus test that cuts to the core of the differences between America and Europe. It's called "the precautionary principle" and it has become the centerpiece of EU regulatory policy governing science and technology in a globalizing world.

The Precautionary Principle

In November 2002, the EU adopted a new policy on the use of the precautionary principle to regulate science and new products derived from technology innovations. According to the EU, reviews occur in "cases where scientific evidence is insufficient, inconclusive or uncertain and preliminary scientific evaluation indicates that there are reasonable grounds for concern that the potentially dangerous effects on the environment, human, animal or plant health may be

---

THE TRANSITION TO ORGANIC AGRICULTURE

Europe is taking the lead in the shift to sustainable farming practices and organic food production. While the organic food sector is soaring in the U.S.—it represents the fastest-growing segment of the food industry—the government has done little to encourage it. Although the U.S. Department of Agriculture fields a small organic food research program, it amounts to only $3 million, less than .004 percent of its $74 billion budget. While American consumers are increasing their purchases of organic food, less than 0.3 percent of total U.S. farmland is currently in organic production.

By contrast, many of the EU member states have made the transition to organic agriculture a critical component of their economic development plans and have even set benchmarks. Germany, which has often been the leader in setting new environmental goals for the continent, has announced its intention to bring 20 percent of its agricultural output into organic production by the year 2020. (Organic agricultural output is now 3.2 percent of all farm output in Germany.)

The Netherlands, Sweden, Great Britain, Finland, Norway, Germany, Switzerland, Denmark, France and Austria also have national programs to promote the transition to organic food production. Denmark and Sweden enjoy the highest consumption of organic vegetables in Europe and both countries project that their domestic markets for organic food will soon reach or exceed 10 percent of domestic consumption.

Sweden has set a goal of having 20 percent of its total cultivated farm area in organic production by 2005. Italy already has 7.2 percent of its farmland under organic production while Denmark is close behind with seven percent.

Great Britain doubled its organic food production in 2002 and now boasts the second-highest sales of organic food in Europe, after Germany. According to a recent survey, nearly 80 percent of British households buy organic food. By comparison, only 33 percent of American consumers buy any organic food.—J.R.

inconsistent with the high level of protection chosen by the EU." The key term is "uncertain." When there is sufficient evidence to suggest a potential negative impact, but not enough to know for sure, the precautionary principle allows regulatory authorities to err on the side of safety. They can suspend the activity altogether, modify it, employ alternative scenarios, monitor the activity or create experimental protocols to better understand its effects.

The precautionary principle allows governments to respond with a lower threshold of scientific certainty than in the past. "Scientific certainty" has been tempered by the notion of "reasonable grounds for concern." The precautionary principle gives authorities the flexibility to respond to events in real time, either before or while they are unfolding.

Advocates of the precautionary principle cite the introduction of halocarbons and the tear in the ozone hole in the Earth's upper atmosphere, the outbreak of mad cow disease in cattle, growing antibiotic resistant strains of bacteria caused by the over-administering of antibiotics to farm animals and the widespread deaths caused by asbestos, benzene and polychlorinated biphenyls (PCBs).

The precautionary principle has been finding its way into international treaties and covenants. It was first recognized in 1982 when the United Nations General Assembly incorporated it into the World Charter for Nature. The precautionary principle was subsequently included in the Rio Declaration on Environment and Development in 1992, the Framework Convention on Climate Change in 1992, the Treaty on EU (Maastricht Treaty) in 1992, the Cartagena Protocol on Biosafety in 2000 and the Stockholm Convention on Persistent Organic Pollutants (POPs) in 2001.

Valuing Nature

Americans, by and large, view nature as a treasure trove of useful resources waiting to be harnessed for productive ends. While Europeans share America's utilitarian perspective, they also have a love for the intrinsic value of nature. One can see it in Europeans' regard for the countryside and their determination to maintain natural landscapes, even if it means providing government assistance in the way of special subsidies, or foregoing commercial development. Nature figures prominently in Europeans' dream of a quality of life. Europeans spend far more time visiting the countryside on weekends and during their vacations than Americans.

The balancing of urban and rural time is less of a priority for most Americans, many of whom are just as likely to spend their weekends at a shopping mall, while their European peers are hiking along country trails. Anyone who spends significant time among Europeans knows that they have a great affinity for rural getaways. Almost everyone I know in Europe—among the professional and business class—has some small second home in the country somewhere—a dacha usually belonging to the family for generations. While working people may not be as fortunate, on any given weekend they can be seen exiting the cities en masse, motoring their way into the nearest rural enclave or country village for a respite from urban pressures.

The strongly held values about rural life and nature is one reason why Europe has been able to support green parties across the continent, with substantial representation in national parliaments as well as in the European Parliament. By contrast, not a single legislator at the federal level in the U.S. is a member of the Green Party.

There is another dimension to the European psyche that makes Europeans supportive of the precautionary principle—their sense of "connectedness."

Because we Americans place such a high premium on autonomy, we are far less likely to see the deep connectedness of things. We tend to see the world in terms of containers, each isolated from the whole and capable of standing alone. We like everything around us to be neatly bundled, autonomous, and self contained. The new view of science that is emerging in the wake of globalization is quite different. Nature is viewed as a myriad of symbiotic relationships, all embedded in a larger whole, of which they are an integral part. In this new vision of nature, nothing is autonomous, everything is connected.

By championing a host of global environmental treaties and accords taking the precautionary approach to regulation, the EU has shown a willingness to act on its commitment to sustainable development and global environmental stewardship. The fact that its commitments in most areas remain weak and are often vacillating is duly noted. But, at least Europe has established a new agenda for conducting science and technology that, if followed, could begin to wean the world from the old ways and toward a second Enlightenment.

Ronald Bailey

 **NO**

Wilting Greens

It's clear that we've suffered a number of major defeats," declared Andrew Hewett, executive director of Oxfam Community Aid, at the conclusion of the World Summit on Sustainable Development, held in Johannesburg, South Africa, in September. Greenpeace climate director Steve Sawyer complained, "What we've come up with is absolute zero, absolutely nothing." The head of an alliance of European green groups proclaimed, "We barely kept our heads above water."

It wasn't supposed to be this way. Environmental activists hoped the summit would set the international agenda for sweeping environmental reform over the next 15 years. Indeed, they hoped to do nothing less than revolutionize how the world's economy operates. Such fundamental change was necessary, said the summiteers, because a profligate humanity consumes too much, breeds too much, and pollutes too much, setting the stage for a global ecological catastrophe.

But the greens' disappointment was inevitable because their major goals—preserving the environment, eradicating poverty, and limiting economic growth—are incompatible. Economic growth is a prerequisite for lessening poverty, and it's also the best way to improve the environment. Poor people cannot afford to worry much about improving outdoor air quality, let alone afford to pay for it. Rather than face that reality, environmentalists increasingly invoke "sustainable development." The most common definition of the phrase comes from the 1987 United Nations report *Our Common Future:* development that "meets the needs of the present without compromising the ability of future generations to meet their own needs."

For radical greens, sustainable development means economic stagnation. The Earth Island Institute's Gar Smith told Cybercast News, "I have seen villages in Africa . . . that were disrupted and destroyed by the introduction of electricity." Apparently, the natives no longer sang community songs or sewed together in the evenings. "I don't think a lot of electricity is a good thing," Smith added. "It is the fuel that powers a lot of multinational imagery." He doesn't want poor Africans and Asians "corrupted" by ads for Toyota and McDonald's, or by Jackie Chan movies.

From *Reason*, vol. 34, no. 7, December 2002, pp. Full text. Copyright © 2002 by Reason Foundation, 3415 S. Sepulveda Blvd., Suite 400, Los Angeles, CA 90034. Reprinted by permission. www.reason.com

Indian environmentalist Sunita Narain decried the "pernicious introduction of the flush toilet" during a recent PBS/BBC television debate hosted by Bill Moyers. Luckily, most other summiteers disagreed with Narain's curious disdain for sanitation. One of the few firm goals set at the confab was that adequate sanitation should be supplied by 2015 to half of the 2.2 billion people now lacking it.

Sustainable development boils down to the old-fashioned "limits to growth" model popularized in the 1970s. Hence Daniel Mittler of Friends of the Earth International moaned that "the summit failed to set the necessary economic and ecological limits to globalization." The *Jo'burg Memo*, issued by the radical green Heinrich Böll Foundation before the summit, summed it up this way: "Poverty alleviation cannot be separated from wealth alleviation."

The greens are right about one thing: The extent of global poverty is stark. Some 1.1 billion people lack safe drinking water, 2.2 billion are without adequate sanitation, 2.5 billion have no access to modern energy services, 11 million children under the age of 5 die each year in developing countries from preventable diseases, and 800 million people are still malnourished, despite a global abundance of food. Poverty eradication is clearly crucial to preventing environmental degradation, too, since there is nothing more environmentally destructive than a hungry human.

Most summit participants from the developing world understood this. They may be egalitarian, but unlike their Western counterparts they do not aim to make everyone equally poor. Instead, they want the good things that people living in industrialized societies enjoy.

That explains why the largest demonstration during the summit, consisting of more than 10,000 poor and landless people, featured virtually no banners or chants about conventional environmentalist issues such as climate change, population control, renewable resources, or biodiversity. Instead, the issues were land reform, job creation, and privatization.

The anti-globalization stance of rich activists widens this rift. Environmentalists claim trade harms the environment and further impoverishes people in the developing world. They were outraged by the dominance of trade issues at the summit.

"The leaders of the world have proved that they work as employees for the transnational corporations," asserted Friends of the Earth Chairman Ricardo Navarro. Indian eco-feminist Vandana Shiva added, "This summit has become a trade summit, it has become a trade show." Yet the U.N.'s own data underscore how trade helps the developing world. As fact sheets issued by the U.N. put it, "During the 1990s the economies of developing countries that were integrated into the world economy grew more than twice as fast as the rich countries. The 'non-globalizers' grew only half as fast and continue to lag further behind."

By invoking a zero sum version of sustainable development, environmentalists not only put themselves at odds with the developing world; they ignore the way in which economic growth helps protect the environment. The real commons from which we all draw is the growing pool of scientific, technological, and institutional concepts, and the capital they create. Past generations have left us far more than they took, and the result has been an explosion in human well-being, longer life spans, less disease, more and cheaper food, and expanding political freedom.

Such progress is accompanied by environmental improvement. Wealthier is healthier for both people and the environment. As societies become richer and more technologically adept, their air and water become cleaner, they set aside more land for nature, their forests expand, they use less land for agriculture, and more people cherish wild species. All indications suggest that the 21st century will be the century of ecological restoration, as humanity uses physical resources ever more efficiently, disturbing the natural world less and less.

In their quest to impose a reactionary vision of sustainable development, the disappointed global greens will turn next to the World Trade Organization, the body that oversees international trade rules. During the summit, the WTO emerged as the greens' bête noire. As Friends of the Earth International's Daniel Mittler carped, "Instead of using the [summit] to respond to global concerns over deregulation and liberalization, governments are pushing the World Trade Organization's agenda." "See you in Cancun!" promised Greenpeace's Steve Sawyer, referring to the location of the next WTO ministerial meeting in September 2003. That confab will build on the WTO's Doha Trade Round, launched last year, which is aimed at reducing the barriers to trade for the world's least developed countries.

The WTO may achieve worthy goals that eluded the Johannesburg summit, such as eliminating economically and ecologically ruinous farm and energy subsidies and opening developed country markets to the products of developing nations. Free marketeers and greens might even form an alliance on those issues.

But environmentalists want to use the WTO to implement their sustainable development agenda: global renewable energy targets, regulation based on the precautionary principle, a "sustainable consumption and production project," a worldwide eco-labeling scheme. According to Greenpeace's Sawyer, nearly everyone at the Johannesburg summit agreed "there is something wrong with unbridled neoliberal capitalism."

Let's hope the greens fail at the WTO just as they did at the U.N. summit. Their sustainable development agenda, supposedly aimed at improving environmental health, instead will harm the natural world, along with the economic prospects of the world's poorest people. The conflicting goals on display at the summit show that at least some of the world's poor are wise to that fact.

POSTSCRIPT

Is Sustainable Development Compatible with Human Welfare?

The first of the Rio Declaration's 22 principles states, "Human beings are at the centre of concerns for sustainable development. They are entitled to a healthy and productive life in harmony with nature." Any solution to the sustainability problem therefore should not infringe human welfare. This makes any solution that involves limiting or reducing human population or blocking improvements in standard of living very difficult to sell. Yet solutions may be possible. David Malin Roodman suggests in *The Natural Wealth of Nations: Harnessing the Market for the Environment* (W. W. Norton, 1998) that taxing polluting activities instead of profit or income would stimulate corporations and individuals to reduce such activities or to discover nonpolluting alternatives. In "Building a Sustainable Society," *State of the World 1999* (W. W. Norton, 1999), he adds recommendations for citizen participation in decision making, education efforts, and global cooperation, without which we are heading for "a world order [that] almost no one wants." (He is referring to a future of environmental crises, not the "new world order" feared by many conservatives, in which national policies are dictated by international [UN] regulators.) Roodman's recommendations may actually be on the way to reality. Arun Agrawal and Maria Carmen Lemos, "A Greener Revolution in the Making? Environmental Governance in the 21st Century," *Environment* (June 2007), argue that budget cuts and globalization are eroding the power of the state in favor of international "hybrid" arrangements that stress public-private partnerships, markets, and community and local participation.

Julie Davidson, in "Sustainable Development: Business as Usual or a New Way of Living?" *Environmental Ethics* (Spring 2000), notes that efforts to achieve sustainability cannot by themselves save the world. But such efforts may give us time to achieve new and more suitable values. It is thus heartening to see that the UN World Summit on Sustainable Development was held in Johannesburg, South Africa, in August 2002. Its aim was to strengthen partnerships between governments, business, nongovernmental organizations, and other stakeholders and to seek to eradicate poverty and make more equal the distribution of the benefits of globalization. See Gary Gardner, "The Challenge for Johannesburg: Creating a More Secure World," *State of the World 2002* (W. W. Norton, 2002), and the United Nations Environmental Programme's Global Environmental Outlook 3 (Earthscan, 2002), prepared as a "global state of the environment report" in preparation for the Johannesburg Summit.

The World Council of Churches brought to the Johannesburg Summit an emphasis on social justice. Martin Robra, in "Justice—The Heart of Sustainability," *Ecumenical Review* (July 2002), writes that the dominant stress on economic growth "has served, first and foremost, the interests of the powerful economic players. It has further marginalized the poor sectors of society, simultaneously undermining their basic security in terms of access to land, water, food, employment, and other basic services and a healthy environment."

Is social justice or equity worth this emphasis? Or is sustainability more a matter of population control, of shielding the natural environment from human impacts, or of economics? A. J. McMichael, C. D. Butler, and Carl Folke, in "New Visions for Addressing Sustainability," *Science* (December 12, 2003), argue that it is wrong to separate—as did the Johannesburg Summit—achieving sustainability from other goals such as reducing fertility and poverty and improving social equity, living conditions, and health. They observe that human population and lifestyle affect ecosystems, ecosystem health affects human health, human health affects population and lifestyle. "A more integrated . . . approach to sustainability is urgently needed," they say, calling for more collaboration among researchers and other fields. Yet there remains reason to focus on single threats. In February 2007, Sigma Xi and the United Nations Foundation released the Scientific Expert Group Report on Climate Change and Sustainable Development, *Confronting Climate Change: Avoiding the Unmanageable and Managing the Unavoidable* (http://www.sigmaxi. org/about/news/UNSEGReport.shtml) (Executive Summary, *American Scientist*, February 2007). Among its many points is that climate change from global warming is a huge threat to sustainability. Even in spite of feasible attempts at mitigation and adaptation, there is a serious risk of "intolerable impacts on human well-being."

Anthony R. Leiserowitz, Robert W. Kates, and Thomas M. Parris, "Do Global Attitudes and Behaviors Support Sustainable Development?" *Environment* (November 2005), find that though the world's people appear to support the component concepts of sustainable development, there is a mismatch between that support and their behavior. In the long term, they say, what is needed is a "shift from materialist to post-materialist values, from anthropocentric to ecological worldview, and a redefinition of the good life." Unfortunately, that shift remains in the future. P. Aarne Vesilind, Lauren Heine, and Jamie Hendry, "The Moral Challenge of Green Technology," *TRAMES: A Journal of the Humanities & Social Sciences* (No.1, 2006), "conclude that the unregulated free market system is incompatible with our search for sustainability. Experience has shown that if green technology threatens profits, green technology loses and profitability wins."

ISSUE 3

Should a Price Be Put on the Goods and Services Provided by the World's Ecosystems?

YES: John E. Losey and Mace Vaughan, from "The Economic Value of Ecological Services Provided by Insects," *BioScience* (April 2006)

NO: Marino Gatto and Giulio A. De Leo, from "Pricing Biodiversity and Ecosystem Services: The Never-Ending Story," *BioScience* (April 2000)

ISSUE SUMMARY

YES: John E. Losey and Mace Vaughan argue that even conservative estimates of the value of the services provided by wild insects are enough to justify increased conservation efforts. They say that "everyone would benefit from the facilitation of the vital services these insects provide."

NO: Professors of applied ecology Marino Gatto and Giulio A. De Leo contend that the pricing approach to valuing nature's services is misleading because it falsely implies that only economic values matter.

Human activities frequently involve trading a swamp or forest or mountainside for a parking lot or housing development or farm. People generally agree that these developments are worthwhile projects, for they have obvious benefits. But are there costs as well? Construction costs, labor costs, and material costs can easily be calculated, but what about the swamp? The forest? The species living there?

How much is a species worth? One approach to answering this question is to ask people how much they would be willing to pay to keep a species alive. If the question is asked when there are a million species in existence, few people will likely be willing to pay much. But if the species is the last one remaining, they might be willing to pay a great deal. Most people would agree that both answers fail to get at the true value of a species, for nature is not expressible solely in terms of cash values. Yet some way must be found to weigh the effects of human activities on nature against the benefits gained from those activities. If it is not, we will continue to degrade the world's ecosystems and threaten our own continued well-being.

Traditional economics views nature as a "free good." That is, forests generate oxygen and wood, clouds bring rain, and the sun provides warmth, all without charge to the humans who benefit. At the same time, nature has provided ways for people to dispose of wastes—such as dumping raw sewage into rivers or emitting smoke into the air—without paying for the privilege. This "free" waste disposal has turned out to have hidden costs in the form of the health effects of pollution (among other things), but it has been up to individuals and governments to bear the costs associated with those effects. The costs are real, but in general, they have not been borne by the businesses and other organizations that produced them. They have thus come to be known as "external" costs.

Environmental economists have recognized the problem of external costs, and government regulators have devised a number of ways to make those who are responsible accept the bill, such as instituting requirements for pollution control and fining those who exceed permitted emissions. Yet some would say that this approach does not help enough.

The *ecosystem services* approach recognizes that undisturbed ecosystems do many things that benefit us. A forest, for instance, slows the movement of rain and snowmelt into streams and rivers; if the forest is removed, floods may follow (a connection that recently forced China to deemphasize forest exploitation). Swamps filter the water that seeps through them. Food chains cycle nutrients necessary for the production of wood and fish and other harvests. Bees pollinate crops and make food production possible. These services are valuable—even essential—to us, and anything that interferes with them must be seen as imposing costs just as significant as the illnesses associated with pollution.

How can those costs be assessed? In 1997 Robert Costanza and his colleagues published an influential paper entitled "The Value of the World's Ecosystem Services and Natural Capital" in the May 15 issue of the journal *Nature*. In it, the authors listed a variety of ecosystem services and attempted to estimate what it would cost to replace those services if they were somehow lost. The total bill for the entire biosphere came to $33 trillion (the middle of a $16–54 trillion range), compared to a global gross national product of $25 trillion. Costanza et al. stated that this was surely an underestimate. Janet N. Abramovitz, "Putting a Value on Natures 'Free' Services," *WorldWatch* (January/February 1998), argues that nature's services are responsible for the vast bulk of the value in the world's economy and that attaching economic value to those services may encourage their protection.

In the following selections, John E. Losey, associate professor of entomology at Cornell University, and Mace Vaughan, Conservation Director at the Xerces Society for Invertebrate Conservation, argue that even conservative estimates of the value of the services provided by wild insects are enough to justify increased conservation efforts. Ecologists Marino Gatto and Giulio A. De Leo argue that the pricing approach to valuing nature's services is misleading because it ignores equally important "nonmarket" values.

YES

John E. Losey and
Mace Vaughan

The Economic Value of Ecological Services Provided by Insects

Natural systems provide ecological services on which humans depend. Countless organisms are involved in these complex interactions that put food on our tables and remove our waste. Although human life could not persist without these services, it is difficult to assign them even an approximate economic value, which can lead to their conservation being assigned a lower priority for funding or action than other needs for which values (economic or otherwise) are more readily calculated. Estimating even a minimum value for a subset of the services that functioning ecosystems provide may help establish a higher priority for their conservation.

In this article we focus on the vital ecological services provided by insects. Several authors have reviewed the economic value of ecological services in general, but none of these reviews focused specifically on insects. Insects comprise the most diverse and successful group of multicellular organisms on the planet, and they contribute significantly to vital ecological functions such as pollination, pest control, decomposition, and maintenance of wildlife species . . . Our twofold goal is to provide well-documented, conservative estimates for the value of these services and to establish a transparent, quantitative framework that will allow the recalculation of the estimates as new data become available. We also should clarify that by "value" we mean documented financial transactions—mostly the purchase of goods or services—that rely on these insect-mediated services.

We restrict our focus to services provided by "wild" and primarily by native insects; we do not include services from domesticated species (e.g., pollination from domesticated honey bees) or pest control from mass-reared insect biological-control agents (e.g., *Trichogramma* wasps). We also exclude the value of commercially produced insect-derived products, such as honey, wax, silk, or shellac, and any value derived from the capture and consumption of insects themselves. The main reasons for these exclusions are that domesticated insects that provide services or products have been covered in many other forums, and they generally do not require the active conservation that we believe is warranted by those undomesticated insects that provide services. Furthermore, in the case of products or food derived directly from wild insects, we simply do not have data to report and therefore wish to maintain a focus on ecological services.

From *BioScience,* by John E. Losey and Mace Vaughan, vol. 56, no. 4, April 2006, pp. 311–316, 318–322. Copyright © 2006 by American Institute of Biological Sciences. Reprinted by permission of American Institute of Biological Sciences via the Copyright Clearance Center.

The four insect services for which we provide value estimates were chosen not because of their importance, but because of the availability of data and an algorithm for their calculation. Three of these services (dung burial, pest control, and pollination) support the production of a commodity that has a quantifiable, published value. To be consistent in our analysis for all three of these commodities, we calculated an estimate for the amount of each commodity that depends on each service or on the amount saved in related expenses (e.g., the cost of fertilizer in our analysis of dung burial). We did not perform an in-depth analysis of how service-dependent changes in the quantity or quality of each commodity may have affected its per-unit price.

One way of looking at the economic implications of the removal of a service was provided by Southwick and Southwick, whose study involved crop pollination by honey bees. Because per-unit cost theoretically increases as supplies decrease, thus mitigating monetary losses, the costs of the service removal in the Southwick and Southwick study were lower than those calculated using our approach. However, all reported values are still within an order of magnitude of each other and, although our approach may not reflect what a consumer would pay for a commodity when these ecological services are *not* being performed, our calculations do provide a measure of the value of these crops at current estimated levels of service.

In the case of insect support of wildlife nutrition, we use a different approach to estimate costs. Instead of basing calculations on the money paid to producers for raw commodities, we use census data to find out how US consumers spent their money. By looking at the consumer end of this system, we immediately see an order-of-magnitude increase in the value reported. We believe it is important for this difference to be understood up front, because it both significantly affects our reported results and provides at least a hint of what happens when raw commodities are converted into value-added products. For example, consumers will spend potentially an order of magnitude more on jellies, pasta sauce, or hamburgers than the price paid to producers for blueberries, tomatoes, or beef.

Using the methods we describe in detail in the following sections, we estimate the annual value of four ecological services provided by primarily native insects in the United States to be more than $57 billion ($0.38 billion for dung burial, $3.07 billion for pollination, $4.49 billion for pest control of native herbivores, and $49.96 billion for recreation). We consider this estimate very conservative. If data were available to support more accurate estimates of the true value of these services (e.g., inclusion of value-added products and wages paid to those who produce such products) or to allow estimation of the value of other services, the results of our calculations would be much higher. In addition to the role of insects in the systems we analyze here, other potentially important services that insects provide could not be quantified, including suppression of weeds and exotic herbivorous species, facilitation of dead plant and animal decomposition, and improvement of the soil. Calculating the value of any of these services could add billions of dollars to our overall estimate. Nevertheless, we hope that even this minimum estimate for a subset of services provided by insects will allow these animals to be more correctly factored into land management and legislative decisions. In the following sections, we present a detailed description of how we calculated these estimates and discuss the implications of our results.

Dung Burial

Confining large mammals in small areas creates challenging waste-management problems. Cattle production in the United States provides a particularly pertinent example, because nearly 100 million head of cattle are in production, and each animal can produce over 9000 kilograms (kg), or about 21 cubic meters, of solid waste per year. Fortunately, insects—especially beetles in the family Scarabaeidae—are very efficient at decomposing this waste. In doing so, they enhance forage palatability, recycle nitrogen, and reduce pest habitat, resulting in significant economic value for the cattle industry.

Dung beetles process a substantial amount of the cattle dung accumulated annually in the United States. . . .

The importance of this service is illustrated by the success of dung beetles introduced into Australia to deal with the dung of nonnative cattle brought to that continent in 1788. Before the introduction of dung beetle species that were adapted to feed on cattle dung, Australia had no insect fauna to process cattle feces. Consequently, rangeland across the country was fouled by slowly decomposing dung. In addition, this dung provided fodder for pest species. Recent research in western Australia has revealed that populations of the pestiferous bush fly (*Musca vetustissima*) have been reduced by 80% following dung beetle introductions. . . .

Using data from Anderson and colleagues, we calculate that the average persistence—or time until complete decomposition—of an untreated dung pat on rangeland in California is 22.74 ± 0.64 months, while the average persistence of a pat treated with insecticides is 28.14 ± 0.71 months. This indicates that dung beetle activity results in a 19% decrease in the amount of time the average pat of dung makes forage unpalatable, which translates into substantial monetary savings. Note that, for the sake of this analysis, we must assume that the 19% decrease applies broadly across the United States, even though the rate of dung burial by beetles probably varies greatly depending upon the location.

Forage fouling.   Fincher estimated a potential value for enhanced palatability based on the concept that cattle will not consume plant material that is fouled with dung. If dung beetles were totally absent, forage fouling by dung would cause estimated annual losses of 7.63 kg of beef per head of cattle (L_{nb}). This level of loss is in comparison with the theoretical zero loss of production if no forage were ever fouled by dung. Fortunately, the cattle industry is not saddled with the full force of this potential loss because range fouling is reduced by the current action of dung beetles.

If we assume that the 19% decrease in dung persistence translates into a 19% decrease in lost beef, then, for cattle whose dung is processed by dung beetles, the per-animal loss would be 6.18 kg (L_b) each year as a result of forage fouling. This assumption seems justified, since for each increment of time a given patch of forage remains fouled, it also remains unavailable for grazing. By applying these estimated losses to the 32 million head that are untreated and on pasture or rangeland, we estimate that in the absence of dung beetles, beef losses due to forage fouling would be 244 million kg of beef per year ($C_p \times L_{nb}$), whereas losses at current levels of dung beetle function would be 198 million kg ($C_p \times L_b$). With an average price over 34 years (1970–2003, corrected for inflation) of live beef cattle at $2.65 per kg ($V_c$), losses would be $647 million ($V_c \times [C_p \times L_{nb}]$) in the absence of dung beetles and $525 million ($V_c \times [C_p \times L_b]$) in the presence of dung beetles. Subtracting the estimated value at current levels of dung beetle activity from the theoretical value if no dung beetles were active, we estimate the value of the reduced forage fouling (V_{rf}) to be approximately $122 million.

Nitrogen volatilization.   Another important service provided by dung beetles is promoting decomposition of dung into labile forms of nitrogen that can be assimilated by plants and thus function as fertilizer when the dung is buried. In the absence of dung beetles, cattle feces that remain on the pasture surface until they are dry lose a large proportion of their inorganic nitrogen to the atmosphere. Experiments in South Africa and the United States have shown that approximately 2% of cattle dung is composed of nitrogen, and that 80% of this nitrogen is lost if the pats dry in the sun before they are buried.

Using Gillard's estimate of 27 kg of nitrogen produced annually per animal and assuming that 80% of this nitrogen is lost in the absence of dung beetle activity, we estimate that 21.6 kg would be lost per animal each year if dung beetles were not functioning (L_{nb}). On the basis of our interpretation of decomposition rates, we assume that these losses will be reduced 19% by the current level of dung beetle activity, compared with the estimate for no beetle activity. Thus, we estimate a loss of 17.5 kg per year (L_b) at current activity levels. Multiplying these per-animal values by the total number of cattle whose dung can potentially be buried by dung beetles (C_p, or 32 million), 691 million kg of nitrogen would be lost annually in the United States in the absence of dung beetle activity, compared with the 560 million kg lost at current levels of activity. With nitrogen valued at $0.44 per kg ($V_n$), we estimate the value of nitrogen lost in the absence of dung beetles to be $304 million and the value of nitrogen lost at current levels of dung beetle activity to be $246 million. Subtracting the estimated value at current levels of dung beetle activity from the theoretical value if no dung beetles were active, the value of the reduction in nitrogen loss is approximately $58 million.

Parasites.   Many cattle parasites and pest flies require a moist environment such as dung to complete their development. Burying dung and removing this habitat can reduce the density of these pests. From field observations that reflected current levels of removal, Fincher estimated the annual losses due to mortality, morbidity, and medication of beef cattle, dairy cattle, and other livestock with internal parasites. To estimate the value of dung burial for reducing these losses, we will use only the losses associated with beef cattle,

FORMULAS USED TO ESTIMATE INSECT SERVICES

Formula used to estimate the number of cattle in the United States whose dung can be processed by dung beetles:

$$C_p = (C_t \times P_r) \times P_{nt},$$

where

C_p = head of cattle producing dung that can be processed by dung beetles,

C_t = total head of cattle produced annually in the United States,

P_r = the proportion of cattle that are raised on range or pasture, and

P_{nt} = the proportion of cattle not treated with avermectins.

Formula used to estimate the value of beef saved because of reduced range fouling resulting from dung burial by dung beetles:

$$V_{rf} = [V_c \times (C_p \times L_{nb})] - [V_c \times (C_p \times L_b)],$$

where

V_{rf} = value of reduced forage fouling,

V_c = value of cattle (per kilogram),

C_p = head of cattle producing dung that can be processed by dung beetles,

L_{nb} = losses (per animal) with no dung beetle activity, and

L_b = losses (per animal) at current levels of dung beetle activity.

Formula used to estimate the value of native insects for suppressing populations of potentially pestiferous native herbivorous insects:

$$V_{ni} = (NC_{ni} - CC_{ni}) \times P_i,$$

where

V_{ni} = the value of suppression of native insect pests by other insects,

NC_{ni} = the cost of damage from native insect pests with no natural control,

CC_{ni} = the cost of damage from native insect pests at current levels of natural control, and

P_i = the proportion of herbivorous insects controlled primarily by other insects.

because we do not have a good estimate for the proportion of dairy cattle or other livestock that live on open pasture or rangeland. Fincher reported that beef cattle ranchers lost $428 million annually because of parasites and pests. Corrected for inflation, this is equal to $912 million in 2003 dollars. Given that 85% of beef cattle are on range or pasture and 44% of these cattle are not treated with insecticides, we calculate that 37% of the beef cattle in the United States have fewer parasites because of the facilitation of dung decomposition by dung beetles.

We go on to assume that cattle whose dung is processed by dung beetles suffer 19% fewer losses because of parasites, on the basis of our previous calculation that dung beetles accelerate decomposition by 19%. We also assume that cattle on rangeland, pasture, and feedlots all face the same level of loss from parasites in the absence of dung beetles. Following this logic, we estimate that damage from parasites is only 93% (100% − [37% × 19%]) of what it would be if dung beetles were not providing this service. In the absence of dung beetle activity, estimated losses would be $981 million instead of the current $912 million, and thus this service saves the cattle industry an estimated $70 million per year.

Pest flies.   Using a similar algorithm, we can calculate a value for the reduction in losses due to pest flies. Fincher estimated that losses due to horn flies and face flies cost ranchers $365 million and $150 million, respectively, for a total of $515 million. Corrected for inflation, this is the equivalent of $1.7 billion in 2003. Using the calculation described above for parasites, we assume that, as a result of the processing of dung by insects, damage from parasites is only 93% of what it would have been if the service were not being provided. We estimate that losses in the absence of dung beetle activity would be $1.83 billion instead of the current $1.7 billion, and thus this service is saving the cattle industry an estimated $130 million per year.

Adding the individual values of increased forage, nitrogen recycling, and reduced parasite and fly densities due to dung processing by beetles, we arrive at a combined annual total of $380 million. This is certainly an underestimate, since these same services are being provided to an unknown proportion of pasture-raised dairy cows, horses, sheep, goats, and pigs. Furthermore, what is said for dung recycling can also be said for burying beetles and flies that decompose carcasses. While the density of carcasses is much lower than the density of dung pats, their removal is important in rangeland, natural areas, and other public areas for returning nutrients to the soil, reducing potential spread of diseases, and increasing site utility.

Pollination by Native Insects

Pollination, especially crop pollination, is perhaps the best-known ecosystem service performed by insects. McGregor estimates that 15% to 30% of the US diet is a result, either directly or indirectly, of animal-mediated pollination. These products include many fruits, nuts, vegetables, and oils, as well as meat and dairy products produced by animals raised on insect-pollinated forage. While this estimate is probably high, it presents one of the best published measures of pollinator-dependant food in the US diet.

Here we attempt to calculate an estimate of the value of crops produced as a result of pollination by wild (i.e., unmanaged) native insects. The US government keeps records of the production of crops and, because of their value, their insect pollinators have been given some attention, especially pollination by managed insects such as the European honey bee (*Apis mellifera L.*). From these studies and personal accounts of crop scientists and entomologists, several authors make generalizations about the proportion of pollination attributed to various insect groups, mostly honey bees. These generalizations are essentially educated guesses of the percentage of necessary pollination provided by insects, and as such, they are likely to be inaccurate. The proportions that could be attributed to native, as opposed to managed, pollinators will vary widely for each crop, depending on geographic location, availability of natural habitat, and use of pesticides. In addition, cultivars of the same species can have drastically different dependencies on insect pollinators, further complicating any calculation of the value of pollinator insects.

To conduct a truly accurate economic analysis of the role of native insects in crop pollination, we would need a much better accounting of current levels of pollination by different species of managed bees (e.g., honey bee [*A. mellifera*], alfalfa leaf-cutter bee [*Megachile rotundata*], blue orchard bee [*Osmia lignaria*], alkali bee [*Nomia melanderi*]), and wild bees (e.g., bumble bees [*Bombus* spp.], southeastern blueberry bee [*Habropoda laboriosa*], squash bee [*Peponapis pruinosa*]) in crop pollination. . . . Although we still lack much of this information, the estimate we provide here for the value of crops produced as a result of wild native bee–mediated pollination is informative.

Several scientists have estimated the value of insect-pollinated crops that are dependent on honey bees, or the financial loss to society that could be expected if managed honey bees were removed from cropping systems. These authors make a variety of assumptions and take different approaches to calculating a value for honey bees. For example, Southwick and Southwick take into account the reduced crop output stemming from a lack of managed honey bees, adjusting their figures for the changes in value of each commodity as demand increases because of reduced supply. They also present a range of possible values based on assumptions of the pollination redundancy of managed honey bees and other bee pollinators, including feral honey bees and other native and normative bees. Taking all of this into account, they give a range of $1.6 billion ($2.1 billion when adjusted for inflation to represent 2003 dollars) to $5.2 billion ($6.8 billion in 2003 dollars) for the value of honey-bee pollinators. The lower estimate included effective pollination by other bees, making the managed honey bees redundant in some localities and thereby reducing their absolute value. On the high end, Southwick and Southwick estimate that honey bees are worth $5.2 billion if few or no other bees visit insect-pollinated crops.

Robinson and colleagues and Morse and Calderone take a simpler approach, summing the value of each commodity that they estimate is dependant on honey-bee pollinators. From this they generate a portion of the overall value of each crop that they attribute to pollination by honey bees and report values of $8.3 billion and $14.6 billion ($12.3 billion and $16.4 billion, respectively, when adjusted for inflation to represent 2003 dollars). This approach is more consistent with our other calculations of the value of eco-system services, and so we choose to use it here to calculate the value of crop production that relies on native insect pollinators.

When we sum the average value of pollinator-dependent commodities reported in Morse and Calderone, we find that native pollinators—almost exclusively bees—may be responsible for almost $3.07 billion of fruits and vegetables produced in the United States. Here we must incorrectly assume that the proportion of honey bees to native species is constant in all settings. In some systems, such as agriculturally diverse, organic farms with nearby pockets of natural or seminatural habitat, native bees may be able to provide all of the pollination needs for certain crops. For example, Morse and Calderone assume that 90% of the insect pollinators of watermelon are honey bees. While this is probably true in most farms, some organic growers can rely on native bees for 100% of their melon pollination.

Pest Control

The best estimate available suggests that insect pests and their control mea-sures cost the US economy billions of dollars every year, but this is only a frac-tion of the costs that would accrue if beneficial insects such as predators and parasitoids, among other forces, did not keep most pests below economically damaging levels. We calculate the value (V) of these natural forces by first esti-mating the cost of damage caused by insect pests at current levels of control (CC) and then subtracting this value from the estimated higher cost that would be caused by the greater damage from these insect pests if no controls were functioning (NC). Finally, we calculate a value for the specific action of insect natural enemies by multiplying the value of these natural forces by an estimate of the proportion (P_i) of pests that are controlled by beneficial insects as opposed to other mechanisms (e.g., pathogens or climate).

Because of data limitations, we restrict our estimate to the value derived from the suppression of insect pests that attack crop plants. Beneficial insects certainly suppress populations of both weeds and insects that attack humans and livestock, but the data were not available to calculate the value of these services. As with the rest of our analysis, we also limit our calculations to pest and beneficial insects native to the United States.

Our first step was to calculate the cost of damage due to insect pests at current levels of control from natural enemies. Drawing on previously pub-lished estimates, Yudelman and colleagues presented monetary values for total production of eight major crops and for the losses to these crops attributable to insects. Using these values, we calculated a ratio of insect loss to actual yield that allowed estimation of losses due to insects for any period for which yield values have been published. Assuming $50.5 billion for total production and $7.5 billion for losses due to insects in North America from 1988 through 1990, we calculated a ratio of 0.1485. . . .

Calkins found that only 35% of the exotic pests in the United States are pests in their home range. Extending this finding, we assume that the same relationship holds true in the United States, and thus only 35% of potential insect pest species that are native to the United States reach damaging levels. In other words, we assume that 65% of the potential damage from native pest species is being suppressed, and that 65% of the potential financial cost of this damage is being saved. We make this assumption based on (a) the abundant evidence of a strong correlation between pest density and the magnitude of loss due to pest damage, and (b) the lack of evidence of a correlation between the destructiveness of a pest and the probability that it will be suppressed.

To clarify, the pool of potential pest species—from which we assume 35% actually reach pest levels—is significantly smaller than the 90,000 described insect species in the United States, because many of the described species are not herbivores, and many of those that are herbivores do not feed on cultivated plants. Only 6000 (7%) of the described species in the United States and Canada cause any damage. For our estimate, we assume that these 6000 species, although they make up only 7% of the total species, account for 35% of the species that would be pests if they were not controlled. Following this logic, we assume that the pool of potential pests would be about 17,000 species, 11,000 of which (65%) are being kept below damage levels by biological or climatic controls.

These native species are estimated to comprise 39% of all pest species in the United States. Since native pests vary greatly in the amount of damage they cause, and include some of the most damaging pests in the United States (e.g., corn rootworm, Colorado potato beetle, and potato leafhopper), we assume that they are responsible for 39% of the cost of damage from all pests in the United States. Hence, we estimate that the cost associated with native pest species at current levels of suppression by natural enemies is 39% of $18.77 billion, or $7.32 billion. We designate this value current control by native insects (CC_{ni}).

On the basis of these assumptions, we estimate that the $7.32 billion lost annually to native insect pests (CC_{ni}) is 35% of what would be lost if natural controls were not functioning. If no natural forces were functioning to control native insect pests, we estimate that they would cause $20.92 billion in damage in the United States each year (NC_{ni}). By subtraction, the value of pest control by our native ecosystems is approximately $13.60 billion.

However, not all of this value for natural control of insect pests is attributable to beneficial insects. Some pest suppression comes from other causes, such as pathogens, climatic conditions, and host-plant resistance. One review of the factors responsible for suppression of 68 herbivore species reported that insects (e.g., predators and parasitoids) were primarily responsible for natural control in 33% of cultivated systems. On the basis of these findings, we estimate that insects are responsible for control of 33% of pests that are suppressed by natural controls, while pathogens or bottom-up forces control the rest. Using this average, we estimate the value of natural control attributable to insects to be $4.5 billion annually (33% of $13.6 billion).

Recreation and Commercial Fisheries

US citizens spend over $60 billion a year on hunting, fishing, and observing wildlife. Insects are a critical food source for much of this wildlife, including many birds, fish, and small mammals. Using 1996 US census data on the spending habits of Americans, adjusted for inflation to 2003 dollars, we estimated the amount of money spent on recreational activities that is dependent on services provided by insects. In this case, the predominant service is concentrating and moving nutrients through the food web.

Small game hunting.   Since most large game are either obligate herbivores or omnivores that are not substantially dependent on insects as a source of nutrition, we restrict our estimate of the value of insects for hunting to small game species. In 1996, expenditures for small game hunting totaled $2.5 billion ($2.9 billion in 2003 dollars). To calculate the proportion of this expenditure that is dependent on insects, we use the proportion of days spent hunting for each insectivorous small game species and the dependence of these birds on insects for food.

On the basis of published reports that most galliform chicks rely on insects as a source of protein and that many cannot even digest plant material, we assume that quail, grouse, and pheasant could not survive without insects as a nutritional resource. Therefore, multiplying the proportion of hunting days spent on each of these small game birds (0.15, 0.13, and 0.23, respectively, for a total of 0.51) by the total value for small game ($2.9 billion), we estimate that insects are required for $1.48 billion in expenditures.

Migratory bird hunting.   Insectivory in migratory birds—primarily waterfowl such as ducks and geese in the order Anseriformes—is not as predominant as in the primarily terrestrial galliform birds discussed above. According to Ehrlich and colleagues, 19 (43%) of the 44 species in this order are primarily insectivorous. Multiplying the total money spent on migratory bird hunting ($1.3 billion) by the 43% of species that are primarily insectivorous, we estimate the value of insects as food for hunted migratory birds at $0.56 billion in hunter expenditures.

Sport and commercial fishing.   The census also provides values for sport or recreational fishing. Since most recreational fishing is in fresh water and a majority of freshwater sport fish are insectivorous, we assume that the entire value of recreational fishing ($27.9 billion) is dependent on insects. In contrast to recreational fishing, the target of most commercial fishing is saltwater fish. There are very few marine insect species, but many fish that are caught in marine systems spend part of their life cycle in fresh water, and insects are often critical sources of nutrition during these periods. Commercial fishing is not covered by the census, but data are available on the number and value of fish landed annually in the United States by commercial operations. Twenty-five of these fish species are primarily insectivorous during at least one life stage. Summing their individual values, we estimate the total value of insects for commercial fishing to be approximately $225 million. Insectivorous fish account for more than 15% of the overall value of commercial fish.

Wildlife observation (bird watching).   The 1996 census reports that Americans spent $33.8 billion on wildlife observation. The census also asked respondents to note which types of wildlife they were watching (e.g., birds, mammals, reptiles, amphibians, insects). Because respondents were allowed to choose more than one category of wildlife, it was impossible to separate out observed groups of organisms that were dependent on insects from those that were not. Bird watching is the most inclusive category, with 96% of respondents indicating that they included birds in their observations. Thus, we assume that 96% of the budget for wildlife observation stems directly from birds, many of which are at least partly dependent on insects as a source of nutrition. Thus, we assume that bird watching accounts for 96% of $33.8 billion spent, or $32.4 billion a year, providing a conservative starting point for calculating the dependency of wildlife observation expenditures on insects.

Our next step is to estimate what proportion of this figure for bird observation was dependent on and attributable to insects. Using data from Ehrlich and colleagues, we calculate that 61% of the bird species known to breed in the United States are primarily insectivorous, and another 28% are at least partially insectivorous. To be conservative, we consider only bird species that are primarily insectivorous. This probably underestimates the importance of insectivory for birds, since many passerine and galliform birds that are listed as partially insectivorous could not survive without the vital protein that insects provide young chicks. This estimate is conservative also because it is based on bird species numbers rather than population numbers, and the passerines, which are overwhelmingly insectivorous, have relatively high population densities. Taking these factors into account, we estimate that insects are responsible for $19.8 billion, which is 61% of the $32.4 billion spent on bird observation annually in the United States.

Discussion

We estimate the value of those insect services we address to be almost $60 billion a year in the United States, which is only a fraction of the value for all the services insects provide. The implication of this estimate is that an annual investment of tens of billions of dollars would be justified to maintain these service-providing insects, were they threatened. And indeed, these beneficial insects are under ever increasing threat from a combination of forces, including habitat destruction, invasion of foreign species, and overuse of toxic chemicals.

Fortunately, no evidence suggests a short-term drastic decline in the insects that provide these services. What the evidence does indicate, however, is a steady decline in these beneficial insects, associated with an overall decline in biodiversity, accompanied by localized, severe declines in environments heavily degraded by human impacts. New evidence indicates that in some situations, the most important species for providing ecosystem services are lost first. The overall, gradual decline in species, coupled with nonlinear changes in service levels, makes it difficult to pinpoint an optimal level of annual investment to conserve beneficial insects and maintain the services they provide. . . .

[E]ven though we provide an estimate of the total value of certain insect services, the complications of redundancy and nonlinearity make it impossible to quantitatively gauge the level of resources that are justified for efforts aimed at conserving the services that insects provide. However, our findings lead us to espouse three qualitative guidelines. First, cost-free or relatively inexpensive measures are almost certainly justified to maintain and increase current service levels. Examples include volunteer construction of nest boxes for wild pollinators and the inclusion of a diverse variety of native plant species in plantings for bank or soil stabilization and site restoration. Second, actions or investments that are estimated to have an economic return at or slightly below the break-even point, such as the use of less toxic pesticides, are probably justified because of their nontarget benefits. Third, actions that lead to substantial decreases in biodiversity should be avoided because of the high probability of a major disruption in essential services.

Finally, although we cannot provide a quantitative formula to determine the optimal level of investment in the conservation of beneficial insects that provide essential services, we do feel justified, on the basis of our estimates, in making some specific recommendations. First, we recommend that conservation funding allocated via Farm Bill programs—such as the Conservation Security Program, Conservation Reserve Program, Wetlands Reserve Program, and Environmental Quality Incentives Program—pay specific attention to insects and the role they play in ecosystems. In particular, funding to provide habitat for beneficial insects such as predators, parasitoids, and pollinators in natural, seminatural, unproductive, or fallow areas in agricultural landscapes not only provides direct benefits to growers but, by focusing on the ecological needs of insects, results in habitat that supports a great diversity of wildlife.

Second, we recommend that ecosystem services performed by insects be taken into account in land-management decisions. Specifically, maintaining ecosystem services should be a goal of land management. With this goal in mind, specific practices such as grazing, burning, and pesticide use should be tailored to protect insect biodiversity. . . .

We believe it is imperative that some federal and local funds be directed toward the study of these beneficial insects and the vital services they provide so that conservation efforts can be optimally allocated, either through the agricultural programs listed above or through other means.

These steps are just a beginning. With greater attention, research, and conservation, the valuable services that insects provide can not only be sustained but increase in capacity. As a result, growers will be able to practice a more sustainable form of agriculture while spending less on managing pest insects or acquiring managed pollinators; ranchers will get more productivity out of their land; and wildlife lovers will find that the birds and fish they hunt occur in greater abundance than in the past few decades. In less direct but no less important ways, everyone would benefit from the facilitation of the vital services that insects provide. Judging from our estimate of the value of these four services, increased investment in the conservation of these services is justified.

**Marino Gatto and
Giulio A. De Leo**

Pricing Biodiversity and Ecosystem Services: The Never-Ending Story

In 1844, the French engineer Jules Juvénal Dupuit introduced cost–benefit analysis to evaluate investment projects. . . . The application of cost–benefit analysis to ecological issues fell out of favor three decades ago, and it was gradually replaced by multicriteria analysis in the decision-making process for projects that have an impact on the environment. Although multicriteria analysis is currently used for environmental impact assessments [EIA] in many nations, [recently] the concept of cost–benefit analysis has again become fashionable, along with the various pricing techniques associated with it, such as contingent valuation methods, hedonic prices, and costs of replacement of ecological services. . . . Economists have generated a wealth of virtuosic variations on the theme of assessing the societal value of biodiversity, but most of these techniques are invariably based on price—that is, on a single scale of values, that of goods currently traded on world markets.

Perhaps the most famous recent study on the issue of pricing biodiversity and ecological services is that by Costanza et al., who argued that if the importance of nature's free benefits could be adequately quantified in economic terms, then policy decisions would better reflect the value of ecosystem services and natural capital. Drawing on earlier studies aimed at estimating the value of a wide variety of ecosystem goods and services, Costanza et al. estimated the current economic value of the entire biosphere at $16–54 trillion per year, with an average value of approximately $33 trillion per year. By contrast, the gross national product of the United States totals approximately $18 trillion per year. The paper, as its authors intended, stimulated much discussion, media attention, and debate. A special issue of *Ecological Economics* (April 1998) was devoted to commentaries on the paper, which, with few exceptions, were laudatory. Some economists have questioned the actual numbers, but many scientists have praised the attempt to value biodiversity and ecosystem functions.

Although Costanza et al. acknowledged that their estimates were crude and imperfect, they also pointed the way to improved assessments. In particular, they noted the need to develop comprehensive ecological economic models that could adequately incorporate the complex interdependencies between

From *BioScience*, vol. 50, no. 4, April 2000, by Marino Gatto and Giulio A. De Leo, pp. 347–354.

ecosystems and economic systems, as well as the complex individual dynamics of both types of systems. Despite the authors' caveats and the fact that many economists have been circumspect in applying their own tools to decisions regarding natural systems, the monetary approach is perceived by scientists, policymakers, and the general public as extremely appealing; a number of biologists are also of the opinion that attaching economic values to ecological services is of paramount importance for preserving the biosphere and for effective decision-making in all cases where the environment is concerned.

In this article, we espouse a contrary view, stressing that, for most of the values that humans attach to biodiversity and ecosystem services, the pricing approach is inadequate—if not misleading and obsolete—because it implies erroneously that complex decisions with important environmental impacts can be based on a single scale of values. We contend that the use of cost–benefit analysis as the exclusive tool for decision-making about environmental policy represents a setback relative to the existing legislation of the United States, Canada, the European Union, and Australia on environmental impact assessment, which explicitly incorporates multiple criteria (technical, economic, environmental, and social) in the process of evaluating different alternatives. We show that there are sound methodologies, mainly developed in business and administration schools by regional economists and by urban planners, that can assist decision-makers in evaluating projects and drafting policies while accounting for the nonmarket values of environmental services.

The Limitations of Cost–Benefit Analysis and Contingent Valuation Methods

Historically, the first important implementation of cost–benefit analysis at the political level came in 1936, with passage of the US Flood Control Act. This legislation stated that a public project can be given a green light if the benefits, to whomsoever they accrue, are in excess of estimated costs. This concept implies that all benefits and costs are to be considered, not just actual cash flows from and to government coffers. However, public agencies (e.g., the US Army Corps of Engineers) quickly ran into a problem: They were not able to give a monetary value to many environmental effects, even those that were predictable in quantitative terms. For instance, engineers could calculate the reduction of downstream water flow resulting from construction of a dam, and biologists could predict the river species most likely to become extinct as a consequence of this flow reduction. However, public agencies were not able to calculate the cost of each lost species. Therefore, many ingenious techniques for the monetary valuation of environmental goods and services have been devised since the 1940s. These techniques fall into four basic categories.

- **Conventional market approaches.** These approaches, such as the replacement cost technique, use market prices for the environmental service that is affected. For example, degradation of vegetation in developing countries leads to a decrease in available fuelwood. Consequently, animal dung has to be used as a fuel instead of a fertilizer, and farmers must therefore replace dung with chemical fertilizers. By computing the cost of these chemical fertilizers, a monetary value for the degradation of vegetation can then be calculated.

- **Household production functions.** These approaches, such as the travel cost method, use expenditures on commodities that are substitutes or complements for the environmental service that is affected. The travel cost method was first proposed in 1947 by the economist Harold Hotelling, who, in a letter to the director of the US National Park Service, suggested that the actual traveling costs incurred by visitors could be used to develop a measure of the recreation value of the sites visited.
- **Hedonic pricing.** This form of pricing occurs when a price is imputed for an environmental good by examining the effect that its presence has on a relevant market-priced good. For instance, the cost of air and noise pollution is reflected in the price of plots of land that are characterized by different levels of pollution, because people are willing to pay more to build their houses in places with good air quality and little noise. . . .
- **Experimental methods.** These methods include contingent valuation methods, which were devised by the resource economist Siegfried V. Ciriacy-Wantrup. Contingent valuation methods require that individuals express their preferences for some environmental resources by answering questions about hypothetical choices. In particular, respondents to a contingent valuation methods questionnaire will be asked how much they would be willing to pay to ensure a welfare gain from a change in the provision of a nonmarket environmental commodity, or how much they would be willing to accept in compensation to endure a welfare loss from a reduced provision of the commodity.

Among these pricing techniques, the contingent valuation methods approach is the only one that is capable of providing an estimate of existence values, in which biologists have a special interest. Existence value was first defined by Krutilla as the value that individuals may attach to the mere knowledge that rare and diverse species, unique natural environments, or other "goods" exist, even if these individuals do not contemplate ever making active use of or benefiting in a more direct way from them. The name "contingent valuation" comes from the fact that the procedure is contingent on a constructed or simulated market, in which people are asked to manifest, through questionnaires and interviews, their demand function for a certain environmental good (i.e., the price they would pay for one extra unit of the good versus the availability of the good). . . .

The limits of cost–benefit analysis were discussed in the 1960s, after more than two decades of experimentation. In particular, many authors pointed out that cost–benefit analysis encouraged policymakers to focus on things that can be measured and quantified, especially in cash terms, and to disregard problems that are too large to be assessed easily. Therefore, the associated price might not reflect the "true" value of social equity, environmental services, natural capital, or human health. In particular, economists themselves recognize that the increasingly popular contingent valuation methods are undermined by several conceptual problems, such as free-riding, overbidding, and preference reversal.

When it comes to monetary valuation of the goods and services provided by natural ecosystems and landscapes specifically, a number of additional problems undermine the effectiveness of pricing techniques and cost–benefit analysis. These problems include the very definition of "existence" value, the dependence of pricing techniques on the composition of the reference group, and the significance of the simulated market used in contingent valuation.

The definition of "existence" value   A classic example of contingent valuation methods is to ask for the amount of money individuals are willing to pay to ensure the continued existence of a species such as the blue whale. However, the existence value of whales does not take into account potential indirect services and benefits provided by these mammals. It is just the value of the existence of whales for humans, that is, the satisfaction that the existence of blue whales provides to people who want them to continue to exist. Therefore, there is a real risk that species with very low or no aesthetic appeal or whose biological role has not been properly advertised will be given a low value, even if they play a fundamental ecological function. Without adequate information, most people do not understand the extent, importance, and gravity of most environmental problems. As a consequence, people may react emotionally and either underestimate or overestimate risks and effects.

Therefore, it is not surprising that five of the seven guidelines issued by the National Oceanic and Atmospheric Administration [NOAA] about how to conduct contingent valuation discuss how to properly inform and question respondents to produce reliable estimates (e.g., in-person interviews are preferred to telephone surveys to elicit values). Of course, acquisition of reliable and complete information is always possible in theory, but in practice strict adherence to NOAA guidelines makes contingent valuation methods expensive and time consuming.

Difficulties with the reference group for pricing   Pricing techniques such as contingent valuation methods provide information about individual willingness to pay or willingness to accept, which must be summed up in the final balance of cost–benefit analysis. Therefore, the outcome of cost–benefit analysis depends strongly on the group of people that is taken as a reference for valuation—particularly on their income. Van der Straaten noted that the Exxon *Valdez* oil spill in 1989 provides a good example of this dependence. The population of the United States was used as a reference group to calculate the damage to the existence value of the affected species and ecosystems using contingent valuation methods. Exxon was ultimately ordered to pay $5 billion to compensate the people of Alaska for their losses. This huge figure was a consequence of the high income of the US population. If the same accident had occurred in Siberia, where salaries are lower, the outcome would certainly have been different.

This example shows that contingent valuation methods simply provide information about the preferences of a particular group of people but do not necessarily reflect the ecological importance of ecosystem goods and services. Moreover, the outcome of cost–benefit analysis depends on which individual willingness to pay or willingness to accept are included in the cost–benefit analysis. If the quality of the Mississippi River is at issue, should the analysis be restricted to US citizens living close to the river, or should the willingness to pay of Californians and New Yorkers be included too? According to Krutilla's definition of existence value, for many environmental goods and ecological services that may ultimately affect ecosystem integrity at the global level, the preferences of the entire human population should potentially be considered in the analysis. Because practical reasons obviously preclude doing so, contingent valuation methods will inevitably only provide information about the preferences of specific groups of people. For many of the ecological services that may be considered the heritage of humanity, contingent valuation methods analyses performed locally in a particular economic situation should be extrapolated only with great caution to other areas. The process of placing a monetary value on biodiversity and ecosystem functioning through nonuser willingness to pay is performed in the same way as for user willingness to pay, but the identification of people who do not use an environmental good directly and still have a legitimate interest in its preservation is problematic.

Significance of the simulated market   Contingent valuation methods are contingent on a market that is constructed or simulated, not real. It is difficult to believe in the efficiency of what Adam Smith called the "invisible hand" of the market for a process that is the artificial production of economic advisors and does not possess the dynamic feedback that characterizes real competitive markets. Is it even possible to simulate a market where units of biodiversity are bought and sold? As Friend stated, "these contingency evaluation methods (CVM) tend to create an illusion of choice based on psychology (willingness) and ideology (the need to pay) which is supposed, somewhat mysteriously, to reflect an equilibrium between the consumer demand for and producer supply of environmental goods and services."

Many additional criticisms of pricing ecological services are more familiar to biologists. For many ecological services, there is simply no possibility of technological substitution. Moreover, the precise contribution of many species is not known, and it may not be known until the species is close to extinction. . . . In addition, specific ecosystem services, as evaluated by Costanza et al., should not be separated from one another and valued individually because the importance of any piece of biodiversity cannot be determined without considering the value of biodiversity in the aggregate. And finally, the use of marginal value theory may be invalidated by the erratic and catastrophic behavior of many ecological systems, resulting in potentially detrimental effects on the health of humans, the productivity of renewable resources, and the vitality and stability of societies themselves.

Despite the efforts of many economists, we believe that some goods and services, especially those related to ecosystems, cannot reasonably be given a monetary value, although they are of great value to humans. Economists coined the term "intangibles" to define these goods. Cost–benefit analysis cannot easily deal with intangibles. As Nijkamp wrote, more than 20 years ago, "the only reasonable way to take account of intangibles in the traditional cost–benefit analysis seems to be the use of a balance with a debit and a credit side in which all intangible project effects (both positive and negative) are represented in their own (qualitative or quantitative) dimensions" as secondary information. In other words, the result of cost–benefit analysis is primarily a single number, the net monetary benefit that comprises all the effects that can be sensibly converted into monetary returns and costs.

Commensurability of Different Objectives and Multicriteria Analysis

Cost–benefit analysis includes intangibles in the decision-making process only as ancillary information, with the main focus being on those effects that can be converted to monetary value. This approach is not a balanced solution to the problem of making political decisions that are acceptable to a wide number of social groups with a range of legitimate interests. . . .

However, even if the attempt to put a price on everything is abandoned, it is not necessary to give up the attempt to reconcile economic issues with social and environmental ones. Social scientists long ago developed multicriteria techniques to reach a decision in the face of multiple different and structurally incommensurable goals. The most important concept in multicriteria analysis was actually conceived by an Italian economist, Vilfredo Pareto, at the end of the nineteenth century. It is best explained by a simple example. Suppose that a natural area hosting several rare species is a target for the development of a mining activity. Alternative mining projects can have different effects in terms of profits from mining (measured in dollars) and in terms of sustained biodiversity (measured in suitable units, for instance, through the Shannon index). Profit from mining can be corrected using welfare economics to include those environmental and social effects that can be priced (e.g., the benefit of providing jobs to otherwise unemployed people, the cost of treating lung disease of miners, and the cost of the loss of the tourists who used to visit the natural area). . . .

The methods of multicriteria analysis are intended to assist the decision-maker in choosing among . . . alternatives . . . (a task that is particularly difficult when there are several incommensurable objectives, not just two). Nevertheless, the initial step of determining [these] alternatives is of enormous importance, for three reasons. First, [doing so] makes perfect sense even if there is no way of pricing a certain environmental good because each objective can be expressed in its own proper units without reduction to a common scale. Second, the determination of all the feasible alternatives . . . requires the joint effort of a multidisciplinary team that includes, for example, economists, engineers, and biologists and that must predict the effects of alternative decisions on all of the different environmental and social components to which humans are sensitive and which, therefore, deserve consideration. Third, the determination of [feasible alternatives] allows the objective elimination of inadequate alternatives because [they are] independent of the subjective perception of welfare . . . [and] in essence describe the tradeoff between the various incommensurable objectives when every effort is made to achieve the best results in all respects; the attention of the authority that must make the final decision is thus directed toward genuine potential solutions because nonoptimal decisions have already been discarded.

It should be noted that a cost–benefit analysis does not elicit tradeoffs between incommensurable goods because it also gives a green light to projects . . . , provided that the benefits that can be converted into a monetary scale exceed the costs. . . . Cost–benefit analysis, however, is not useful for eliciting the tradeoffs between two incommensurable goods, neither of which is monetary. For instance, there might be a conflict between the goals of preserving wildlife within a populated area and minimizing the risk that wild animals are vectors of dangerous diseases. A multicriteria analysis can describe this tradeoff, whereas a cost–benefit analysis cannot.

Another philosophical point concerning the issue of commensurability is the question of implicit pricing. Economists often argue that to make a decision is to put an implicit price on such intangibles as human life or aesthetics and, therefore, to reduce their value to a common scale (as pointed out also by Costanza et al.). . . .

Environmental Impact Assessment and Multiattribute Decision-Making

Because of the flaws of cost–benefit analysis, many countries have taken a different approach to decision-making through the use of environmental impact assessment legislation (e.g., the United States in 1970, with the signing of the National Environmental Policy Act, NEPA; France in 1976, with the act 76/629; the European Union in 1985, with the directive 85/337). Environmental impact assessment procedures, if properly carried out, represent a wiser approach than setting an a priori value of biodiversity and ecosystem services because these procedures explicitly recognize that each situation, and every regulatory decision, responds to different ethical, economic, political, historical, and other conditions and that the final decision must be reached by giving appropriate consideration to several different objectives. As Canter noted, all projects, plans, and policies that are expected to have a significant environmental impact would ideally be subject to environmental impact assessment.

The breadth of goals embraced by environmental impact assessment is much wider than that of cost–benefit analysis. Environmental impact assessment provides a conceptual framework and formal procedures for comparing different alternatives to a proposed project (including the possibilities of not development a site, employing different management rules, or using mitigation measures); for fostering interdisciplinary team formation to investigate all possible environmental, social, and economic consequences of a proposed activity; for enhancing administrative review procedures and coordination among the agencies involved in the process; for producing the necessary documentation to enhance transparency in the decision-making process and the possibility of reviewing all the objective and subjective steps that resulted in a given conclusion; for encouraging broad public participation and the input of different interest groups; and for including monitoring and feedback procedures. Classical multiattribute analysis can be used to rank different alternatives. . . . Ranking usually requires the use of value functions to transform environmental and other indicators (e.g., biological oxygen demand or animal density) to levels of satisfaction on a normalized scale, and the weighting of factors to combine value functions and to rank the alternatives. These weights explicitly reflect the relative importance of the different environmental, social, and economic compartments and indicators.

A wide range of software packages for decision support can assist experts in organizing the collected information; in documenting the various phases of EIA; in guiding the assignment of importance weights; in scaling, rating, and ranking alternatives; and in conducting sensitivity analysis for the overall decision-making process. This last step, of testing the robustness and consistency of multiattribute analysis results, is especially important because it shows how sensitive the final ranking is to small or large changes in the set of weights and value functions, which often reflect different and subjective perspectives. It is important to stress that, although the majority of environmental impact assessments have been conducted on specific projects, such as road construction or the location of chemical plants, there is no conceptual barrier to extending the procedure to evaluation of plans, programs, policies, and regulations. In fact, according to NEPA, the procedure is mandatory for any federal action with an important impact on the environment. The extension of environmental impact assessment to a level higher than a single project is termed "strategic environmental assessment" and has received considerable attention.

Conclusions

An impressive literature is available on environmental impact assessment and multiattribute analysis that documents the experience gained through 30 years of study and application. Nevertheless, these studies seem to be confined to the area of urban planning and are almost completely ignored by present-day economists as well as by many ecologists. Somewhere between the assignment of a zero value to biodiversity (the old-fashioned but still used practice, in which environmental impacts are viewed as externalities to be discarded from the balance sheet) and the assignment of an infinite value (as advocated by some radical environmentalists), lie more sensible methods to assign value to biodiversity than the price tag techniques suggested by the new wave of environmental economists. Rather than collapsing every measure of social and environmental value onto a monetary axis, environmental impact assessment and multiattribute analysis allow for explicit consideration of intangible nonmonetary values along with classical economic assessment, which, of course, remains important. It is, in fact, possible to assess ecosystem values and the ecological impact of human activity without using prices.

Concepts such as Odum's eMergy [the available energy of one kind previously required to be used up directly and indirectly to make the product or service] and Rees' ecological footprint [the area of land and water required to support a defined economy or population at a specified standard of living], although perceived by some as naive, may aid both ecologists and economists in addressing this important need.

To summarize our viewpoint, economists should recognize that cost–benefit analysis is only part of the decision-making process and that it lies at the same level as other considerations. Ecologists should accept that monetary valuation of biodiversity and ecosystem services is possible (and even helpful) for part of its value, typically its use value. We contend that the realistic substitute for markets, when they fail, is a transparent decision-making process, not old-style cost–benefit analysis. The idea that, if one could get the price right, the best and most effective decisions at both the individual and public levels would automatically follow is, for many scientists, a sort of Panglossian obsession. In reality, there is no simple solution to complex problems. We fear that putting an a priori monetary value on biodiversity and ecosystem services will prevent humans from valuing the environment other than as a commodity to be exploited, thus reinvigoraing the old economic paradigm that assumes a perfect substitution between natural and human-made capital. As Rees wrote, "for all its theoretical attractiveness, ascribing money values to nature's services is only a partial solution to the present dilemma and, if relied on exclusively, may actually be counterproductive."

POSTSCRIPT

Should a Price Be Put on the Goods and Services Provided by the World's Ecosystems?

In "Can We Put a Price on Nature's Services?" *Report From the Institute for Philosophy and Public Policy* (Summer 1997), Mark Sagoff objects that trying to attach a price to ecosystem services is futile because it legitimizes the accepted cost-benefit approach and thereby undermines efforts to protect the environment from exploitation. The March 1998 issue of *Environment* contains environmental economics professor David Pearce's detailed critique of the 1997 Costanza et al. study. Pearce objects chiefly to the methodology, not the overall goal of attaching economic value to ecosystem services. Costanza et al. reply to Pearce's objections in the same issue. Pearce and Edward B. Barbier have published *Blueprint for a Sustainable Economy* (Earthscan, 2000), in which they discuss how governments worldwide are now applying economics to environmental policy.

Despite the controversy over the worth of assigning economic values to various aspects of nature, researchers continue the effort. Gretchen C. Daily et al., in "The Value of Nature and the Nature of Value," *Science* (July 21, 2000), discuss valuation as an essential step in all decision making and argue that efforts "to capture the value of ecosystem assets . . . can lead to profoundly favorable effects." Daily and Katherine Ellison continue the theme in *The New Economy of Nature: The Quest to Make Conservation Profitable* (Island Press, 2002). In "What Price Biodiversity?" *Ecos* (January 2000), Steve Davidson describes an ambitious program funded by the Commonwealth Scientific and Industrial Research Organization (CSIRO) and the Myer Foundation that is aimed at developing principles and methods for objectively valuing "ecosystem services—the conditions and processes by which natural ecosystems sustain and fulfil human life—and which we too often take for granted. These include such services as flood and erosion control, purification of air and water, pest control, nutrient cycling, climate regulation, pollination, and waste disposal." Jim Morrison, "How Much Is Clean Water Worth?" *National Wildlife* (February/March 2005), argues that ecosystem services such as cleaning water, controlling floods, and pollinating crops have sufficient economic value to make it profitable to spend millions of dollars to protect natural systems.

Stephen Farber, et al., "Linking Ecology and Economics for Ecosystem Management," *Bioscience* (February 2006), find "the valuation of ecosystem services . . . necessary for the accurate assessment of the trade-offs involved in different management options." Ecosystem valuation is currently being used to justify restoration efforts, "linking the science to human welfare," as shown (for example) in Chungfu Tong, et al., "Ecosystem Service Values and Restoration in the Urban Sanyang Wetland of Wenzhou, China," *Ecological Engineering* (March 2007).

Internet References . . .

ECOLEX: A Gateway to Environmental Law

This site, sponsored by the United Nations and the World Conservation Union, is a comprehensive resource for environmental treaties, national legislation, and court decisions.

http://www.ecolex.org/index.php

Environmental Defense

Environment Defense (once The Environmental Defense Fund) is dedicated to "protecting the environmental rights of all people, including future generations." Guided by science, Environmental Defense evaluates environmental problems and works "to create solutions that win lasting economic and social support because they are nonpartisan, cost-efficient, and fair."

http://www.environmentaldefense.org/home.cfm

Office of Environmental Justice

The U.S. Environmental Protection Agency (EPA) pursues environmental justice under the Office of Enforcement and Compliance Assurance as part of its "firm commitment to the issue of environmental justice and its integration into all programs, policies, and activities, consistent with existing environmental laws and their implementing regulations."

http://www.epa.gov/compliance/environmentaljustice/index.html

The Heritage Foundation

The Heritage Foundation is a think-tank whose mission is to formulate and promote conservative public policies on many issues, including environmental ones. It bases its work on the principles of free enterprise, limited government, individual freedom, and traditional American values.

http://www.heritage.org/

SourceWatch

SourceWatch is a collaborative project of the Center for Media and Democracy. Its primary focus is on documenting the interconnections and agendas of public relations firms, think tanks, industry-funded organizations and industry-friendly experts that work to influence public opinion and public policy on behalf of corporations, governments, and special interests.

http://www.sourcewatch.org

The National Endangered Species Act Reform Coalition

The National Endangered Species Act Reform Coalition is a broad based coalition of roughly 150 member organizations, representing millions of individuals across the United States, that is dedicated to improving and updating the Endangered Species Act. Our membership includes rural irrigators, municipalities, farmers, electric utilities, and many other individuals and organizations that are directly affected by the ESA.

http://www.nesarc.org/

Principles Versus Politics

*I*n many environmental issues, it is easy to tell what basic principles apply and therefore determine what is the right thing to do. Ecology is clear on the value of species to ecosystem health. Sociology and politics have agreed that racism is an evil to be avoided. Medicine makes no bones about the ill effects of pollution. But are the environmental problems so bad that we must act immediately? How much of the "right thing" should we do? How should we do it? Such questions arise in connection with every environmental issue, not just the three in this part of the book, but these three will serve to introduce the theme of principles versus politics.

- Should the Endangered Species Act Be Strengthened?

- Should the EPA Be Doing More to Fight Environmental Injustice?

- Can Pollution Rights Trading Effectively Control Environmental Problems?

ISSUE 4

Should the Endangered Species Act Be Strengthened?

YES: John Kostyack, from Testimony before the Oversight Hearing on the Endangered Species Act, U.S. Senate Committee on Environment & Public Works (May 19, 2005)

NO: Monita Fontaine, from Testimony before the Oversight Hearing on the Endangered Species Act, U.S. Senate Committee on Environment & Public Works (May 19, 2005)

ISSUE SUMMARY

YES: Representing the National Wildlife Foundation, John Kostyack argues that the Endangered Species Act has been so successful that it should not be weakened but strengthened.

NO: Speaking for the National Endangered Species Act Reform Coalition, a group that represents those affected by the Endangered Species Act, Monita Fontaine argues that federal regulation under the ESA should be replaced by a system that relies more on voluntary and state species conservation efforts.

Extinction is normal. Indeed, 99.9 percent of all the species that have ever lived are extinct, according to some estimates. But the process is normally spread out over time, with the formation of new species by mutation and selection balancing out the loss of old ones to disease, new predators, climate change, habitat loss, and other factors. Today, human activities are an important cause of species loss mostly because humans destroy or alter habitat but also because of hunting (including commercial fishing), the introduction of foreign species as novel competitors, and the introduction of diseases. According to Martin Jenkins, "Prospects for Biodiversity," *Science* (November 14, 2003), some 350 (3.5 percent) of the world's bird species may vanish by 2050. Other categories of living things may suffer greater losses, leading to a "biologically impoverished" world; indeed, according to Sarah DeWeerdt, "Bye Bye, Birdie," *World Watch* (July/August 2006), over a third of all species may be on their way to extinction by 2050. Julia Whitty says "By the End of the Century Half of All Species Will Be Gone. Who Will Survive?" *Mother Jones* (May/June 2007). Jenkins states that the consequences for human life are "unforeseeable but potentially catastrophic."

Awareness of the problem has been growing. When the United States adopted the Endangered Species Act (ESA) in 1973, the goal was to protect species that were so reduced in numbers or restricted in habitat that a single untoward event could wipe them out. Both environmental groups and politicians were concerned over declining populations of some birds and plants. According to Ted Williams, "Law of Salvation," *Audubon* (November/December 2005), "Protecting the planet's genetic wealth made sense morally and economically. It was considered, rightly enough, what decent, civilized people do." The ESA therefore barred construction projects that would further threaten endangered species. In one famous case, construction on the Tellico Dam on the Little Tennessee River in Loudon County, Tennessee, was halted because it threatened the snail darter, a small fish. Another case involved the spotted owl, which was threatened by logging in the Northwest. Those in favor of the dam or the timber industry felt that the value of the endangered species was trivial compared to the human benefits at stake. Those in favor of the act argued that the loss of a single species might not matter to the world, but where one species went, others would follow. Protecting one species also protects others.

The ESA has had some notable successes despite lengthy legal battles, which have used up funds intended for protecting species, and pressures to ease restrictions on activities that might damage species or their habitat. These pressures have made it difficult to maintain the ESA. It was last authorized in 1988. Efforts to reauthorize it have repeatedly stalled. According to Peter Uimonen and John Kostyack, "Unsound Economics: The Bush Administration's New Strategy for Undermining the Endangered Species Act," National Wildlife Federation (June 2004) (http://www.nwf.org/wildlife/pdfs/UnsoundEconomics.pdf), the current administration has reduced critical habitat protection, suppressed and distorted information on the economic benefits to local economies of habitat conservation, exaggerated costs, and reduced funding. Erik Stokstad, "What's Wrong with the Endangered Species Act?" *Science* (September 30, 2005), describes the Republican-sponsored attempt to "reform" the ESA with the Threatened and Endangered Species Recovery Act as further restricting the ESA budget by requiring that landowners be compensated "for the fair market value of any development or other activity that the government vetoes because it would impact endangered species." Critics charge that the legislation, which passed the House late in 2005, is clearly more favorable to private interests than to endangered species. Proponents of reform argue that improving protection of private property rights is crucial.

In the following selections, John Kostyack argues that the Endangered Species Act has been so successful that it should not be weakened. What is needed is increased funding, earlier, speedier, and more effective recovery planning, reinforced habitat protection, and financial assistance for cooperative landowners. Monita Fontaine argues that federal regulation under the ESA should be replaced by a system that relies more on voluntary and state species conservation efforts.

YES

John Kostyack

Testimony

Good morning, Senator Chafee and members of the subcommittee. My name is John Kostyack, and I am Senior Counsel and Director of Wildlife Conservation Campaigns with the National Wildlife Federation. I appreciate your invitation for me to testify here today on the Endangered Species Act. I have been working on Endangered Species Act law and policy, both here in Washington, D.C., and in various regions around the country, for the past 12 years. Over this time my appreciation for the value and wisdom of this law has grown continuously.

I'd like to talk today about how Congress could update the law to deal with the wildlife conservation challenges of the coming decades. The challenges are many. Consider, for example, the following threats, each of which is accelerating over time:

- *Invasive Species.* According to the USDA, 133 million acres of land in the U.S. are already covered by invasive plants, and each year another 1.7 million acres are invaded. Invasive species threaten the survival of nearly half of all listed species.
- *Sprawling Development Patterns.* The amount of land covered by urban and suburban development in the U.S. has quadrupled since 1950, with the rate of land consumption greatly outpacing population growth and increasing every decade. According to Endangered by Sprawl (2005), a study recently completed by National Wildlife Federation, Smart Growth America, and Nature Serve, over 1,200 plant and animal species will be threatened with extinction by sprawl in just the next two decades.
- *Global Warming.* According to the U.S. State Department's recent Climate Action Report (2002), global warming poses serious risks to species and habitat types throughout the United States, threatening, among other things, alpine meadows across the West, prairie potholes in the Great Plains, and salmon spawning habitats in the Pacific Northwest.

If we truly want to pass on this nation's wildlife heritage to our children and grandchildren, we are going to need a strong Endangered Species Act to address these threats.

Before moving to some suggested updates to the Endangered Species Act, I would first like to talk about what kind of law we already have. It is crucial that Congress understands the benefits the law is already providing, and the

United States Senate Committee on Environment & Public Works Hearing on the National Wildlife Federation Oversight on the Endangered Species Act, May 19, 2005.

law's many on-the-ground success stories, before it proceeds to reauthorization. The positive accomplishments of the past 32 years are the foundation that future changes to the Act must be built upon.

The Benefits of the Endangered Species Act

The Endangered Species Act represents the only effort by this nation to grapple in a comprehensive way with the problem of human-caused extinctions. For the many animal and plant species at risk of extinction, it is the only safety net that our nation provides.

Fortunately, the Endangered Species Act has been quite successful in rescuing plants and animals from extinction.

- Over 98% of species ever protected by the Act remain on the planet today.
- Of the listed species whose condition is known, 68% are stable or improving and 32% are declining.
- The longer a species enjoys the ESA's protection, the more likely its condition will stabilize or improve.

This is the most important thing for Congress to understand about the Endangered Species Act. **It has worked to keep species from disappearing forever into extinction and, over time, it has generally stabilized and improved the condition of species.** As a result, we have a fighting chance of achieving recovery, and more importantly, we are passing on to future generations the practical and aesthetic benefits of wildlife diversity that we have enjoyed.

The other key benefit provided by the Endangered Species Act, besides stopping extinction, is that it **protects the habitats that species depend upon for their survival.** The habitats protected by the Act are not only essential for wildlife, they are oftentimes the very natural areas that people count on to filter drinking water, prevent flooding, provide healthy conditions for hunting, fishing and other outdoor recreation, and provide a quiet and peaceful respite from our noisy and frenetic everyday lives.

To this date, no one has come up with a better way to protect our wildlife and wild places for future generations. So, when our children peer into the eyes of a manatee swimming by their canoe in a clear cool Florida river, or listen to a wolf howl in Yellowstone, or watch a condor soar majestically over the Grand Canyon, our generation and the one before ours should take pride in what we have done for them in the past 32 years. As a result of the commitment Congress made in enacting the Endangered Species Act in 1973, and as a result of the efforts of many people working with the law ever since, we still have a rich and wonderful wildlife legacy to pass along.

Measuring Success—A Lesson from the Ivory-Billed Woodpecker

In the past few years, opponents of the Endangered Species Act have repeatedly tried to persuade the American people that despite the law's success in stopping extinction, the law is broken and needs a radical overhaul. Their argument boils down to a single statistic: only 13 or so species have been removed from the endangered species list due to recovery.

Recovery and delisting are certainly goals that the National Wildlife Federation shares, and I will speak in a moment about how to improve the odds of achieving them. However, I must first challenge the premise of the ESA's opponents that recovery and delisting should be the only measure of the success of the Endangered Species Act. Because it is not the only measure of success—it is not even the best measure—the entire case for a radical overhaul of the Act evaporates.

The story of the ivory-billed woodpecker highlights three reasons why the Endangered Species Act cannot be evaluated based upon the number of species fully recovered and delisted. Although the ESA has not yet been applied to the ivory bill, this species symbolizes the challenges facing wildlife agencies today. It shows that some of the biggest obstacles to recovery and delisting are largely beyond the influence of the Endangered Species Act.

First, Restoring Species and Habitats Requires Funding

Although the ivory-billed woodpecker has been listed as endangered under the ESA and predecessor laws since 1967, it has been presumed extinct since the 1940s. In perhaps one of the most exciting wildlife stories in our nation's history, a single bird was recently sighted in the Cache River National Wildlife Refuge in eastern Arkansas. We hope and expect that there are more birds in that area, but in any case, the bird's numbers are extremely low.

The ivory bill historically inhabited swampy bottomland hardwood forests. It prefers older trees, where it finds its primary food source, beetle larvae, living under the bark. In the southeastern U.S. where the bird once ranged, the vast majority of these old-growth forests are now gone, cleared for farms and pine plantations, and it will take decades to grow them back.

Restoring the habitats that the ivory bill needs to recover is going to take a lot more than the Endangered Species Act. Although safe harbor agreements under the ESA can remove disincentives, substantial public and private dollars will be needed to create positive incentives for private landowners to plant bottomland hardwood trees and protect them until they reach the stage where they are suitable habitat for the ivory bill. The fact that the ivory bill is listed as endangered under the Endangered Species Act will help concentrate everyone's attention on this task. However, if sufficient restoration dollars are not raised, it will not be a failure of the Endangered Species Act. Congress and other key actors need to provide funding to make this large-scale restoration project happen.

Second, as a Matter of Biology, Achieving Full Recovery Often Takes a Long Time

The average period of time in which species have been listed under the ESA is 15.5 years. In that amount of time, our best-case scenario is that we will have discovered and begun protecting a few more ivory bills and developed a strategy for accommodating range expansion. As a matter of simple biology—there aren't currently enough old trees around that could sustain a viable meta-population—full recovery of the ivory bill will take many decades.

Although the condition of most other listed species is not as dire as the ivory bill, many have severely depleted population numbers and habitats. As with the ivory bill, bringing their population numbers back and restoring their habitats often takes a long time for reasons of biology alone. Add in economic and political obstacles—such as the fact that many areas that need to be restored as habitat have potentially competing uses—and you can reasonably expect that recovery will not be completed for many species for a long while.

Third, Delisting Requires Putting in Place Non-ESA Regulatory Measures

Once a species' numbers and habitats are restored to the point of long-term viability, delisting still may not be feasible. Under the ESA, the Fish and Wildlife Service or NOAA Fisheries must first ensure that adequate regulatory measures are in place to prevent immediate backsliding after delisting.

For the ivory bill and many other listed species, there are no protections in place to prevent immediate habitat losses after the Endangered Species Act's protections are removed. In addition, many species require continuing management even after their population sizes and habitats have been restored to targeted levels. Conservation agreements with funding, monitoring and enforcement mechanisms must be negotiated with land managers to ensure that this management is carried out over the long run.

In summary, those who claim the ESA is broken due to the absence of a sizable number of delistings are ignoring the facts. The realities that impede quick recovery and delisting—inadequate funding, slow biological processes, and the absence of any alternative safety net—are not the fault of the Endangered Species Act.

The Endangered Species Act is making an essential contribution to recovery by stabilizing and improving the condition of species over time. Thanks to the Act, the ivory bill has a real chance of making it into the next century. But Congress needs to look outside the four corners of the Act to fully understand and address the reasons why so few species are removed from the threatened and endangered list due to recovery each year.

In addition, members of Congress should stop relying a single statistic about delistings as the measure of the Act's success, and instead encourage the wildlife agencies to develop new and better mechanisms for tracking progress. As authors Michael Scott and Dale Goble point out in the April 2005 issue of *BioScience,* the wildlife agencies currently do not maintain a database enabling policymakers and the public to track Endangered Species Act actions. A database that identifies, among other things, how much habitat is being conserved and how much is being authorized for destruction as a result of ESA consultation processes, would greatly inform the debate over the effectiveness of the law.

On-the-Ground Success Stories to Build Upon

The Endangered Species Act has produced numerous on-the-ground successes. The small list of examples below is designed simply to highlight the variety and creativity of the conservation actions that the law has fostered. These examples show that the Endangered Species Act is empowering people to find a place for wildlife in a country that is increasingly crowded with extractive industries, real estate developments, and other human uses of natural resources. Because of the Act's safety net features and its recovery programs, native wildlife still has a place on the American landscape.

1. **Whooping Crane** The whooping crane is a dynamic and charismatic bird that, if it were not for the Endangered Species Act and its predecessors, would probably no longer exist in the wild today. As a result of a recovery program developed under the Act, birds have been bred in captivity, released into the wild, and trained with the help of an aircraft to fly and migrate. Endangered Species Act enforcement action to protect the bird's designated critical habitat led to the creation of the Platte River Critical Habitat Maintenance Trust, which has acquired over 10,000 acres of riparian habitat along the crane's migratory route. Prior to the Endangered Species Act, a mere 16 birds existed in the wild. Today, nearly 200 birds thrive in the wild, attracting birdwatchers from around the world.

2. **Florida Panther** The Florida panther is one of the most endangered large mammals in the world. As recently as fifteen years ago, its numbers had been reduced to somewhere between 30 and 50. Due to the Endangered Species Act, a number of innovative conservation measures have been taken to bring the animal back from the brink. The U.S. Fish and Wildlife Service successfully addressed the panther's inbreeding problem by bringing Texas cougars (a closely related subspecies) into south Florida. Vehicle mortality, one of the leading causes of panther deaths, has been greatly reduced with the construction of highway underpasses. The underpasses created for the Florida panther now serve as a world model for facilitating movement of wildlife in an urbanizing landscape. Today, the number of cats living in the wild approaches 100. The Florida panther is still a long way from full recovery, but it has a fighting chance.

3. **Gray Wolf** Although the gray wolf once ranged across much of the continental United States, several centuries of hunting and predator control programs, reduction of prey, and habitat loss greatly reduced the species' numbers. By the mid-1960s, the only gray wolves in the lower 48 states were the 200 to 500 animals in Minnesota and roughly 20 on Isle Royale, Michigan. Today, thanks to the Endangered Species Act, there are thriving gray wolf populations in the Western Great Lakes and Northern Rockies, a small population in the Southwest, and occasional wolf sightings in the Northeast and Pacific Northwest. The dramatic recovery of the gray wolf in the Northern Rockies was jump-started by an historic reintroduction of wolves to Yellowstone National Park and the central Idaho wilderness—one of the most successful wildlife reintroductions in the nation's history.

4. **Bald Eagle** In the 1960s, the bald eagle, our Nation's symbol, had fewer than 500 breeding pairs remaining in the continental U.S. Widespread use of the pesticide DDT in the post-World War II period had contaminated the majestic bird's food supply, causing its populations across the country to plummet. Although the federal ban on DDT in 1972 was a major factor in turning around the bald eagle's decline, the Endangered Species Act also played an essential role in its recovery. The Act protected the bird's key habitat and facilitated translocations of eaglets from areas where the bird was numerous to states where it had been eliminated or severely depleted. Today, the number of bald eagles in the lower 48 states exceeds 7,600 breeding pairs.

5. **Puget Sound Chinook Salmon** Chinook salmon have long been a symbol of the Pacific Northwest, providing important cultural values for Native American tribes and sustenance and recreation for all residents. The Puget Sound population of the Chinook was listed in 1999 after declining steadily due to logging, mining, dam-building and suburban development in its habitat, and interbreeding of hatchery fish. Recently, in response to the Endangered Species Act, Seattle City Light improved prospects for the fish by modifying its dam operations on the Skagit, the Puget Sound's largest river. Prospects for the fish and habitats also have improved due to the emergence of Shared Strategy, a groundbreaking collaborative effort by a diverse array of citizens and organizations to build an ESA recovery plan for the Puget Sound chinook from the ground up, watershed by watershed. This effort will ensure broad public support for the array of recovery actions that will ultimately be needed to bring the chinook back to full recovery.

6. **Robbins' Cinquefoil** The Robbins' cinquefoil is a species of the rose family, found at just two locations on the slopes of the White Mountains in New Hampshire. In the 1970s, its numbers were reduced to roughly 1,800 plants due to trampling by horses and hikers and harvesting by commercial plant collectors. After listing and critical habitat designation pursuant to the Endangered Species Act, the Appalachian Mountain Club and New England Wild Flower Society teamed up with federal agencies to relocate a hiking trail, educate the public and reestablish healthy populations. By 2002, the species' numbers had rebounded to over 14,000 plants in two populations, and the species was removed from the endangered list. A cooperative agreement with the U.S. Forest Service helps ensure the continuation of the Robbins' cinquefoil's success story through management and monitoring.

Opportunities for Updating and Improving the Act

Many lessons can be learned from the successes described above and from the numerous other positive experiences implementing the Endangered Species Act. The following are some ideas for updating and improving the Act that are drawn from these experiences.

- **Implement Recovery Plans and Encourage Proactive Conservation.** Any effort to update the Endangered Species Act must begin with steps to promote greater and earlier progress toward recovery. As discussed above, due to Act's flexibility the Nation has benefited in recent years from numerous collaborative initiatives to restore species and habitats. Wildlife agencies should build recovery plans around these proactive recovery initiatives, and Congress should support them with funding so long as they are consistent with recovery plans. If such an approach were taken, ESA conflicts would be reduced because there would be greater buy-in to the Act's implementation. Because greater amounts of habitats would be restored, wildlife agencies would have greater management flexibility.

The Endangered Species Act already provides a solid foundation for this approach. Section 4(f) calls for one of the two wildlife agencies to develop a recovery plan with objective measurable criteria for success and to implement it. However, recovery plans oftentimes are not completed for many years after listing, and thus there is no early blueprint to guide management and restoration actions. A simple solution to this problem would be to require that recovery plans be finalized within a specified time after listing (e.g., 3 years).

A related problem is that the two wildlife agencies are typically not in the position to carry out many of the actions that are needed to bring about recovery. Section 7(a)(1) of the Act requires all federal agencies to utilize their authorities in furtherance of species recovery, but it does not link this duty to the recovery plan. As a result, agencies have often chosen recovery actions in an arbitrary manner.

A solution to this problem would be for federal agencies to be required to develop and implement Recovery Implementation Plans to set forth the specific actions, timetables, and funding needed for that agency to help achieve the recovery goals set forth in the Recovery Plan. The Western Governors Association developed a variation of this idea when it adopted its ESA legislative proposal in the 1990s. "Implementation agreements" for federal and state agencies to help carry out recovery plans remains part of WGA policy to this day.

Another problem related to implementation of recovery plans is that federal agencies oftentimes carry out actions that are at odds with those plans. For example, the Corps of Engineers has issued dredge-and-fill permits for development in Florida panther habitat despite the fact that the habitat is deemed essential for the species in the recovery plan. Congress could easily fix this problem by clarifying that federal agencies must ensure that their actions do not undermine the recovery needs of listed species. The recovery needs of the species would be identified in the recovery plan, and updated by the latest scientific data. If Congress were to adopt this approach, agency decisions would more likely to contribute to the Act's recovery goal. They would also be easier to defend in court, and less likely to attract litigation, because they would be tied to a larger strategic framework, the recovery plan.

- **Provide incentives for private landowners to contribute to recovery.** According to the GAO, roughly 80 percent of all listed species have at least some of their habitat on non-federal land; about 50 percent have the majority of their habitat on non-federal land. Much of this non-federal land is private land, and yet the current Endangered Species Act does not provide many incentives for private landowners to carry out the management measures that are often needed for listed species to thrive. Although ESA regulatory programs such as Safe Harbor remove disincentives, they do not provide incentives. Technical assistance programs can help, but by far the most meaningful incentive that Congress can provide is financial assistance. To ensure a reliable source of funding, this assistance should be provided through the tax code. In return for conservation agreements in which private landowners commit to actively manage habitats for the benefit of listed species, Congress should defer indefinitely federal estate taxes or provide immediate income tax credits for expenses incurred.

- **Protect critical habitat.** The Administration has attempted to justify its efforts to weaken the Act's critical habitat protections by claiming that these protections are redundant with other ESA protections and therefore without value to listed species. At the same time, the Administration contradicts itself by generating cost-benefit analyses claiming that critical habitat protections are imposing enormous costs on the private sector. None of this rhetoric is supported by any meaningful analysis of data. The only quantitative studies on critical habitat have shown that critical habitat indeed provides benefits to many listed species. Species with critical habitat designations tend to do better than species without such designations.

  Critical habitat is particularly important when it comes to protecting unoccupied habitat, because the other protections in the Endangered Species Act generally do not adequately protect such habitat. Most species will never recover unless they can return to some part of their historic range that is currently unoccupied.

  Because of the hostility shown by the current Administration toward critical habitat, it will be essential for Congress, when it reauthorizes the ESA, to strongly reaffirm the importance of critical habitat protection. Congress should push back the deadlines to three years after listing, thereby giving the wildlife agencies the time they need to get the science right. It also should encourage the wildlife agencies to integrate recovery plan and critical habitat designation decisions. Congress also should develop a schedule, and authorize the funding, for cleaning up the backlog of species awaiting critical habitat designations. When the late Senator Chafee took these steps in S. 1100 back in 1999, they attracted broad public support.

- **Provide adequate funding.** Finally, there perhaps can be no more important step that Congress can take to improve implementation of the Endangered Species Act than to increase funding to reasonable levels. At a bare minimum, Congress must provide the funding that the wildlife agencies need to carry out their mandatory duties. For example, the U.S. Fish and Wildlife Service has estimated that it would take approximately $153 million over 10 years to eliminate the current backlog of listings and critical habitat designations. Congress could immediately eliminate dozens of lawsuits simply by providing these funds and other funds needed for the basic implementation steps of the Act. In addition, many of the concerns about the Act's impact on states, local governments and private landowners could be alleviated if Congress were to expand its Section 6 and other grant funding for recovery actions.

Testimony

The Endangered Species Act (ESA) was enacted in 1973 with the promise that we can do better in the job of protecting and conserving our nation's resident species and the ecosystems that support them. Today, over thirty years later, I bring that same message back to this Committee—*we can, and must, do better.* We have learned many lessons over the past three decades about how and what can be done to protect endangered and threatened species, and it is time to update and improve the ESA to reflect those lessons.

I am here before you today on behalf of the National Endangered Species Act Reform Coalition (NESARC), an organization of 110 national associations, businesses and individuals that are working to develop bipartisan legislation that updates and improves the ESA. Personally, my organization, the National Marine Manufacturers Association (NMMA), joined NESARC in 2003 largely due to our members' experiences with listed marine species such as the manatee population in Florida, as well for as the opportunity to join a diverse group of interests working on this matter. I have the pleasure of sitting on the NESARC Board of Directors. On behalf of the NESARC Board of Directors and, all of the NESARC members, I want to commend the efforts being undertaken by members of this Committee, other members of the Senate and in the House of Representatives to develop a bipartisan bill that updates and improves the ESA. We look forward to working with the Committee, its able staff, and other members of the Senate to find common ground.

NESARC members come from a wide range of backgrounds. Among our ranks are farmers, ranchers, cities and counties, rural irrigators, electric utilities, forest and paper operators, mining, homebuilders and other businesses and individuals throughout the United States. What our members have in common is that they have been impacted by the operation of the ESA. Frankly speaking, the burdens and rewards of protecting listed species are borne, in a very large part, by the members of NESARC. NESARC members are actively involved in a broad range of species conservation efforts including:

- The development of State management plans for wolf populations in the Rocky Mountains and in Minnesota, Michigan and Wisconsin;
- Recovery implementation programs such as the Upper Colorado and San Juan Rivers Endangered Fish Recovery Implementation Program and Platte River Endangered Species Recovery Program;

United States Senate Committee on Environment & Public Works Hearing on the National Wildlife Federation Oversight on the Endangered Species Act, May 19, 2005.

- Numerous habitat conservation plans ranging from county-wide HCPs in Southern California to single parcel plans for covering agricultural operations; and
- Observation, research and monitoring programs for listed and candidate species.

Many environmental groups (including some of those who are testifying today) have recognized the need for on-the-ground partnerships. The reality is that, without the support and active commitment to the protection of listed species by the private landowners, businesses and communities where the species reside, the chances of success are slim. We need to learn from the experiences of those who are faced with the real-world decisions on how to make a living and still protect species if we are to make the Act work better.

If we are to do a better job protecting endangered and threatened species, we need an ESA that can fully accommodate the range of efforts that are necessary. As detailed later in my testimony, NESARC has developed a number of recommendations for ways to improve the ESA. These recommendations are the product of an extensive reassessment by NESARC members as to what improvements to the ESA would be useful for the future implementation of the Act.

At the end of 2003, NESARC decided to look inward, to reassess the state of the ESA's implementation on the ground and to identify the success stories of its members in protecting endangered and threatened species as well as those roadblocks that had to be overcome. What we learned was that, more often than not, our members have succeeded in protecting endangered and threatened species *in spite of, rather than because of,* the ESA.

When we asked our members to share their success stories and positive experiences, what we received were very personal observations from the ground reporting that success is occurring—but not easily.

"Our HCP process has had some very beneficial elements, but it's been painfully slow and costly to get there. Given the experience, [it is] hard to endorse it for others to pursue. Yet an HCP embodies concepts for species protection which are very good and could be more effective. [We] advocate moving to a system with more incentives and much greater penalties for abuses." **Carol Rische, Humboldt Bay Metropolitan Water District.**

"Some of the regulators that we deal with are very results-oriented. Their practical approach has been beneficial to our operations and beneficial to species recovery. Working together with practical regulators to the benefit of the species has been a positive experience." **Tom Squeri, Granite Rock Company.**

The experience of my own members within NMMA is similar—with the hope of cooperative efforts between federal and state agencies limited by the realities of working within an Act that was enacted more than thirty years ago and does not provide the necessary flexibility and tools to effectively and efficiently develop workable solutions. As many of you know, Florida has a long history of protecting its endangered manatee population—in which NMMA members have actively participated. As a result of efforts led by the State of Florida and stakeholders, the manatee population has grown from an estimated 1,465 manatees in 1991 to at least 3,142 (as documented by a 2005 aerial survey)—more than a doubling of the population in approximately 14 years. Further, the U.S. Fish & Wildlife Service has joined with the Florida Fish & Wildlife Conservation Commission to begin a "manatee forum" which is aimed at developing a consensus, science-based approach to continuing to protect and enhance manatee populations in balance with marine activities. However, such cooperative efforts remain the exception, not the norm.

As I am here today representing NESARC, I do not wish to dwell on the particular problems facing boaters, marina operators and other marine services; however, to the extent that the Committee wishes to hear more about the personal experiences of any of our individual NESARC members, including NMMA, we are happy to provide that information and brief you or your staff on particular issues of interest.

Drawing from our members' experiences and observations, NESARC identified a series of guideposts from which to consider future improvements to the ESA, which include the following:

- Encourage Sound Decision-making
- Promote Innovation
- Promote Certainty
- Increase Funding
- Reduce Economic Impacts
- Increase Roles for State, Local Governments
- Provide Greater Public Participation
- Limit Litigation

After developing these initial guideposts, over the latter half of 2004, NESARC worked to draft a white paper which was publicly released in November 2004. This white paper is attached to my testimony and provides an outline of a new approach to ESA legislation that we hope the Members of this Committee will take into consideration.

In sum, a new approach is needed to change the focus of the debate from a clash over existing terms and programs to the development of new tools that improve the Act. We need *new provisions* of the Act that encourage recovery of listed species through voluntary species conservation efforts and the active involvement of States. This new approach can and should maintain the goal of species conservation. Simultaneously, we must recognize that species conservation and recovery will only be accomplished if we can find ways to provide stakeholders the tools and flexibility to take action and, most importantly, certainty that quantifiable success will be rewarded by the lifting of the ESA restrictions.

As this Committee reviews ways to improve the ESA, we would ask that you take into consideration the following proposals:

- **Expand and Encourage Voluntary Conservation Efforts**—A universal concern with the Act is that it does not fully promote and accommodate voluntary conservation efforts. Many landowners want to help listed species, but the ESA doesn't let them. A critical element of updating and improving the Act must be the development of additional voluntary conservation programs. These efforts should include: (1) creating a habitat reserve program, (2) tax incentives, (3) loan or grant programs and (4) other initiatives that encourage landowners to voluntarily participate in species conservation efforts. Further, existing programs like the Safe Harbor Agreements should be codified.

- **Give the States the Option of Being On the Front Line of Species Conservation**—In 1973, the National Wildlife Federation testified before Congress that "[s]tates should continue to exercise the prime responsibility for endangered species" and "should be given the opportunity to prepare and manage recovery plans and retain jurisdiction over resident species." Thirty-plus years later, the Western Governors' Association, in a February 25, 2005 letter (attached) noted that "[t]he [ESA] can be effectively implemented only through a full partnership between the states and the federal government" and asked Congress to "give us the tools and authority to make state and local conservation efforts meaningful."

    NESARC agrees that States should have a wider role in facilitating landowner/operator compliance with the Act and, ultimately, the recovery of species. States have significant resources, research capabilities and coordination abilities that can allow for better planning of species management activities. Further, States know their lands and are often better situated to work with stakeholders to protect and manage the local resources and species.

- **Increase Funding of Voluntary and State Programs for Species Conservation**—A significant amount of federal funding for ESA activities is presently tied up in addressing multiple lawsuits and the review of existing and new listing and critical habitat proposals. In contrast, actual funding for on-the ground projects that will recover species is limited.

    Federal funding priorities need to be re-focused to active conservation measures that ultimately serve to achieve the objectives of the Act. Further, we need to financially support the voluntary, community-based programs that are critical to ensuring species recovery.

- **Encourage Prelisting Measures**—Recently, a nationwide coalition of state and local governments, stakeholders and conservation organizations worked together to develop a comprehensive sage grouse conservation program that has been able to stand in the place of a listing of that species under the ESA. Those efforts were supported by many members of this committee including Senator Harry Reid of Nevada who stated that, ". . . I have advocated using the Farm Security and Rural Investment Act of 2002 (Farm Bill) conservation programs to help local communities like Elko, Nevada, engage in voluntary conservation efforts for species like sage grouse. In fact, the Farm Bill's Wildlife Habitat Incentives Program (WHIP) encourages private and public agencies to develop wildlife habitat on their properties, and specifically has directed funds to enhance habitats for sage grouse. I know more can be done, and I am committed to improving local conservation efforts." *Statement of Senator Harry Reid, September 24, 2004.*

    Private landowners, State and local governmental agencies should be encouraged to develop and implement programs for species that are being considered for listing. The protections afforded by all such programs (including existing activities) should be considered in determining whether a listing is warranted or whether such voluntary programs, other federal agency programs and State/local conservation efforts already provide sufficient protections and enhance species populations so that application of the ESA is not necessary.

- **Establish Recovery Objectives**—We need to be able to identify and establish recovery objectives. Knowing what ultimately must be achieved is a critical first step in understanding what must be done. Since the goal of the ESA is to assure recovery of endangered and threatened species, implementation of the ESA should reward progress when it is made toward recovery. There must be a determination of specific recovery goals necessary to reach the point where a species can and will be downlisted or delisted—and there must be certainty in such a goal so that the goal is not continually shifted to perpetuate a listing.

- **Strengthen the Critical Habitat Designation Process**—We need to strengthen the critical habitat designation process by ensuring that these designations are supported by sound decision-making procedures, do not overlap with existing habitat protection measures (such as habitat conservation plans, safe harbor agreements or candidate conservation agreements, and other state and federal land conservation or species management programs) and rely on timely field survey data.

- **Improve Habitat Conservation Planning Procedures and Codify "No Surprises"**—The HCP process has the potential to be a success story, but too often private property owners are stymied by the delays and costs of getting HCP approval. HCP approval should be streamlined, and the HCP process must be adapted so that it is practical for the smaller landowner. Further, landowners involved in conservation efforts need to be certain that a "deal is a deal." The "No Surprises" policy must be codified under the Act and cover all commitments by private parties to voluntary protection and enhancement of species and habitat—not just HCPs.

- **Ensure an Open and Sound Decision-Making Process**—The ESA must be open to new ideas and data. A good example of this principle is the emerging data regarding the effect of boat speeds on manatees and their avoidance mechanisms. Because the principal threat to manatees is impact from boat propellers, federal and state manatee-protection policies historically have focused on slowing boats passing through manatee habitats. However, research by Dr. Edmund Gerstein of Florida Atlantic University and Joseph E. Blue, retired director of the Naval Undersea Warfare Center and the Naval Research Laboratory's Underwater Sound Reference Detachment challenges some of the existing protection measures. This new research shows that while manatees have good hearing abilities at high frequencies, they have relatively poor sensitivity in the low frequency ranges associated with boat noise, which means that manatees may be least able to hear the propellers of boats that have slowed down in compliance with boat speed regulations designed to reduce collisions. My point is not to suggest that there should not be speed limits in areas occupied by manatees, but rather that we need to make sure that our policy decisions (like setting boat speed limits) are informed by up-to-date research. By providing for better data collection and independent scientific review, we can ensure that the necessary and appropriate data is available.

In addition to making sure we have better information upon which to act, we need a decision-making process that allows for full public participation in the listing, critical habitat and recovery decisions. It has been my experience that providing full and open access to the decision-making processes—beyond simply the submission of letter comments—through mechanisms like stakeholder representatives and data collection programs provides a much more diverse and ultimately stronger record from which to act.

For more than a decade, Congress has struggled with the question of what, if any, changes to the ESA should be made. In the interim, stakeholders like NESARC members, have had to take the existing Act and make it work. It has been time-consuming, expensive and often frustrating—and the successes have been limited. Today, less than 1% of all listed species in the United States have been recovered.

The Congressional history on ESA legislation has had its ebbs and flows over the past thirteen years with at least two distinct sets of legislative efforts—both of which ultimately failed. NESARC is not interested in going down that same path again where stakeholders (on both sides) re-open old battles and try to right perceived wrongs from past court decisions. NESARC urges this Committee to take stock of the lessons we have learned and successes that have been achieved in order to identify the improvements that are necessary to make this Act work better in the future.

POSTSCRIPT

Should the Endangered Species Act Be Strengthened?

After the Threatened and Endangered Species Recovery Act passed the House of Representatives in September 2005 (see "House Passes Bill Reforming Endangered Species Act," *Human Events* [October 31, 2005]), Senator Mike Enzi (R-Wyo) introduced in December similar legislation in the Senate, calling it a Christmas present for "Wyoming farmers and ranchers. . . . We have a bill that will both help recover species and preserve landowner livelihood." Six U.S. Senators, both Republican and Democrat, asked the Keystone Center to prepare a report on changes needed in the Endangered Species Act. The Keystone Center is a Colorado think tank that helps "leaders from governmental, non-governmental, industrial, and academic organizations to find productive solutions to controversial and complex public policy issues." Its report, released in April 2006 (http://www.keystone.org/spp/env-esa.html), concluded that the ESA's effectiveness for species recovery could be improved in several ways and that the burdens imposed by the ESA on private interests could be relieved with incentives, but reconciling the two goals is difficult and the House bill is not the way to do it. Senator Lincoln Chafee (R-RI), one of the six who requested the Keystone report, said in March 2006 that Senate action was unlikely for that year. In May 2007, at a hearing of the House Natural Resources Committee, committee chairman Nick Rahall called for recommitting the Endangered Species Act to being based on science, even as republicans insisted the Act interfered with private property rights (see Mike Deehan, "Democrats Say Science Will Guide Endangered Species Act," *CongressDaily* [May 9, 2007].

Michael J. Bean, "The Endangered Species Act under Threat," *Bioscience* (February 2006), cautioned that the House act undermines "the government's long-standing trust responsibility to safeguard wildlife. The Senate should think long and hard before embracing the House's radical proposals."

As a contribution to that thinking process, The Ecological Society of America, together with other scientific societies, published the "Scientific Societies's Statement on the Endangered Species Act" (http://www.esa.org/pao/policyStatements/pdfDocuments/2-2006_finalStatement_Scientific%20Societies%20ESA.pdf) on February 27, 2006. Among other things, the statement objects to bureaucratic attempts to limit the definition of useful scientific data (included in the House legislation), calls for eliminating delays in evaluating rare species for listing as threatened or endangered, improving funding for research, and restoring protections for critical habitat. It does recognize that "parties experiencing economic and social impacts from recovery activities should be included" in planning.

ISSUE 5

Should the EPA Be Doing More to Fight Environmental Injustice?

YES: Robert D. Bullard, from Statement on "Environmental Justice Programs" before the Senate Committee on Environment and Public Works, Subcommittee on Superfund and Environmental Health, July 25, 2007

NO: Granta Y. Nakayama, from Statement on "Environmental Justice Programs" before the Senate Committee on Environment and Public Works, Subcommittee on Superfund and Environmental Health, July 25, 2007

ISSUE SUMMARY

YES: Professor Robert D. Bullard argues that despite a 1994 Executive Order directing the Environmental Protection Agency (EPA) to ensure that minority and poor communities not bear a disproportionate burden of pollution and other environmental ills, environmental justice still eludes many people.

NO: Granta Y. Nakayama, Assistant Administrator of the EPA's Office of Enforcement and Compliance, argues that the EPA is a "trailblazer" in the government's effort to achieve environmental justice, and is planning to review its programs.

Archeologists delight in our forebears' habit of dumping their trash behind the house or barn. Today, however, most people try to arrange for their junk to be disposed of as far away from home as possible. Landfills, junkyards, recycling centers, and other operations with large negative environmental impacts tend to be sited in low-income and minority areas. Boston's Great Molasses Flood (see http://www.mv.com/ipusers/arcade/molasses.htm and Stephen Puleo, *Dark Tide: The Great Boston Molasses Flood of 1919* [Beacon Press, 2003]) happened when a two-million-gallon molasses storage tank burst; the tank had been built in a neighborhood crowded with immigrant laborers who lacked the political influence to say "Not in My Back Yard." Was such a location mere coincidence? Or was it deliberate?

The environmental movement has, in fact, been charged with having been created to serve the interests of white middle- and upper-income people. Native Americans, blacks, Hispanics, and poor whites were not well represented among early environmental activists. It has been suggested that the reason for this is that these people were more concerned with more basic needs, such as jobs, food, health, and safety. However, the situation has been changing. In 1982, for example, in Warren County, North Carolina, poor black and Native American communities held demonstrations in protest of a poorly planned PCB (polychlorinated biphenyl) disposal site. This incident kicked off the environmental justice movement, which has since grown to include numerous local, regional, national, and international groups. The movement's target is systematic discrimination in the setting of environmental goals and in the siting of polluting industries and waste disposal facilities—also known as environmental racism. The global reach of the problem is discussed by Jan Marie Fritz in "Searching for Environmental Justice: National Stories, Global Possibilities," *Social Justice* (Fall 1999).

In 1990 the Environmental Protection Agency (EPA) published "Environmental Equity: Reducing Risks for All Communities," a report that acknowledged the need to pay attention to many of the concerns raised by environmental justice activists. At the 1992 United Nations Earth Summit in Rio de Janeiro, a set of "Principles of Environmental Justice" was widely discussed. In 1993 the EPA opened an Office of Environmental Equity (now the Office of Environmental Justice) with plans for cleaning up sites in several poor communities. In February 1994 President Bill Clinton made environmental justice a national priority with an executive order. Since then, many complaints of environmental discrimination have been filed with the EPA under Title VI of the federal Civil Rights Act of 1964; and in March 1998 the EPA issued guidelines for investigating those complaints. However, in April 2001 the U.S. Supreme Court ruled that individuals cannot sue states by charging that federally funded policies unintentionally violate the Civil Rights Act of 1964. . . . See the Environmental Justice Coalition's home page at http:// groups.msn.com/environmentaljusticecoalition/homepage.msnw.

Critics of the environmental justice movement contend that inequities in the siting of sources of pollution are the natural consequence of market forces that make poor neighborhoods (whether occupied by whites or minorities) the economically logical choice for locating such facilities, as well as a source of needed jobs (see David Friedman, "The 'Environmental Racism' Hoax," *The American Enterprise* [November/December 1998]). In the following selections, Professor Robert D. Bullard argues that despite the 1994 Executive Order directing the Environmental Protection Agency (EPA) to ensure that minority and poor communities not bear a disproportionate burden of pollution and other environmental ills, environmental justice still eludes many people. Part of the reason is that the EPA has weakened its Strategic Plan and the regulatory protections inherent in the Toxic Release Inventory program. It needs to do much more. Granta Y. Nakayama, Assistant Administrator of the EPA's Office of Enforcement and Compliance, argues that the EPA is a "trailblazer" in the government's effort to achieve environmental justice. It builds partnerships with other government agencies and communities, provides grants to promote community empowerment, shares information, and is planning to review its programs.

YES

Robert D. Bullard

Subcommittee on Superfund and Environmental Health of the Senate Environment and Public Works Committee Regarding Environmental Justice

The environmental justice movement has come a long way from its humble beginnings in rural and mostly African American Warren County, North Carolina. It has now been twenty-five years since the controversial 1982 decision to dump 40,000 cubic yards (or 60,000 tons) of soil in the mostly black county. The soil was contaminated with the highly toxic polychlorinated biphenyls (PCB) illegally dumped along 210 miles of roadways in fourteen North Carolina counties in 1978. The roadways were cleaned up in 1982.

Warren County won the dubious prize of hosting the toxic dump. The landfill decision became the shot heard around the world and put environmental racism on the map and the catalyst for mass mobilization against environmental injustice. Over 500 protesters were arrested, marking the first time any Americans had been jailed protesting the placement of a waste facility.

After waiting more than two decades for justice, victory finally came to the residents of predominately black Warren County when detoxification work ended the latter part of December 2003. State and federal sources spent $18 million to detoxify contaminated soil stored at the PCB landfill.

After mounting scientific evidence and much prodding from environmental justice advocates, the EPA created the Office of Environmental Justice in 1992 and produced its own study, *Environmental Equity: Reducing Risks for All Communities*, a report that finally acknowledged the fact that low-income and minority populations shouldered greater environmental health risks than others.

In 1992, staff writers from the *National Law Journal* uncovered glaring inequities in the way the federal EPA enforces its laws. The authors found a "racial divide in the way the U.S. government cleans up toxic waste sites and punishes polluters. White communities see faster action, better results and stiffer penalties than communities where blacks, Hispanics and other minorities

From U.S. Senate Committee on Environment and Public Works, Subcommittee on Superfund and Environmental Health, July 25, 2007.

live. This unequal protection often occurs whether the community is wealthy or poor." These findings suggest that unequal protection is placing communities of color at special risk and that their residents who are differentially impacted by industrial pollution can also expect different treatment from the government.

On February 11, 1994, environmental justice reached the White House when President Clinton signed Executive Order 12898, *"Federal Actions to Address Environmental Justice in Minority Populations and Low-Income Populations."* The EPA defines environmental justice as: "The fair treatment and meaningful involvement of all people regardless of race, color, national origin, or income with respect to the development, implementation, and enforcement of environmental laws, regulations and policies. Fair treatment means that no group of people, including racial, ethnic, or socioeconomic groups should bear a disproportionate share of the negative environmental consequences resulting from industrial, municipal, and commercial operations or the execution of federal, state, local, and tribal programs and policies."

Numerous studies have documented that people of color in the United States are disproportionately impacted by environmental hazards in their homes, neighborhoods, and workplace. A 1999 Institute of Medicine study, *Toward Environmental Justice: Research, Education, and Health Policy Needs,* concluded that low-income and people of color communities are exposed to higher levels of pollution than the rest of the nation and that these same populations experience certain diseases in greater number than more affluent white communities.

A 2000 study by *The Dallas Morning News* and the University of Texas-Dallas found that 870,000 of the 1.9 million (46 percent) housing units for the poor, mostly minorities, sit within about a mile of factories that reported toxic emissions to the U.S. Environmental Protection Agency.

Even schools are not safe from environmental assaults. A 2001 Center for Health, Environment, and Justice study, *Poisoned Schools: Invisible Threats, Visible Action,* reports that more than 600,000 students in Massachusetts, New York, New Jersey, Michigan and California were attending nearly 1,200 public schools, mostly populated by low-income and people of color students, that are located within a half mile of federal Superfund or state-identified contaminated sites.

EPA Response to Environmental Justice Needs

Thirteen years after the signing of Executive Order 12898, environmental justice still eludes many communities across this nation. In its 2003 report, *Not in My Backyard: Executive Order and Title VI as Tools for Achieving Environmental Justice,* the U.S. Commission on Civil Rights (USCCR) concluded that "Minority and low-income communities are most often exposed to multiple pollutants and from multiple sources. . . . There is no presumption of adverse health risk from multiple exposures, and no policy on cumulative risk assessment that considers the roles of social, economic and behavioral factors when assessing risk."

A March 2004 EPA Inspector General report, *EPA Needs to Conduct Environmental Justice Reviews of Its Programs, Policies, and Activities,* concluded that the agency "has not developed a clear vision or a comprehensive strategic plan, and has not established values, goals, expectations, and performance measurements" for integrating environmental justice into its day-to-day operations.

In July 2005, the U.S. Government Accountability Office (GAO) criticized EPA for its handling of environmental justice issues when drafting clean air rules. That same month, EPA proposed major changes to its Environmental Justice Strategic Plan. This proposal outraged EJ leaders from coast to coast. The agency's Environmental Justice Strategic Plan was described as a "giant step backward." The changes would clearly allow EPA to shirk its responsibility for addressing environmental justice problems in minority populations and low-income populations and divert resources away from implementing Executive Order 12898.

The agency then attacked community right-to-know by announcing plans to modify the Toxic Release Inventory (TRI) program—widely credited with reducing toxic chemical releases by 65 percent. In December 2006, the EPA announced final rules that undermine this critical program by eliminating detailed reports from more than 5,000 facilities that release up to 2,000 pounds of chemicals every year; and eliminating detailed reports from nearly 2,000 facilities that manage up to 500 pounds of chemicals known to pose some of the worst threats to human health, including lead and mercury.

In September 2006, EPA's Inspector General issued another report chastising the agency for falling to "conduct environmental justice reviews of its programs, policies, and activities."

And in June 2007, the U.S. General Accountability Office (GAO) issued yet another report, *Hurricane Katrina: EPA's Current and Future Environmental Protection Efforts Could Be Enhanced by Addressing Issues and Challenges Faced on the Gulf Coast,* that criticized EPA's handling of contamination in post-Katrina New Orleans and the Gulf Coast. The GAO found inadequate monitoring for asbestos around demolition and renovation sites. Additionally, the GAO investigation uncovered that "key" information released to the public about environmental contamination was neither timely nor adequate, and in some cases, easily misinterpreted to the public's detriment."

In December 2005, the Associated Press released results from its study, *More Blacks Live with Pollution,* showing African Americans are 79 percent more likely than whites to live in neighborhoods where industrial pollution is suspected of posing the greatest health danger. Using EPA's own data and government scientists, the AP study found blacks in 19 states were more than twice as likely as whites to live in neighborhoods with high pollution; a similar pattern was discovered for Hispanics in 12 states and Asians in seven states.

The AP analyzed the health risk posed by industrial air pollution using toxic chemical air releases reported by factories to calculate a health risk score for each square kilometer of the United States. The scores can be used to compare risks from long-term exposure to factory pollution from one area to another. The scores are based on the amount of toxic pollution released by each factory, the path the pollution takes as it spreads through the air, the level of danger to humans posed by each different chemical released, and the number of males and females of different ages who live in the exposure paths.

Toxic Wastes and Race at Twenty

This year represents the twentieth anniversary of *Toxic Wastes and Race*. To commemorate this milestone, the United Church of Christ (UCC) asked me to assemble a team of researchers to complete a new study, *Toxic Wastes and Race at Twenty 1987–2007*. The Executive Summary of the new study was released at the 2007 American Association for the Advancement of Science (AAAS) in San Francisco. . . . The full report was released in March 2007 at the National Press Club in Washington, DC. In addition to myself, the principal authors of new UCC report are Professors Paul Mohai (University of Michigan), Beverly Wright (Dillard University of New Orleans), and Robin Saha (University of Montana).

Toxic Wastes and Race at Twenty is the first national-level study to employ 2000 Census data and distance-based methods to a current database of commercial hazardous waste facilities to assess the extent of racial and socio-economic disparities in facility locations. Disparities are examined by region and state, and separate analyses are conducted for metropolitan areas, where most hazardous waste facilities are located.

The new report also includes two detailed case studies: one on environmental cleanup in post-Katrina New Orleans and the other on toxic contamination in the mostly African American Eno Road community in Dickson, Tennessee.

Study Findings

- People of color make up the majority (56%) of those living in neighborhoods within 3 kilometers (1.8 miles) of the nation's commercial hazardous waste facilities, nearly double the percentage in areas beyond 3 kilometers (30%).
- People of color make up a much larger (over two-thirds) majority (69%) in neighborhoods with clustered facilities.
- Percentages of African Americans, Hispanics/Latinos, and Asians/Pacific Islanders in host neighborhoods are 1.7, 2.3, and 1.8 times greater in host neighborhoods than non-host areas (20% vs. 12%, 27% vs. 12%, and 6.7% vs. 3.6%), respectively.
- 9 out of 10 EPA regions have racial disparities in the location of hazardous waste sites.
- Forty of 44 states (90%) with hazardous waste facilities have disproportionately high percentages of people of color in host neighborhoods— on average about two times greater than the percentages in non-host areas (44% vs. 23%).
- Host neighborhoods in an overwhelming majority of the 44 states with hazardous waste sites have disproportionately high percentages of Hispanics (35 states), African Americans (38 states), and Asians/Pacific Islanders (27 states).

- Host neighborhoods of 105 of 149 metropolitan areas with hazardous waste sites (70%) have disproportionately high percentages of people of color, and 46 of these metro areas (31%) have majority people of color host neighborhoods.

Study Conclusions

- Environmental injustice in people of color communities is as much or more prevalent today than two decades ago.
- Racial and socioeconomic disparities in the location of the nation's hazardous waste facilities are geographically widespread throughout the country.
- People of color are concentrated in neighborhoods and communities with the greatest number of facilities; and people of color in 2007 are more concentrated in areas with commercial hazardous sites than in 1987.
- Race continues to be a significant independent predictor of commercial hazardous waste facility locations when socioeconomic and other non-racial factors are taken into account. . . .

Policy Recommendations

The Toxic Wastes and Race at Twenty report gives more than three dozen recommendations for action at the Congressional, state and local levels to help eliminate the disparities. The report also makes recommendations for nongovernmental agencies and the commercial hazardous waste industry. Based on these findings, I along with my colleagues and more than a hundred environmental justice, civil rights and human rights, and health allies are calling for steps to reverse this downward spiral. . . . We recommend the following policy actions:

1. **Hold Congressional Hearings on EPA Response to Contamination in EJ Communities.** We urge the U.S. Congress to hold hearings on the U.S. Environmental Protection Agency's (EPA's) response to toxic contamination in EJ communities, including post-Katrina New Orleans, the Dickson County (Tennessee) Landfill water contamination problem and similar problems throughout the United States.
2. **Pass a *National Environmental Justice Act* Codifying the Environmental Justice Executive Order 12898.** Executive Order 12898 "Federal Actions to Address Environmental Justice in Minority Populations and Low-Income Populations" provides significant impetus to advance environmental justice at the federal level and in the states. Congress should codify Executive Order 12898 into law. Congress will thereby establish an unequivocal legal mandate and impose federal responsibility in ways that advance equal protection under law in communities of color and low-income communities.
3. **Provide a Legislative "Fix" for Title VI of the Civil Rights Act of 1964.** Work toward a legislative "fix" of Title VI of the Civil Rights Act of 1964 that was gutted by the 2001 *Alexander v. Sandoval* U.S. Supreme Court decision that requires intent, rather than disparate

impact, to prove discrimination. Congress should act to reestablish that there is a private right of action for disparate impact discrimination under Title VI.

4. **Require Assessments of Cumulative Pollution Burdens in Facility Permitting.** EPA should require assessments of multiple, cumulative and synergistic exposures, unique exposure pathways, and impacts to sensitive populations in issuing environmental permits and regulations.

5. **Require Safety Buffers in Facility Permitting.** The EPA (states and local governments too) should adopt site location standards requiring a safe distance between a residential population and an industrial facility. It should also require locally administered Fenceline Community Performance Bonds to provide for the recovery of residents impacted by chemical accidents.

6. **Protect and Enhance Community and Worker Right-to-Know.** Reinstate the reporting of emissions and lower reporting thresholds to the Toxic Release Inventory (TRI) database on an annual basis to protect communities' right to know.

7. **Enact Legislation Promoting Clean Production and Waste Reduction.** State and local governments can show leadership in reducing the demand for products produced using unsustainable technologies that harm human health and the environment. Government must use its buying power and tax dollars ethically by supporting clean production systems.

8. **Adopt Green Procurement Policies and Clean Production Tax Policies.** Require industry to use clean production technologies and support necessary R&D for toxic use reduction and closed loop production systems. Create incentives and buy-back programs to achieve full recovery, reuse and recycling of waste and product design that enhances waste material recovery and reduction.

9. **Reinstate the Superfund Tax.** Congress should act immediately to reinstate the Superfund Tax, re-examine the National Priorities List (NPL) hazardous site ranking system and reinvigorate Federal Relocation Policy in communities of color to move those communities that are directly in harms way.

10. **Establish Tax Increment Finance (TIP) Funds to Promote Environmental Justice-Driven Community Development.** Environmental justice organizations should become involved in redevelopment processes in their neighborhoods to integrate brownfields priorities into long-range neighborhood redevelopment plans. This will allow for the use of Tax Increment Finance funds for cleanup and redevelopment of brownfields sites expressly for community-determined uses.

Getting government to respond to the environmental and health concerns of low-income and people of color communities has been an uphill struggle long before the world witnessed the disastrous Hurricane Katrina response nearly two years ago. The time to act is now. Our communities cannot wait another twenty years. Achieving environmental justice for all makes us a much healthier, stronger, and more secure nation as a whole.

 **NO**

U.S. Environmental Protection Agency before the Committee on Environment and Public Works Subcommittee on Superfund and Environmental Health United States Senate, July 25, 2007

EPA is a trailblazer in Federal government implementation of environmental justice programs. No other Federal agency has attempted to incorporate environmental justice into its programs, policies, and activities as comprehensively as the EPA. EPA is the lead for implementing Executive Order 12898, "Federal Actions to Address Environmental Justice in Minority Populations and Low-Income Populations. This Executive Order directs each Federal Agency to "make achieving environmental justice part of its mission." EPA works to comply with this Executive Order, and has taken significant and meaningful steps to integrate environmental justice into its mission."

In its role as lead agency for the Executive Order, EPA provides technical assistance to other Federal agencies on integrating environmental justice. For example, EPA has been working with the Centers for Disease Control and Prevention (CDC) in developing an environmental justice policy. EPA also is working with the National Center for Environmental Health/Agency for Toxic Substances and Disease Registry (ATSDR) to develop a strategy for integrating environmental justice goals within its programs and operations. Last week, EPA, CDC and ATSDR announced a memorandum of understanding (MOU) to collaborate on data gathering and sharing, and to find solutions for community health problems that could be linked to environmental hazards. Environmental justice was an important consideration in the development of this MOU.

Under the leadership of Administrator Johnson, EPA maintains an ongoing commitment to protect the environment for all people, regardless of race, color, national origin, or income, so that all people have the clean environment they deserve. We recognize that minority and/or low-income communities

From U.S. Senate Committee on Environment and Public Works, Subcommittee on Superfund and Environmental Health, July 25, 2007.

may be exposed disproportionately to environmental harms and risks. EPA works to protect these and other communities from adverse human health and environmental effects. Ensuring environmental justice means not only protecting human health and the environment for everyone, but also ensuring that all people are treated fairly and are given the opportunity to participate meaningfully in the development, implementation, and enforcement of environmental laws, regulations, and policies.

Integrating Environmental Justice into EPA's Mission

On November 4, 2005, Administrator Johnson reaffirmed EPA's commitment to environmental justice. He directed the Agency's managers and staff to integrate environmental justice considerations into EPA's core planning and budgeting processes. As a result, EPA has made transparent, measurable, and accountable environmental justice commitments and targets in all five goals of EPA's Strategic Plan for 2006–2011. Administrator Johnson identified eight national environmental justice priorities. Specifically, he directed the Agency to work with our partners to:

- Reduce asthma attacks;
- Reduce exposure to air toxics;
- Reduce incidences of elevated blood lead levels (ASTDR and the Department of Housing and Urban Development);
- Ensure that companies meet environmental laws;
- Ensure that fish and shellfish are safe to eat (Federal Drug Administration);
- Ensure water is safe to drink;
- Revitalize brownfields and contaminated sites; and
- Foster collaborative problem-solving.

EPA's Program Offices and Regions each implement an Environmental Justice Action Plan (Action Plan) to support EPA national priorities. These Action Plans are prospective planning documents that identify measurable commitments from each organization.

EPA's Chief Financial Officer directed the Agency's National Program Managers (NPMs) to include language in their FY2008 National Program Guidance that addresses the use of Action Plans and the Agency's 2006–2011 Strategic Plan to identify activities, initiatives, and/or strategies for the integration of environmental justice and incorporate them into planning and budgeting documents and program agreements. By instituting these types of programmatic requirements, EPA is building a stronger foundation to successfully integrate environmental justice into its programs for the long-term.

In addition, EPA's Inspector General recently identified the need for environmental justice program reviews. EPA agreed, and we have embarked on an extensive effort to develop and conduct those reviews. We are developing and piloting environmental justice review protocols for the Agency's core function areas—rule-making/standard setting, permitting, enforcement, and remediation/cleanup. Once these protocols are complete, the Agency will begin conducting the reviews in March 2008.

Lastly, the Office of Environmental Justice was made an *ex officio* member of the Agency's Regulatory Steering Committee. Its most important contribution in this role so far has been to develop environmental justice template language that assists rule writers in developing their Federal Register publications. The template ensures that the Agency's environmental justice considerations are accurately described to the public when proposed and final regulations are published after January 2007.

Obtaining the Best Available Environmental Justice Advice

EPA is taking actions to obtain the best available environmental justice advice and to impart any lessons learned to those who can work with us to address environmental justice issues at the federal, state and local levels.

Importantly, in 2006, EPA renewed the charter for the National Environmental Justice Advisory Council (NEJAC) thereby ensuring that EPA will continue to receive valuable advice and recommendations on national environmental justice policy issues from its stakeholders. The NEJAC is comprised of prominent representatives of local communities, academia, industry, and environmental, indigenous, as well as state, local, and tribal governments that can identify and recommend solutions to environmental justice problems. It is essential that EPA provide an opportunity for such discussions and for ideas to be aired, and that the NEJAC's advice and recommendations be appropriately integrated into EPA's environmental justice priorities and initiatives.

During the response to Hurricanes Katrina and Rita, EPA worked closely with NEJAC to ensure that environmental justice issues were addressed in a timely manner. Among a number of new initiatives, EPA has modified its Incident Command System to ensure an environmental justice function is incorporated into future responses. As part of this initiative, the Incident Commander is responsible for assuring that adequate resources are devoted to environmental justice issues. In addition, EPA Region 6's environmental justice team now participates in the Regional Incident Command Team. EPA also provided $300,000 in grant funding to encourage community-based organizations in EPA Regions 4 and 6 to participate in the decision-making (at all levels of government) related to cleanup, recovery, and rebuilding the hurricane-impacted areas in the Gulf Coast.

Imparting Lessons Learned

During the past 13 years and through the course of our more recent efforts, EPA has experienced first-hand the complexities of integrating environmental justice into the programs, policies, and activities of an agency as large and diverse as EPA.

Partnering for Maximum Effect

Most importantly, EPA has learned that addressing environmental justice issues is everyone's shared responsibility. Most environmental justice issues are local or site-specific—resolving these issues requires the concerted efforts of many stakeholders—Federal, state, local and tribal governments, community organizations, NGOs, academic institutions, business/industry, and even the community residents themselves. Since 1993, EPA has awarded more than $31 million in grants to more than 1,100 community-based organizations and others to take on an active role in our nation's environmental stewardship.

These environmental justice grants promote community empowerment and capacity-building—essential ingredients to maximize meaningful participation in the regulatory process. This year, EPA awarded $1 million in environmental justice grants to 10 community-based organizations, and will award an additional $1 million later this month to 20 community-based organizations to raise awareness and build their capacity to solve local environmental and public health issues.

The Power of Collaborative Problem Solving

EPA is proud of the progress that our many programs have made in environmental justice since President Clinton signed Executive Order 12898 in 1994. I would be remiss not to highlight a particular example that demonstrates not only EPA's success, but the success of other Federal, state, and local partners, and community groups.

EPA's relationship with ReGenesis, a community-based organization in Spartanburg, South Carolina, began in 1999 with a $20,000 grant award to address local environmental, health, economic and social issues. In 2003, EPA developed a Collaborative Problem-Solving (CPS) Model as a framework for others to follow. The model has worked well with amazing results. The ReGenesis Environmental Justice Partnership used elements of the CPS Model to leverage the initial grant from EPA to generate more than $166 million in funding, including over $1 million from EPA Region 4. ReGenesis marshaled the collaboration of more than 200 partner agencies, and local residents, industry, and a university to revitalize two Superfund sites and six Brownfields sites into new housing developments, an emergency access road, recreation areas, green space, and job training that are vital to the community's economic growth and well-being. This result was beyond anyone's expectation.

ReGenesis proved to be such an excellent example of what can be accomplished with EPA's funding, training and partnerships that we created a documentary film about it as a training tool to put thousands of other communities on the path of collaborative-problem solving. The DVD is being distributed across the country.

With the ongoing efforts in collaborative problem-solving and the grant programs, EPA is creating new opportunities to effectively target and address local environmental justice issues. By working together, everyone can benefit from the results.

Sharing Information

Since 2002, EPA has provided environmental justice training nationwide through the *Fundamentals of Environmental Justice* workshop, to almost 4,000 people, including staff in EPA and other government agencies. It is a long-term investment to ensure our workforce knows how to integrate environmental justice into their daily responsibilities. Some EPA offices have customized the training for their own organizations. For example, Region 1 has trained 98% of its workforce on environmental justice and has made it a training requirement for all new employees.

Drawing on the success of its classroom-based training, the Office of Environmental Justice introduced three Web-based courses during FY 2006: (1) *Introduction to Environmental Justice*, (2) *Introduction to the Toolkit for Assessing Potential Allegations of Environmental Injustice*, and (3) *Incorporating Environmental Justice Considerations into RCRA Permitting*. By using the latest on-line technology, EPA's training has become more cost effective and reaches a greater audience.

In addition to the importance of training, we also have learned that we must have a consistent approach to identify potential areas for environmental justice concern. My Office is developing a prototype tool, the Environmental Justice Strategic Enforcement Assessment Tool (EJSEAT), to enhance OECA's ability to consistently identify potential environmental justice areas, and assist us in making fair and efficient enforcement and compliance resource deployment decisions. Although we may have a tool and a process for ensuring consistency, variations in data availability may affect the tool's usefulness.

Future EPA Environmental Justice Efforts

The EPA successes I have highlighted today demonstrate that we are making significant headway on the road to environmental justice. To fully integrate and implement these concerns, the EPA and its Federal, state, tribal, local and community partners continue to work together to build a better model for the future. We are on that path today, and will continue to address all issues that come our way.

In moving forward, we will complete the environmental justice program reviews so that we can appropriately evaluate the effectiveness of EPA's actions for environmental justice. A number of successes thus far have been the result of innovative outreach rather than traditional EPA regulatory activity. That has to be factored into our plans for the future. We will focus on leveraging resources so that we can broaden our reach and replicate successes in encouraging collaborative problem-solving.

We will also finalize the Environmental Justice Strategic Enforcement Assessment Tool (EJSEAT) to enhance EPA's ability to consistently identify potential environmental justice areas of concern and assist EPA in making fair and efficient enforcement and compliance resource deployment decisions. We will evaluate the potential for applying the tool in other EPA programs and activities.

Based on the lessons we have learned and our efforts over the past 13 years, we are on a path forward with EPA's environmental justice programs. EPA will continue to integrate environmental justice considerations into the Agency's core programs, policies and activities and to engage others in collaborative problem-solving to address environmental justice concerns at every turn. Whenever and wherever we address environmental justice issues, we strive to build staying power in those communities and share any lessons learned with others.

POSTSCRIPT

Should the EPA Be Doing More to Fight Environmental Injustice?

The problems that led to the environmental justice movement have been documented in many reports. For example, in "Who Gets Polluted? The Movement for Environmental Justice," *Dissent* (Spring 1994), Ruth Rosen presents a history of the environmental justice movement, stressing how the movement has woven together strands of the civil rights and environmental struggles. Rosen argues that racial discrimination plays a significant role in the unusually intense exposure to industrial pollutants experienced by disadvantaged minorities, and she expresses the hope that "greening the ghetto will be the first step in greening our entire society." In addition, Robert D. Bullard's *Dumping in Dixie: Race, Class and Environmental Quality* (Westview Press, 1990, 1994, 2000) has become a standard text in the environmental justice field. In 2005, he edited *The Quest for Environmental Justice: Human Rights and the Politics of Pollution* (Sierra Club Books). Also see his *Unequal Protection: Environmental Justice and Communities of Color* (Sierra Club Books, 1994); Michael Heiman's "Waste Management and Risk Assessment: Environmental Discrimination Through Regulation," *Urban Geography* (vol. 17, no. 5, 1996); and Luke W. Cole and Sheila R. Foster's *From the Ground Up: Environmental Racism and the Rise of the Environmental Justice Movement* (New York University Press, 2000). David W. Allen, in "Social Class, Race, and Toxic Releases in American Counties, 1995," *Social Science Journal* (vol. 38, no. 1, 2001), finds that the data support the existence of environmental racism but that the effect is strongest in the southern portion of the United States (the Sun Belt). Julian Agyeman, "Where Justice and Sustainability Meet," *Environment* (July/August 2005), argues that improving justice, equity, and human rights, and reducing poverty, are essential components of sustainable development. Robert J. Brulle and David N. Pellow, "Environmental Justice: Human Health and Environmental Inequalities," *Annual Review of Public Health* (No. 1, 2006), review unequal exposures to environmental pollution and call for more research. Bradley C. Parks and J. Timmons Roberts, "Globalization, Vulnerability to Climate Change, and Perceived Injustice," *Society & Natural Resources* (April 2006), link environmental injustice to vulnerability to global climate change. Michael Eric Dyson, *Come Hell or High Water: Hurricane Katrina and the Color of Disaster* (Basic Books, 2006), argues that racism influenced the disastrous attempts to cope and recover when Hurricane Katrina struck New Orleans. It is perhaps not unreasonable to say that it is environmental racism when minority groups are forced to bear a disproportionate burden of the impact of environmental disasters, whether they are natural or human-caused.

Elisabeth Jeffries, "E-Wasted," *World Watch* (July/August 2006), notes that an enormous amount of toxic waste is discarded electronic products such as computers and cell phones, and most is shipped to Africa and Asia where poor people—including children—employed to extract valuable metals are exposed to toxic dusts and fumes. See also Ted Smith, David A. Sonnenfeld, and David Naguib Pellow, eds., *Challenging the Chip: Labor Rights and Environmental Justice in the Global Electronics Industry* (Temple University Press, 2006). The Silicon Valley Toxics Coalition website (http://svtc.etoxics.org/site/PageServer) offers information and access to numerous reports on the topic of e-waste.

Those who criticize the environmental justice movement tend to focus on other studies. In "Green Redlining: How Rules Against 'Environmental Racism' Hurt Poor Minorities Most of All," *Reason* (October 1998), Henry Payne labels the Environmental Protection Agency's efforts to impose environmental equity "redlining" and, like Friedman, argues that the practice reduces job opportunities and economic benefits for minorities.

There is great contrast in the sides to this debate. In such cases, the reader must not ignore the social values and political commitments of the debaters. The reader must also be careful to consider the data relied on by the debaters and to watch for unsupported claims and simplistic explanations for events whose causes are likely to be more complicated.

Where is government policy going? Jim Motavalli, in "Toxic Targets: Polluters That Dump on Communities of Color Are Finally Being Brought to Justice," *E: The Environmental Magazine* (July–August 1998), states that although minorities and the poor have been forced to bear a disproportionate share of the burden of industrial pollution, changes in environmental policy and law are finally offering remedies. The Environmental Protection Agency has affirmed and reaffirmed its commitment to improving environmental justice for all. However, what it can do is limited, perhaps largely because recent federal budgets have slashed EPA funding. See Chris Mooney, "Earth Last," *The American Prospect* (May 2004).

ISSUE 6

Can Pollution Rights Trading Effectively Control Environmental Problems?

YES: James Allen and Anthony White, from "Carbon Trading," *Electric Perspectives* (September/October 2005)

NO: Brian Tokar, from "Trading Away the Earth: Pollution Credits and the Perils of 'Free Market Environmentalism,'" *Dollars & Sense* (March/April 1996)

ISSUE SUMMARY

YES: James Allen and Anthony White describe the European Union's Greenhouse Gas Emissions Trading Scheme and argue that it encourages investment in carbon-abatement technologies and depends on governmental commitments to reducing emissions despite possible adverse economic effects.

NO: Author, college teacher, and environmental activist Brian Tokar maintains that pollution credits and other market-oriented environmental protection policies do nothing to reduce pollution while transferring the power to protect the environment from the public to large corporate polluters.

F ollowing World War II the United States and other developed nations experienced an explosive period of industrialization accompanied by an enormous increase in the use of fossil fuel energy sources and a rapid growth in the manufacture and use of new synthetic chemicals. In response to growing public concern about the pollution and other forms of environmental deterioration resulting from this largely unregulated activity, the U.S. Congress passed the National Environmental Policy Act of 1969. This legislation included a commitment on the part of the government to take an active and aggressive role in protecting the environment. The next year the Environmental Protection Agency (EPA) was established to coordinate and oversee this effort. During the next two decades an unprecedented series of legislative acts and administrative rules were promulgated, placing numerous restrictions on industrial and commercial activities that might result in the pollution, degradation, or contamination of land, air, water, food, and the workplace.

Such forms of regulatory control have always been opposed by the affected industrial corporations and developers as well as by advocates of a free-market policy. More moderate critics of the government's regulatory program recognize that adequate environmental protection will not result from completely voluntary policies. They suggest that a new set of strategies is needed. Arguing that "top down, federal, command and control legislation" is not an appropriate or effective means of preventing ecological degradation, they propose a wide range of alternative tactics, many of which are designed to operate through the economic marketplace. The first significant congressional response to these proposals was the incorporation of tradable pollution emission rights into the 1990 Clean Air Act amendments as a means for achieving the set goals for reducing acid rain-causing sulfur dioxide emissions. More recently, the 1997 international negotiations on controlling global warming in Kyoto, Japan, resulted in a protocol that includes emissions trading as one of the key elements in the plan to limit the atmospheric buildup of greenhouse gases.

Charles W. Schmidt, "The Market for Pollution," *Environmental Health Perspectives* (August 2001), argues that emissions trading schemes represent "the most significant developments" in the use of economic incentives to motivate corporations to reduce pollution. . . . Many environmentalists, however, continue to oppose the idea of allowing anyone to pay to pollute, either on moral grounds or because they doubt that these tactics will actually achieve the goal of controlling pollution. Diminishment of the acid rain problem is often cited as an example of how well emission rights trading can work, but in "Dispelling the Myths of the Acid Rain Story," *Environment* (July–August 1998), Don Munton argues that other control measures, such as switching to low-sulfur fuels, deserve much more of the credit for reducing sulfur dioxide emissions.

In "A Low-Cost Way to Control Climate Change," *Issues in Science and Technology* (Spring 1998), Byron Swift argues that the "cap-and-trade" feature of the U.S. Acid Rain Program has been so successful that a similar system for implementing the Kyoto Protocol's emissions trading mandate as a cost-effective means of controlling greenhouse gases should work. In March 2001 the U.S. Senate Committee on Agriculture, Nutrition, and Forestry held a "Hearing on Biomass and Environmental Trading: Opportunities for Agriculture and Forestry," in which witnesses urged Congress to encourage trading for both its economic and its environmental benefits.

According to Bret Schulte, "Putting a Price on Pollution," *U.S. News and World Report* (May 14, 2007), the Bush administration continues to oppose emissions trading but support in the United States is growing. The European Union has actually established a Greenhouse Gas Emissions Trading Scheme, which James Allen and Anthony White describe in the following selections, arguing that it encourages investment in carbon-abatement technologies and depends on governmental commitments to reducing emissions despite possible adverse economic effects. "The carbon market in Europe is here to stay," they say. Brian Tokar has a much more negative assessment of pollution credit trading. Focusing on its application to sulfur dioxide, not carbon dioxide, he argues that such "free-market environmentalism" tactics fail to reduce pollution while turning environmental protection into a commodity that corporate powers can manipulate for private profit.

YES

James Allen and
Anthony White

Carbon Trading

The European Union Greenhouse Gas Emissions Trading Scheme (EUETS) rests on a simple concept: The right to emit greenhouse gases (GHGs) can be allocated and traded. The scheme's purpose is relatively straightforward, as well: to help the 25 member states of the European Union achieve an 8-percent collective reduction below 1990 levels of six greenhouse gases by 2008–12.

More than 12,000 installations, totaling nearly half of all emissions in the European Union, must retire these rights (EU emissions allowances, or EUAs) corresponding to their actual emissions over the compliance period. Failure to do so will result in a fine of €40 (around $48) per ton of CO_2 in excess during Phase I (2005–07), rising to €100 ($120) per ton (equivalent) in Phase II (2008–12). And paying the fine does not remove the obligation to retire the missing certificates.

Also, the European Commission environmental directorate designed EUETS with the Kyoto Protocol in mind. In place of EUAs, installations can retire "clean development mechanism" [CDM] credits (generated through GHG reduction projects in developing countries) in Phase I and, in Phase II, CDM and joint implementation (JI) credits (the latter being credits generated in developed countries that have ratified the protocol). Companies may trade all these credits among themselves—the market mechanism is to ensure that abatement be achieved more efficiently than through a command-and-control regimen. Trading is currently limited primarily to futures contracts, traded over-the-counter, but increasingly it is mediated via exchanges, such as the European Climate Exchange. EUETS provides a price signal to companies that operate and own emissions-intensive installations, which encourages investment in carbon-abatement technologies, as well as measures to increase efficiency. This kind of cap-and-trade system is fundamentally policy-driven—that is, it depends entirely on the will of governments to impose emission reduction obligations, despite the possibility that these may create adverse economic effects.

The European carbon market is in a critical phase. Already, fundamental changes in the European economy—particularly a switch from coal- to gas-fired power generation—are taking place. Carbon funds are beginning to attract private capital to invest in renewable technologies, often in developing countries. And even at this early stage, when only half of all installations have received allowances, fully 1 million tons of CO_2 are traded each day.

From *Electric Perspectives,* September/October 2005, pp. 50–59. Copyright © 2005 by Edison Electric Institute. Reprinted by permission.

Still, the price of emitting CO_2 has risen beyond many predictions, mostly due to unexpected growth in natural gas prices. Meanwhile, member states are debating whether to expand EUETS to gases other than CO_2 and to additional sectors of the economy, such as transportation, responsible for a great deal of all EU emissions. Moreover, bureaucracy at the United Nations (which oversees Kyoto and the CDM and JI credits) and a lack of experience in this new market are limiting investment opportunities for carbon speculators.

But benefits and challenges aside, one thing is clear: The world's largest emissions trading scheme has arrived.

Long-Term Commitment or Unstable Pact?

From a European perspective, policy on climate change in the United States is uneven, though certainly there are efforts to push markets. The federal government is investing large sums of money to promote new renewable energy, nuclear, and clean coal technologies; at the same time the Bush administration, opposed to mandatory CO_2 emission reductions, has focused on voluntary efforts. Meanwhile at the state and regional levels, efforts are moving forward—but they face many hurdles.

In contrast, European governments have settled on a long-term course of action. And, in addition to the targets agreed at Kyoto and implemented via the EUETS, some EU nations have set mandatory emission targets for dates beyond 2012. Some of these targets may be aspirational, but the United Kingdom, for example, has set a long-term target for a 60-percent reduction of CO_2 emissions of 1990 levels by 2050. In any event, it is likely that EUETS will be the linchpin for all efforts to reduce European emissions. Successful implementation of the scheme, along with a demonstration of minimal cost and competitiveness effects, will fortify efforts within the European Union (and possibly abroad) to guarantee a long-term continuation of the current arrangements.

Still, people compare EUETS with the European Stability and Growth Pact, which was negotiated in 1997 by members of the economic and monetary union (the pact included adoption of the euro) within the European Union. Among other things, the pact limits a nation's annual budget deficit to lower than 3 percent of GDP, and a public debt lower than 60 percent of GDP. Several member states have flouted these rules (France and Germany, in particular) but faced little punishment from the European Commission.

Regarding EUETS, the argument goes something like this: If France and Germany could not adhere to their commitment to a fundamental economic component of European monetary union, what is to keep them from doing the same thing with EUETS? Actually, several things, first, the attempt to reduce global GHG emissions does not derive from an arbitrary line drawn in the macro-economic sand—the member states accept the fundamental proposition in the Kyoto Protocol about climate change and its scientific underpinnings. The recent pronouncement by the scientific academies of the G8 nations (together with those of China, Brazil, and India) regarding the validity of the problem, highlights the necessity for addressing it.

There are economic incentives, as well. One way for member states to weaken the scheme would be to issue more allowances. But, since each state has passed national legislation creating the allowances, such action would damage the "property rights" of the various plant operators. Moreover, member states could also be held liable to compensate for economic losses associated with a collapsed market.

And, after 2012 (under the Kyoto Protocol), member states in noncompliance will have to retire allowances corresponding to any excess over their target level, plus an additional 30 percent. (It may be that a larger international trading scheme may not come to fruition, thereby making Kyoto compliance measures after 2012 problematic. But most observers expect such a scheme to be established.) The strength of the legislation agreed to by European governments is therefore a more onerous constraint on their behavior than the debt limits implied by the Stability and Growth Pact. In short, countries should not fail to fulfil their obligations under the EUETS directive. But, because an international emissions trading scheme inevitably creates competitive effects on industries covered by it, any scheme will likely result in protectionist measures. Therefore, the commitment of individual countries, the strength of legislation, and the effectiveness of noncompliance procedures will all be important determinants of its eventual success.

National Plans

For Phase I of EUETS, the member states have set targets, defined right down to the plant level for 2007, in National Allocation Plans (NAPs). (Nations must submit Phase II allocations plans for approval by the end of June in 2006.) In Phase I, each plant will know, from the moment of its allocation, precisely how much it may emit during the period 2005-07. If it needs to emit more than allocated, the company must purchase the right to do so from other installations within the scheme, or from CDM projects in developing countries. The carbon-intensity of an installation within the European Union is now an asset or a liability that can be monetized. It is equal to the value of allowances that must be bought to offset . . . emissions above the allocated level.

The level of allocation, also, defines the scale of potential asset or liability. EUETS prevents member states from auctioning more than 5 percent of EUAs in Phase I and 10 percent in Phase II. Allowances have so far been grandfathered to installations based on historical projections and the "business as usual" case, with a little bit subtracted. It is possible that the allocation methodology will be revised in Phase II to allow for "benchmarking," which defines "standard" levels of emissions for different sectors based on both historical trends and newly available technologies. The benchmark provides the level of emissions the regulator thinks [is] necessary (or appropriate) for a sector or installation.

Many debate whether allocation or auction of allowances is most efficient in a new market. Some see allocations as a market-creation strategy that has a proven track record in other cap-and-trade regimes; and auctions as a way to make a company pay twice for compliance (that is, for compliance technology and the allowance). On the other hand, many consider grandfathering and benchmarking as attempts to "predict" the market, with government allocating allowances where it thinks they are most needed. Historical and benchmarked emissions may not reflect technological feasibility but rather the amount of effort invested in reducing emissions from a particular sector. Allocating rather than auctioning allowances may therefore impose minimal requirements on inefficient (or high-emissions-intensity) economic sectors while punishing efficient ones.

Perhaps changes will take place in the post-2012, Phase III EUETS, but it will likely depend on the structure of the post-2012 international trading scheme.

Another issue is the harmonization of NAPs across the European Union. Member states have the flexibility to implement several aspects of the EUETS Directive in their own ways. Clearly this is important, due to different conditions in different nations. A country with low capacity margin, for example, may wish to allocate more allowances to its power sector (and less elsewhere) than another country, in order to protect domestic consumers against high electricity prices.

Decisions to include or exclude some plants (based on the definition of "small" and "combustion"), and the treatment of new entrants to the scheme during 2005-07, have also varied among states. Whether this is related to economics or simply politics (probably both), such decisions will have an impact on domestic competitiveness and may therefore put some companies at a disadvantage. But other issues, such as the scope of the scheme (relating both to gases—only CO_2 thus far—and sectors—currently excluding transportation), have fallen evenly across member states. Harmonization across the European Union is necessary, because no country wishes to move first and risk being left alone to the detriment of its domestic industry. Because such unanimity is difficult to achieve, no substantial changes appear likely before 2012 (with the possible exception of the inclusion of the aviation sector).

Like aviation, many sectors (transportation, most significantly) that contribute large amounts of GHGs are not included in EUETS. Still, these areas will face measures to reduce emissions in line with the overall national targets. The Directive on Renewable Transport Fuels, for example, obliges member states to devise plans to meet (nonbinding) targets to replace a proportion of gasoline and diesel with ethanol and biodiesel. Other measures include a directive on energy performance of buildings; and a directive that sets targets for the reduction of biodegradable waste sent to landfill. None of these measures has the strength of EUETS, but for all these sectors, there are efforts to create markets to provide reduction incentives.

Business as Usual—Except Power

Let's look at the "business-as-usual" (BAU) allocations. In total they have equaled roughly 97 percent of covered entities projected emissions in a BAU scenario. That makes a shortfall of 200 million tons (plus or minus 50 million tons) of CO_2 during Phase I. The sector that has taken the biggest hit is the power sector. In the United Kingdom, powerplants have received, on average, 72 percent of reference emissions. (These are the emissions upon which allocations are based. The "reference" is the average of the five highest annual emissions totals from 1998-2003.)

For example, AES Kilroot, a coal-fired power station in Northern Ireland with 520 megawatts of capacity, received allowances corresponding to 2.05 million tons of CO_2 emissions per year during Phase I. This is 25 percent below Kilroot's 2.71 million tons of actual emissions in 2003. In comparison, cement manufacturers received 96.5 percent of reference emissions, while chemicals plants received 88.0 percent of theirs. Moreover, a company loses remaining allocations if a plant is closed during the commitment period, a fact that serves to ensure that aging capacity remains available, even if does not operate for much of that time.

All this means that operating costs for coal plants have risen sufficiently to induce coal generators to switch from operating at their traditional baseload or midpeak load to move to the margin, displacing existing marginal gas capacity.

From that method of abatement, approximately 120 million EUAs per year are available within EUETS. In fact, the "switch" from coal to gas is the cheapest form of abatement. Hence the cost of the coal-gas switch drives the cost of EUAs. And what drives the cost of the coal-gas switch? The relative cost of coal and gas within the European Union.

Many analysts have been surprised by the levels EUA prices have reached. Indeed, our own estimates were of the range €5–10 ($6-12) per ton of CO_2 for Phase I and €10–15 ($12-18) for Phase II. But this was based on costs of coal and natural gas at the time, which were in line with the long-run marginal cost of production. Since then gas prices have doubled. Our models show that current prices (as of July) imply a price of roughly €20–25 ($24-30) per ton of CO_2. In both cases, the coal-gas differential provides the best estimate of allowance prices in EUETS. Volumes have reached an all-time high, with 1 million tons of CO_2 being traded per day, on average. But liquidity is still low, given that less than half of all installations have actually received their allowances. A cold winter and hot, dry summer have also helped to drive prices upward.

There should also be a change in allowance prices between Phases I and II because of the tighter allocations expected in the latter. All but two member states restrict the banking of allowances from one period to the other (though EUETS allows banking of CDM credits). Banking, however, would have reduced the volatility in allowance prices by placing an option value on Phase I allowances. It would also have rewarded early abatement in the first phase. However, governments were reluctant to permit banking because they worried it would result in unacceptably high allowance prices during Phase I and give states less control over Phase II allocations.

The Next Phase and the United Nations

As June 30, 2006, approaches, the focus turns to Phase II NAPs. Broadly speaking, if the European Union wants to reach the Kyoto targets, allocations must be far stricter in this phase than the previous one. Current targets put EU emissions at 2 percent below their 1990 levels by the end of 2007; by 2010, this will need to be 8 percent below, requiring roughly 100 million tons of abatement each year (compared to 70 million tons from 2005-07). As for the NAPs, it is unlikely that the member states will add other gases to the trading list or include other sectors in EUETS. The European Commission recently announced that it will not seek to make any changes to the scheme until 2012. While it is possible for governments to agree to changes without Commission involvement, the time this would normally take seems prohibitive given the deadline for Phase II NAPs.

Clouding the picture are difficulties with the CDM mechanism and a lack of visibility over the JI mechanism. The CDM Executive Board, a United Nations institution, oversees the CDM process and registers emission reduction projects in developing countries as being able to generate CDM allowances, known as certified emissions reductions, or CERs. But, so far in Phase I, a lack of adequate funding and an overly bureaucratic registration process have seriously affected the volume of authorized projects. Currently, in fact, only 11 projects are capable of generating CERs. Just 60 million CERs are expected by the end of 2007, compared to a total EU shortfall of roughly 200 million. Volumes up to 2012 are very uncertain but may not reach the originally predicted 500 million.

The JI mechanism is uncertain because it depends heavily on the governments of developed countries, which must create the procedures by which a project can claim JI allowances (known as emission reduction units, or ERUs). ERUs cannot be used for compliance in EUETS until 2008, so there has been little movement. Russia, the last Kyoto signatory, recently created the first JI project, although it awaits government approval, procedures for which are not yet in place.

Russia is, in a sense, the OPEC of Kyoto. In addition to allowances at the installation level, the Kyoto Protocol established "assigned amount units" (AAUs) for national governments. One allowance for each ton of CO_2 equivalent corresponding to the total amount of emissions permitted between 2008 and 2012. And Russia has significantly more allowances than required to meet its Kyoto obligations—due to the collapse of the Soviet Union, emissions have dropped since 1990, the reference year for Kyoto targets. With all that so-called extra hot air, Russia and other former Soviet Union states could probably supply the entire demand for allowances in EUETS Phase II. They can't due to Kyoto's supplementarity principle, which will probably require a 50-percent limit on the use of foreign-sourced AAUs. (The principle was included in the protocol at the behest of environmental groups, which sought to encourage abatement in the industrialized world and let developed nations show leadership in "cleaning up their own backyards.") But, just as the Organization of Petroleum Exporting Countries restricts its supply of oil, Russia may restrict its supply of allowances to keep prices high and to maintain a competitive position for a post-2012 system.

Would Governments Limit the Price of Carbon?

But back to fossil fuels. Driving the current high prices of natural gas in the European Union is the soaring cost of oil, which has recently reached historic highs. What if oil prices were to climb even higher? Under these circumstances, allowance prices could reach sufficient levels to prompt coal plants to close in 2007 or perhaps for carbon capture and sequestration to become economical.

One can reasonably expect governments not to allow prices in EUETS to go too high; and prices in excess of €20–30 ($12–36) per ton of CO_2 sustained over the Phase I period could force governments to limit the reductions for coal plants in Phase II. Governments are likely to be unwilling to let plants close when there are decreasing capacity margins across the European Union and a policy objective to maintain diversity in energy supplies. Also, many coal plants have installed or are in the process of installing flue-gas desulphurization equipment (typically costing €45 million, or $54 million) to comply with the EU directive on large combustion plants. Sectors under EUETS that are most exposed to international competitors (such as the aluminum sector) face potentially damaging competitive effects. Recently, for example, the owners of the HAW aluminum mill in Hamburg, Germany, said high EUA prices contributed to the decision to shut the facility. But most important is the political fall-out that would result from high electricity prices when the full costs of carbon are passed through.

To keep plants in operation and electricity prices down, governments might allocate to them sufficient allowances over Phase II and apply the hit in another industry sector. Indeed, governments could avoid EUETS entirely by purchasing CDMs, JIs, and AAUs. There are currently more than 20 European government procurement funds (including investments in World Bank and other multilateral funds) with a total committed capital of over €1 billion—at a final purchase price of €6 per ton of CO_2, this would yield 170 million allowances. The value of such emission reduction purchase agreements actually signed, however, is unknown, given a lack of market data. But is likely to be less than €1 billion.

The fact that this is a policy-driven market means that governments will take whatever action necessary to minimize impacts on domestic industry and consumers—that is, voters. But because European governments have also shown a willingness to use EUETS to achieve their policy goals, it is likely they will allocate few enough allowances to guarantee a minimum price in Phase II (say, €10 per ton), but not too few to make prices spiral out of control.

After Kyoto

And what about post-2012? These discussions continue. At the G8 summit in Scotland last July, the participants agreed to an ongoing dialogue on climate change, as well as on action to develop low-carbon technologies. But there was no progress about binding emission reduction targets outside the Kyoto Protocol. In December 2005, the members of the United Nations Foundation on Climate Change will meet in Montreal. The best estimate is that the "Conference of the Parties" will hold to the status quo and not wrangle over the future course of international policy.

A recent study published by Climate Change Capital suggests that an agreement may be possible that would incorporate the United States and China, India, and Brazil in emission reduction efforts. The study finds that price caps (to limit abatement costs) and flexible targets (such as emissions-intensity targets, linking reductions to the rate of GDP growth) would be attractive to governments concerned about the economic impacts of reducing GHG emissions. Other mechanisms that could prompt agreement include positively-binding targets on developing countries and an expanded CDM. The positively-binding target in an international emissions trading scheme would reward over-achievement of a target with tradable credits, but would not impose a penalty for under-achievement. The expanded CDM would let companies in developed countries buy a greater number of less expensive credits generated in developing countries, where abatement costs are lower.

Of course, if nations reach an international agreement beyond the Kyoto Protocol, European efforts to reduce GHG emissions will not occur alone. To many, the signs for wider participation are favorable. Recent developments in the United States, for example, seem to make an enhancement of its emissions trading scheme a possibility in the future. Whatever the outcome, the political climate in Europe is currently widely supportive of EUETS; for the most part, skepticism exists not about scientific basis but about competitiveness with nonparticipating countries. In the main, the carbon market in Europe is here to stay.

Brian Tokar  **NO**

Trading Away the Earth: Pollution Credits and the Perils of "Free Market Environmentalism"

The Republican takeover of Congress has unleashed an unprecedented assault on all forms of environmental regulation. From the Endangered Species Act to the Clean Water Act and the Superfund for toxic waste cleanup, laws that may need to be strengthened and expanded to meet the environmental challenges of the next century are instead being targeted for complete evisceration.

For some activists, this is a time to renew the grassroots focus of environmental activism, even to adopt a more aggressively anti-corporate approach that exposes the political and ideological agendas underlying the current backlash. But for many, the current impasse suggests that the movement must adapt to the dominant ideological currents of the time. Some environmentalists have thus shifted their focus toward voluntary programs, economic incentives and the mechanisms of the "free market" as means to advance the cause of environmental protection. Among the most controversial, and widespread, of these proposals are tradeable credits for the right to emit pollutants. These became enshrined in national legislation in 1990 with President George Bush's amendments to the 1970 Clean Air Act.

Even in 1990, "free market environmentalism" was not a new phenomenon. In the closing years of the 1980s, an odd alliance had developed among corporate public relations departments, conservative think tanks such as the American Enterprise Institute, Bill Clinton's Democratic Leadership Council (DLC), and mainstream environmental groups such as the Environmental Defense Fund. The market-oriented environmental policies promoted by this eclectic coalition have received little public attention, but have nonetheless significantly influenced debates over national policy.

Glossy catalogs of "environmental products," television commercials featuring environmental themes, and high profile initiatives to give corporate officials a "greener" image are the hallmarks of corporate environmentalism in the 1990s. But the new market environmentalism goes much further than these showcase efforts. It represents a wholesale effort to recast environmental protection based on a model of commercial transactions within the marketplace. "A

From *Dollars & Sense*, March/April 1996, pp. 24–29. Reprinted by permission of Dollars & Sense, a progressive economics magazine. www.dollarsandsense.org.

new environmentalism has emerged," writes economist Robert Stavins, who has been associated with both the Environmental Defense Fund and the DLC's Progressive Policy Institute, "that embraces . . . market-oriented environmental protection policies."

Today, aided by the anti-regulatory climate in Congress, market schemes such as trading pollution credits are granting corporations new ways to circumvent environmental concerns, even as the same firms try to pose as champions of the environment. While tradeable credits are sometimes presented as a solution to environmental problems, in reality they do nothing to reduce pollution—at best they help businesses reduce the costs of complying with limits on toxic emissions. Ultimately, such schemes abdicate control over critical environmental decisions to the very same corporations that are responsible for the greatest environmental abuses.

How It Works, and Doesn't

A close look at the scheme for nationwide emissions trading reveals a particular cleverness; for true believers in the invisible hand of the market, it may seem positively ingenious. Here is how it works: The 1990 Clean Air Act amendments were designed to halt the spread of acid rain, which has threatened lakes, rivers and forests across the country. The amendments required a reduction in the total sulfur dioxide emissions from fossil fuel burning power plants, from 19 to just under 9 million tons per year by the year 2000. These facilities were targeted as the largest contributors to acid rain, and participation by other industries remains optional. To achieve this relatively modest goal for pollution reduction, utilities were granted transferable allowances to emit sulfur dioxide in proportion to their current emissions. For the first time, the ability of companies to buy and sell the "right" to pollute was enshrined in U.S. law.

Any facility that continued to pollute more than its allocated amount (roughly half of its 1990 rate) would then have to buy allowances from someone who is polluting less. The 110 most polluting facilities (mostly coal burners) were given five years to comply, while all the others would have until the year 2000. Emissions allowances were expected to begin selling for around $500 per ton of sulfur dioxide, and have a theoretical ceiling of $2000 per ton, which is the legal penalty for violating the new rules. Companies that could reduce emissions for less than their credits are worth would be able to sell them at a profit, while those that lag behind would have to keep buying credits at a steadily rising price. For example, before pollution trading every company had to comply with environmental regulations, even if it cost one firm twice as much as another to do so. Under the new system, a firm could instead choose to exceed the mandated levels, purchasing credits from the second firm instead of implementing costly controls. This exchange would save money, but in principle yield the same overall level of pollution as if both companies had complied equally. Thus, it is argued, market forces will assure that the most cost-effective means of reducing acid rain will be implemented first, saving the economy billions of dollars in "excess" pollution control costs.

Defenders of the Bush plan claimed that the ability to profit from pollution credits would encourage companies to invest more in new environmental technologies than before. Innovation in environmental technology, they argued, was being stifled by regulations mandating specific pollution control methods. With the added flexibility of tradeable credits, companies could postpone costly controls—through the purchase of some other company's credits—until new technologies became available. Proponents argued that, as pollution standards are tightened over time, the credits would become more valuable and their owners could reap large profits while fighting pollution.

Yet the program also included many pages of rules for extensions and substitutions. The plan eliminated requirements for backup systems on smokestack scrubbers, and then eased the rules for estimating how much pollution is emitted when monitoring systems fail. With reduced emissions now a marketable commodity, the range of possible abuses may grow considerably, as utilities will have a direct financial incentive to manipulate reporting of their emissions to improve their position in the pollution credits market.

Once the EPA actually began auctioning pollution credits in 1993, it became clear that virtually nothing was going according to their projections. The first pollution credits sold for between $122 and $310, significantly less than the agency's estimated minimum price, and by 1995, bids at the EPA's annual auction of sulfur dioxide allowances averaged around $130 per ton of emissions. As an artificial mechanism superimposed on existing regulatory structures, emissions allowances have failed to reflect the true cost of pollution controls. So, as the value of the credits has fallen, it has become increasingly attractive to buy credits rather than invest in pollution controls. And, in problem areas air quality can continue to decline, as companies in some parts of the country simply buy their way out of pollution reductions.

At least one company has tried to cash in on the confusion by assembling packages of "multi-year streams of pollution rights" specifically designed to defer or supplant purchases of new pollution control technologies. "What a scrubber really is, is a decision to buy a 30-year stream of allowances," John B. Henry of Clean Air Capital Markets told the *New York Times*, with impeccable financial logic. "If the price of allowances declines in future years," paraphrased the *Times*, "the scrubber would look like a bad buy."

Where pollution credits have been traded between companies, the results have often run counter to the program's stated intentions. One of the first highly publicized deals was a sale of credits by the Long Island Lighting Company to an unidentified company located in the Midwest, where much of the pollution that causes acid rain originates. This raised concerns that places suffering from the effects of acid rain were shifting "pollution rights" to the very region it was coming from. One of the first companies to bid for additional credits, the Illinois Power Company, canceled construction of a $350 million scrubber system in the city of Decatur, Illinois. "Our compliance plan is based almost totally on purchase of credits," an Illinois Power spokesperson told the *Wall Street Journal*. The comparison with more traditional forms of commodity trading came full circle in 1991, when the government announced that the entire system for trading and auctioning emissions allowances would be administered by the Chicago Board of Trade, long famous for its ever-frantic markets in everything from grain futures and pork bellies to foreign currencies.

Some companies have chosen not to engage in trading pollution credits, proceeding with pollution control projects, such as the installation of new scrubbers, that were planned before the credits became available. Others have switched to low-sulfur coal and increased their use of natural gas. If the 1990 Clean Air Act amendments are to be credited for any overall improvement in the air quality, it is clearly the result of these efforts and not the market in tradeable allowances.

Yet while some firms opt not to purchase the credits, others, most notably North Carolina-based Duke Power, are aggressively buying allowances. At the 1995 EPA auction, Duke Power alone bought 35% of the short-term "spot" allowances for sulfur dioxide emissions, and 60% of the long-term allowances redeemable in the years 2001 and 2002. Seven companies, including five utilities and two brokerage firms, bought 97% of the short-term allowances that were auctioned in 1995, and 92% of the longer-term allowances, which are redeemable in 2001 and 2002. This gives these companies significant leverage over the future shape of the allowances market.

The remaining credits were purchased by a wide variety of people and organizations, including some who sincerely wished to take pollution allowances out of circulation. Students at several law schools raised hundreds of dollars, and a group at the Glens Falls Middle School on Long Island raised $3,171 to purchase 21 allowances, equivalent to 21 tons of sulfur dioxide emissions over the course of a year. Unfortunately, this represented less than a tenth of one percent of the allowances auctioned off in 1995.

Some of these trends were predicted at the outset. "With a tradeable permit system, technological improvement will normally result in lower control costs and falling permit prices, rather than declining emissions levels," wrote Robert Stavins and Brad Whitehead (a Cleveland-based management consultant with ties to the Rockefeller Foundation) in a 1992 policy paper published by the Progressive Policy Institute. Despite their belief that market-based environmental policies "lead automatically to the cost-effective allocation of the pollution control burden among firms," they are quite willing to concede that a tradeable permit system will not in itself reduce pollution. As the actual pollution levels still need to be set by some form of regulatory mandate, the market in tradeable allowances merely gives some companies greater leverage over how pollution standards are to be implemented.

Without admitting the underlying irrationality of a futures market in pollution, Stavins and Whitehead do acknowledge (albeit in a footnote to an Appendix) that the system can quite easily be compromised by large companies' "strategic behavior." Control of 10% of the market, they suggest, might be enough to allow firms to engage in "price-setting behavior," a goal apparently sought by companies such as Duke Power. To the rest of us, it should be clear that if pollution credits are like any other commodity that can be bought, sold and traded, then the largest "players" will have substantial control over the entire "game." Emissions trading becomes yet another way to assure that large corporate interests will remain free to threaten public health and ecological survival in their unchallenged pursuit of profit.

Trading the Future

Mainstream groups like the Environmental Defense Fund (EDF) continue to throw their full support behind the trading of emissions allowances, including the establishment of a futures market in Chicago. EDF senior economist Daniel Dudek described the trading of acid rain emissions as a "scale model" for a much more ambitious plan to trade emissions of carbon dioxide and other gases responsible for global warming. This plan was unveiled shortly after the passage of the 1990 Clean Air Act amendments, and was endorsed by then-Senator Al Gore as a way to "rationalize investments" in alternatives to carbon dioxide-producing activities.

International emissions trading gained further support via a U.N. Conference on Trade and Development study issued in 1992. The report was coauthored by Kidder and Peabody executive and Chicago Board of Trade director Richard Sandor, who told the *Wall Street Journal*, "Air and water are simply no longer the 'free goods' that economists once assumed. They must be redefined as property rights so that they can be efficiently allocated."

Radical ecologists have long decried the inherent tendency of capitalism to turn everything into a commodity; here we have a rare instance in which the system fully reveals its intentions. There is little doubt that an international market in "pollution rights" would widen existing inequalities among nations. Even within the United States, a single large investor in pollution credits would be able to control the future development of many different industries. Expanded to an international scale, the potential for unaccountable manipulation of industrial policy by a few corporations would easily compound the disruptions already caused by often reckless international traders in stocks, bonds and currencies.

However, as long as public regulation of industry remains under attack, tradeable credits and other such schemes will continue to be promoted as market-savvy alternatives. Along with an acceptance of pollution as "a by-product of modern civilization that can be regulated and reduced, but not eliminated," to quote another Progressive Policy Institute paper, self-proclaimed environmentalists will call for an end to "widespread antagonism toward corporations and a suspicion that anything supported by business was bad for the environment." Market solutions are offered as the only alternative to the "inefficient," "centralized," "command-and-control" regulations of the past, in language closely mirroring the rhetoric of Cold War anti-communism.

While specific technology-based standards can be criticized as inflexible and sometimes even archaic, critics choose to forget that in many cases, they were instituted by Congress as a safeguard against the widespread abuses of the Reagan-era EPA. During the Reagan years, "flexible" regulations opened the door to widely criticized—and often illegal—bending of the rules for the benefit of politically favored corporations, leading to the resignation of EPA administrator Anne Gorsuch Burford and a brief jail sentence for one of her more vocal legal assistants.

The anti-regulatory fervor of the present Congress is bringing a variety of other market-oriented proposals to the fore. Some are genuinely offered to further environmental protection, while others are far more cynical attempts to replace public regulations with virtual blank checks for polluters. Some have proposed a direct charge for pollution, modeled after the comprehensive pollution taxes that have proved popular in Western Europe. Writers as diverse as Supreme Court Justice Stephen Breyer, American Enterprise Institute economist Robert Hahn and environmental business guru Paul Hawken have defended pollution taxes as an ideal market-oriented approach to controlling pollution. Indeed, unlike tradeable credits, taxes might help reduce pollution beyond regulatory levels, as they encourage firms to control emissions as much as possible. With credits, there is no reduction in pollution below the threshold established in legislation. (If many companies were to opt for substantial new emissions controls, the market would soon be glutted and the allowances would rapidly become valueless.) And taxes would work best if combined with vigilant grassroots activism that makes industries accountable to the communities in which they operate. However, given the rapid dismissal of Bill Clinton's early plan for an energy tax, it is most likely that any pollution tax proposal would be immediately dismissed by Congressional ideologues as an outrageous new government intervention into the marketplace.

Air pollution is not the only environmental problem that free marketeers are proposing to solve with the invisible hand. Pro-development interests in Congress have floated various schemes to replace the Endangered Species Act with a system of voluntary incentives, conservation easements and other schemes through which landowners would be compensated by the government to protect critical habitat. While these proposals are being debated in Congress, the Clinton administration has quietly changed the rules for administering the Act in a manner that encourages voluntary compliance and offers some of the very same loopholes that anti-environmental advocates have sought. This, too, is being offered in the name of cooperation and "market environmentalism."

Debates over the management of publicly-owned lands have inspired far more outlandish "free market" schemes. "Nearly all environmental problems are rooted in society's failure to adequately define property rights for some resource," economist Randal O'Toole has written, suggesting a need for "property rights for owls and salmon" developed to "protect them from pollution." O'Toole initially gained the attention of environmentalists in the Pacific Northwest for his detailed studies of the inequities of the U.S. Forest Service's long-term subsidy programs for logging on public lands. Now he has proposed dividing the National Forest system into individual units, each governed by its users and operated on a for-profit basis, with a portion of user fees allocated for such needs as the protection of biological diversity. Environmental values, from clean water to recreation to scenic views, should simply be allocated their proper value in the marketplace, it is argued, and allowed to out-compete unsustainable resource extraction. Other market advocates have suggested far more sweeping transfers of federal lands to the states, an idea seen by many in the West as a first step toward complete privatization.

Market enthusiasts like O'Toole repeatedly overlook the fact that ecological values are far more subjective than the market value of timber and minerals removed from public lands. Efforts to quantify these values are based on various sociological methods, market analysis and psychological studies. People are asked how much they would pay to protect a resource, or how much money they would accept to live without it, and their answers are compared with the prices of everything from wilderness expeditions to vacation homes. Results vary widely depending on how questions are asked, how knowledgeable respondents are, and what assumptions are made in the analysis. Environmentalists are rightfully appalled by such efforts as a recent Resources for the Future study designed to calculate the value of human lives lost due to future toxic exposures. Outlandish absurdities like property rights for owls arouse similar skepticism.

The proliferation of such proposals—and their increasing credibility in Washington—suggest the need for a renewed debate over the relationship between ecological values and those of the free market. For many environmental economists, the processes of capitalism, with a little fine tuning, can be made to serve the needs of environmental protection. For many activists, however, there is a fundamental contradiction between the interconnected nature of ecological processes and an economic system which not only reduces everything to isolated commodities, but seeks to manipulate those commodities to further the single, immutable goal of maximizing individual gain. An ecological economy may need to more closely mirror natural processes in their stability, diversity, long time frame, and the prevalence of cooperative, symbiotic interactions over the more extreme forms of competition that thoroughly dominate today's economy. Ultimately, communities of people need to reestablish social control over economic markets and relationships, restoring an economy which, rather than being seen as the engine of social progress, is instead, in the words of economic historian Karl Polanyi, entirely "submerged in social relationships."

Whatever economic model one proposes for the long-term future, it is clear that the current phase of corporate consolidation is threatening the integrity of the earth's living ecosystems—and communities of people who depend on those ecosystems—as never before. There is little room for consideration of ecological integrity in a global economy where a few ambitious currency traders can trigger the collapse of a nation's currency, its food supply, or a centuries-old forest ecosystem before anyone can even begin to discuss the consequences. In this kind of world, replacing our society's meager attempts to restrain and regulate corporate excesses with market mechanisms can only further the degradation of the natural world and threaten the health and well-being of all the earth's inhabitants.

POSTSCRIPT

Can Pollution Rights Trading Effectively Control Environmental Problems?

Does pollution rights trading give major corporate polluters too much power to control and manipulate the market for emission credits? This is one of the key issues that continues to inspire developing countries to withhold their endorsement of the greenhouse gas emissions trading provisions of the Kyoto Protocol. The evidence that Tokar cites, which is primarily based on short-term experience with trading in sulfur dioxide pollution credits, does not appear to fully justify the broad generalizations he makes about the inherent perils in market-based regulatory plans. Recent assessments of the Acid Rain Program by the EPA and such organizations as the Environmental Defense Fund are more positive. So is the corporate world: In "Economic Man, Cleaner Planet," *The Economist* (September 29, 2001), it is asserted that economic incentives have proved very useful and that "market forces are only just beginning to make inroads into green policymaking." T. H. Tietenberg, *Emissions Trading: Principles and Practice*, 2nd ed. (RFF Press, 2006), notes that emissions trading has a definite niche in pollution control policies. Ruth Greenspan Bell, "The Kyoto Placebo," *Issues in Science and Technology* (Winter 2006), notes that "Though heavily promoted by the World Bank, U.S.-style environmental trading has yet to be tested on a global scale and has never been successfully deployed on a national level in the developing world." There is more to be gained by helping developing nations gain regulatory skills ("What to Do about Climate Change," *Foreign Affairs*, May/June 2006).

The position of those who are ideologically opposed to pollution rights is concisely stated in Michael J. Sandel's op-ed piece "It's Immoral to Buy the Right to Pollute," *The New York Times* (December 15, 1997). In "Selling Air Pollution," *Reason* (May 1996), Brian Doherty supports the concept of pollution rights trading but argues that the kind of emission cap imposed in the case of sulfur dioxide is an inappropriate constraint on what he believes should be a completely free-market program. Richard A. Kerr, in "Acid Rain Control: Success on the Cheap," *Science* (November 6, 1998), contends that emissions trading has greatly reduced acid rain and that the annual cost has been about a tenth of the $10 billion initially forecast. According to Barry D. Solomon and Russell Lee, in "Emissions Trading Systems and Environmental Justice," *Environment* (October 2000), "a significant part of the opposition to emissions trading programs is a perception that they do little to reduce environmental injustice and can even make it worse."

The threat of global warming from continuing emissions of greenhouse gases has prompted the extension of emissions trading to carbon dioxide. Canada has a proposed trading plan (see Danny Bradbury, "The Trading Game," *Alternatives Journal*, October 2005). Europe is implementing its Greenhouse

Gas Emissions Trading Scheme, although so far its effectiveness is in doubt; Marianne Lavelle, "The Carbon Market Has a Dirty Little Secret," *U.S. News and World Report* (May 14, 2007), reports that in Europe the value of tradable emissions allowances has fallen so low, partly because too many allowances were issued, that it has become cheaper to burn more fossil fuel and emit more carbon than to burn and emit less. Future trading schemes will need to be designed to avoid the problem, and there are several bills before Congress that address the issue (see "Support Grows for Capping and Trading Carbon Emissions," *Issues in Science and Technology*, Summer 2007). Meanwhile, there is great interest in what is known as "carbon offsets," by which corporations, governments, and even individuals compensate for carbon dioxide emissions by investing in activities that remove carbon dioxide from the air or reduce emissions from a different source. See Anja Kollmuss, "Carbon Offsets 101," *World Watch* (July/August 2007).

Internet References . . .

The Arctic National Wildlife Refuge: A Special Report

This site offers a cogent review of the debate over exploiting the Arctic National Wildlife Refuge (ANWR) for oil.

http://arcticcircle.uconn.edu/ANWR/anwrindex.html

National Wilderness Preservation System

Operated by representatives of the Arthur Carhart National Wilderness Training Center, the Aldo Leopold Wilderness Research Institute, and the Wilderness Institute at the University of Montana's School of Forestry, the National Wilderness Preservation System provides information, news, and Internet links related to wilderness. It includes a database of information on all 680 wilderness areas.

http://www.wilderness.net/index.cfm?fuse=NWPS

Intergovernmental Panel on Climate Change

The Intergovernmental Panel on Climate Change (IPCC) was formed by the World Meteorological Organization (WMO) and the United Nations Environment Programme (UNEP) to assess the scientific, technical, and socio-economic information relevant for the understanding of the risk of human-induced climate change.

http://www.ipcc.ch/

Climate Change

The United Nations Environmental Program maintains this site as a central source for substantive work and information resources with regard to climate change.

http://www.unep.org/themes/climatechange/

CAFE Overview

The U.S. National Highway Traffic Safety Administration provides an overview of existing CAFE standards here.

http://www.nhtsa.dot.gov/cars/rules/cafe/overview.htm

The National Renewable Energy Laboratory

The National Renewable Energy Laboratory is the nation's primary laboratory for renewable energy and energy efficiency research and development. Among other things, it works on wind power and biofuels.

http://www.nrel.gov/learning/re_basics.html

Nuclear Energy

The U.S. Department of Energy's Office of Nuclear Energy leads U.S. efforts to develop new nuclear energy generation technologies; to develop advanced, proliferation-resistant nuclear fuel technologies that maximize energy from nuclear fuel; and to maintain and enhance the national nuclear technology infrastructure.

http://www.ne.doe.gov/

UNIT 3

Energy Issues

*H*umans cannot live and society cannot exist without producing environmental impacts. The reason is very simple: Humans cannot live and society cannot exist without using resources (e.g., soil, water, ore, wood, space, plants, animals, oil, sunlight), and those resources come from the environment. Many of these resources (e.g., wood, oil, coal, water, wind, sunlight, uranium) have to do with energy. The environmental impacts come from what must be done to obtain these resources and what must be done to dispose of the wastes generated in the process of obtaining and using them. The issues that arise are whether and how we should obtain these resources, whether and how we should deal with the wastes, and whether alternative answers to these questions may be preferable to the answers that experts think they already have.

In 2007, the Intergovernmental Panel on Climate Change released its fourth assessment report, summarizing the scientific consensus as the climate is warming, human activities are responsible for it, and the impact on human well-being and ecosystems will be severe. This brought energy issues to the fore with unprecedented urgency and prompted expansion of this section of this book. The six issues presented here are by no means the only issues related to energy, but they will serve to demonstrate the vigor and variety of the current energy debates.

- Should the Arctic National Wildlife Refuge Be Opened to Oil Drilling?

- Is Global Warming Skepticism Just Smoke and Mirrors?

- Is Wind Power Green?

- Should Cars Be More Efficient?

- Do Biofuels Enhance Energy Security?

- Is It Time to Revive Nuclear Power?

ISSUE 7

Should the Arctic National Wildlife Refuge Be Opened to Oil Drilling?

YES: Dwight R. Lee, from "To Drill or Not to Drill: Let the Environmentalists Decide," *The Independent Review* (Fall 2001)

NO: Jeff Bingaman, et al., from Dissenting Views on ANWR Drilling Senate Energy Committee (October 24, 2005)

ISSUE SUMMARY

YES: Professor of economics Dwight R. Lee argues that the economic and other benefits of Arctic National Wildlife Refuge (ANWR) oil are so great that even environmentalists should agree to permit drilling—and they probably would if they stood to benefit directly.

NO: The Minority Members of the Senate Energy Committee objected when the Committee approved a bill that would authorize oil and gas development in the Arctic National Wildlife Refuge. They argued that though the bill contained serious legal and environmental flaws, the greatest flaw lay in its choice of priorities: Wilderness is to be preserved, not exploited.

$\mathbf{T}$he birth of environmental consciousness in the United States was marked by two strong, opposing views. Late in the nineteenth century, John Muir (1838–1914) called for the preservation of natural wilderness, untouched by human activities. At about the same time, Gifford Pinchot (1865–1946) became a strong voice for conservation (not to be confused with preservation; Gifford's conservation allowed the use of nature but in such a way that it was not destroyed; his aim was "the greatest good of the greatest number in the long run"). Both views agree that nature has value; however, they disagree on the form of that value. The preservationist says that nature has value in its own right and has a right to be left alone, neither developed with houses and roads nor exploited with farms, dams, mines, and oil wells. The conservationist says that nature's value lies chiefly in the benefits it provides to human beings.

The first national parks date back to the 1870s. Parks and the national forests are managed for "multiple use" on the premise that wildlife protection, recreation, timber cutting, and even oil drilling and mining can coexist. The first "primitive areas," where all development is barred, were created by the U.S. Forest Service in the 1920s. However, pressure from commercial interests (the timber and mining industries, among others) led to the reclassification of many such areas and their opening to exploitation. In 1964 the Federal Wilderness Act provided a mechanism for designating "wilderness" areas, defined as areas "where the earth and its community of life are untrammeled by man, where man himself is a visitor who does not remain." Since then it has become clear that pesticides and other man-made chemicals are found everywhere on earth, drifting on winds and ocean currents and traveling in migrant birds even to areas without obvious human presence. Humans might not be present in these places, but their effects are. And commercial interests are just as interested in the wealth that may be extracted from these areas as they ever were. There is continual pressure to expand commercial use of national forests and parks and to open wilderness areas to exploitation.

The Arctic National Wildlife Refuge (ANWR) provides a good illustration. It is not a "wilderness" area, for it was designated a wildlife preserve in 1960 and enlarged and renamed in 1980 with the proviso that its coastal plain be evaluated for its potential value in terms of oil and gas production. In 1987 the Department of the Interior recommended that the coastal plain be opened for oil and gas exploration. In 1995 Congress approved doing so, but President Bill Clinton vetoed the legislation. In 2001, after California experienced electrical blackouts, President George W. Bush declared that opening the ANWR to oil exploitation was essential to national energy security but could not muster enough votes in Congress to make it happen. In 2003, an attempt to link the need for Arctic oil to the war in Iraq failed in the Senate. In 2004, the Bush Administration proposed once more to open the ANWR to oil drilling, and in December 2005, the House of Representatives approved a Defense bill with a provision approving ANWR drilling. The Senate immediately blocked the provision, but with gasoline prices soaring, in March 2006 Bush renewed his call for approval to exploit the ANWR for oil. In May 2006, the House provided that approval. According to Darren Goode, Jill Smallen, and Charlie Mitchell, "House OKs ANWR Drilling, Again," *National Journal* (May 27, 2006), House Resources Committee Chairman Richard Pombo (R-CA) argued that "there is no logical reason" to oppose ANWR drilling. "America needs more American-made energy." As usual (so far), the Senate did not follow suit.

Strict preservationists still remain, but the debate over protecting wilderness areas generally centers on economic arguments. In the following selections, Dwight R. Lee argues that the economic and other benefits of Arctic National Wildlife Refuge oil are so great that drilling should be permitted. When in October 2005 the Senate Energy Committee approved a bill that would authorize oil and gas development in the Arctic National Wildlife Refuge, the Committee's minority members argued that the bill contained serious legal and environmental flaws, but its greatest flaw lay in its choice of priorities: Wilderness is to be preserved, not exploited for its energy resources.

YES

Dwight R. Lee

To Drill or Not to Drill

High prices of gasoline and heating oil have made drilling for oil in Alaska's Arctic National Wildlife Refuge (ANWR) an important issue. ANWR is the largest of Alaska's sixteen national wildlife refuges, containing 19.6 million acres. It also contains significant deposits of petroleum. The question is, Should oil companies be allowed to drill for that petroleum?

The case for drilling is straightforward. Alaskan oil would help to reduce U.S. dependence on foreign sources subject to disruptions caused by the volatile politics of the Middle East. Also, most of the infrastructure necessary for transporting the oil from nearby Prudhoe Bay to major U.S. markets is already in place. Furthermore, because of the experience gained at Prudhoe Bay, much has already been learned about how to mitigate the risks of recovering oil in the Arctic environment.

No one denies the environmental risks of drilling for oil in ANWR. No matter how careful the oil companies are, accidents that damage the environment at least temporarily might happen. Environmental groups consider such risks unacceptable; they argue that the value of the wilderness and natural beauty that would be spoiled by drilling in ANWR far exceeds the value of the oil that would be recovered. For example, the National Audubon Society characterizes opening ANWR to oil drilling as a threat "that will destroy the integrity" of the refuge (see statement at . . .).

So, which is more valuable, drilling for oil in ANWR or protecting it as an untouched wilderness and wildlife refuge? Are the benefits of the additional oil really less than the costs of bearing the environmental risks of recovering that oil? Obviously, answering this question with great confidence is difficult because the answer depends on subjective values. Just how do we compare the convenience value of using more petroleum with the almost spiritual value of maintaining the "integrity" of a remote and pristine wilderness area? Although such comparisons are difficult, we should recognize that they can be made. Indeed, we make them all the time.

We constantly make decisions that sacrifice environmental values for what many consider more mundane values, such as comfort, convenience, and material well-being. There is nothing wrong with making such sacrifices because up to some point the additional benefits we realize from sacrificing a

This article is reprinted with permission from the publisher of *The Independent Review: A Journal of Political Economy* (vol. VI, no. 2, Fall 2001, pp. 217–226). Copyright © 2001 by The Independent Institute, 100 Swan Way, Oakland, CA 94621-1428. info@independent.org; www.independent.org.

little more environmental "integrity" are worth more than the necessary sacrifice. Ideally, we would somehow acquire the information necessary to determine where that point is and then motivate people with different perspectives and preferences to respond appropriately to that information.

Achieving this ideal is not as utopian as it might seem; in fact, such an achievement has been reached in situations very similar to the one at issue in ANWR. In this article, I discuss cases in which the appropriate sacrifice of wilderness protection for petroleum production has been responsibly determined and harmoniously implemented. Based on this discussion, I conclude that we should let the Audubon Society decide whether to allow drilling in ANWR. That conclusion may seem to recommend a foregone decision on the issue because the society has already said that drilling for oil in ANWR is unacceptable. But actions speak louder than words, and under certain conditions I am willing to accept the actions of environmental groups such as the Audubon Society as the best evidence of how they truly prefer to answer the question, To drill or not to drill in ANWR?

Private Property Changes One's Perspective

What a difference private property makes when it comes to managing multiuse resources. When people make decisions about the use of property they own, they take into account many more alternatives than they do when advocating decisions about the use of property owned by others. This straightforward principle explains why environmental groups' statements about oil drilling in ANWR (and in other publicly owned areas) and their actions in wildlife areas they own are two very different things.

For example, the Audubon Society owns the Rainey Wildlife Sanctuary, a 26,000-acre preserve in Louisiana that provides a home for fish, shrimp, crab, deer, ducks, and wading birds, and is a resting and feeding stopover for more than 100,000 migrating snow geese each year. By all accounts, it is a beautiful wilderness area and provides exactly the type of wildlife habitat that the Audubon Society seeks to preserve. But, as elsewhere in our world of scarcity, the use of the Rainey Sanctuary as a wildlife preserve competes with other valuable uses.

Besides being ideally suited for wildlife, the sanctuary contains commercially valuable reserves of natural gas and oil, which attracted the attention of energy companies when they were discovered in the 1940s. Clearly, the interests served by fossil fuels do not have high priority for the Audubon Society. No doubt, the society regards additional petroleum use as a social problem rather than a social benefit. Of course, most people have different priorities: they place a much higher value on keeping down the cost of energy than they do on bird-watching and on protecting what many regard as little more than mosquito-breeding swamps. One might suppose that members of the Audubon Society have no reason to consider such "anti-environmental" values when deciding how to use their own land. Because the society owns the Rainey Sanctuary, it can ignore interests antithetical to its own and refuse to allow drilling. Yet, precisely because the society owns the land, it has been willing to accommodate the interests of those whose priorities are different and has allowed thirty-seven wells to pump gas and oil from the Rainey Sanctuary. In return, it has received royalties of more than $25 million.

One should not conclude that the Audubon Society has acted hypocritically by putting crass monetary considerations above its stated concerns for protecting wilderness and wildlife. In a wider context, one sees that because of its ownership of the Rainey Sanctuary, the Audubon Society is part of an extensive network of market communication and cooperation that allows it to do a better job of promoting its objectives by helping others promote theirs. Consumers communicate the value they receive from additional gas and oil to petroleum companies through the prices they willingly pay for those products, and this communication is transmitted to owners of oil-producing land through the prices the companies are willing to pay to drill on that land. Money really does "talk" when it takes the form of market prices. The money offered for drilling rights in the Rainey Sanctuary can be viewed as the most effective way for millions of people to tell the Audubon Society how much they value the gas and oil its property can provide.

By responding to the price communication from consumers and by allowing the drilling, the Audubon Society has not sacrificed its environmental values in some debased lust for lucre. Instead, allowing the drilling has served to reaffirm and promote those values in a way that helps others, many of whom have different values, achieve their own purposes. Because of private ownership, the valuations of others for the oil and gas in the Rainey Sanctuary create an opportunity for the Audubon Society to purchase additional sanctuaries to be preserved as habitats for the wildlife it values. So the society has a strong incentive to consider the benefits as well as the costs of drilling on its property. Certainly, environmental risks exist, and the society considers them, but if also responsibly weighs the costs of those risks against the benefits as measured by the income derived from drilling. Obviously, the Audubon Society appraises the benefits from drilling as greater than the costs, and it acts in accordance with that appraisal.

Cooperation Between Bird-Watchers and Hot-Rodders

The advantage of private ownership is not just that it allows people with different interests to interact in mutually beneficial ways. It also creates harmony between those whose interests would otherwise be antagonistic. For example, most members of the Audubon Society surely see the large sport utility vehicles and high-powered cars encouraged by abundant petroleum supplies as environmentally harmful. That perception, along with the environmental risks associated with oil recovery, helps explain why the Audubon Society vehemently opposes drilling for oil in the ANWR as well as in the continental shelves in the Atlantic, the Pacific, and the Gulf of Mexico. Although oil companies promise to take extraordinary precautions to prevent oil spills when drilling in these areas, the Audubon Society's position is no off-shore drilling, none. One might expect to find Audubon Society members completely unsympathetic with hot-rodding enthusiasts, NASCAR racing fans, and drivers of Chevy Suburbans. Yet, as we have seen, by allowing drilling for gas and oil in the Rainey Sanctuary, the society is accommodating the interests of those with gas-guzzling lifestyles, risking the "integrity" of its prized wildlife sanctuary to make more gasoline available to those whose energy consumption it verbally condemns as excessive.

The incentives provided by private property and market prices not only motivate the Audubon Society to cooperate with NASCAR racing fans, but also motivate those racing enthusiasts to cooperate with the Audubon Society. Imagine the reaction you would get if you went to a stock-car race and tried to convince the spectators to skip the race and go bird-watching instead. Be prepared for some beer bottles tossed your way. Yet by purchasing tickets to their favorite sport, racing fans contribute to the purchase of gasoline that allows the Audubon Society to obtain additional wildlife habitat and to promote bird-watching. Many members of the Audubon Society may feel contempt for racing fans, and most racing fans may laugh at bird-watchers, but because of private property and market prices, they nevertheless act to promote one another's interests.

The Audubon Society is not the only environmental group that, because of the incentives of private ownership, promotes its environmental objectives by serving the interests of those with different objectives. The Nature Conservancy accepts land and monetary contributions for the purpose of maintaining natural areas for wildlife habitat and ecological preservation. It currently owns thousands of acres and has a well-deserved reputation for preventing development in environmentally sensitive areas. Because it owns the land, it has also a strong incentive to use that land wisely to achieve its objectives, which sometimes means recognizing the value of developing the land.

For example, soon after the Wisconsin chapter received title to 40 acres of beach-front land on St. Croix in the Virgin Islands, it was offered a much larger parcel of land in northern Wisconsin in exchange for its beach land. The Wisconsin chapter made this trade (with some covenants on development of the beach land) because owning the Wisconsin land allowed it to protect an entire watershed containing endangered plants that it considered of greater environmental value than what was sacrificed by allowing the beach to be developed.

Thanks to a gift from the Mobil Oil Company, the Nature Conservancy of Texas owns the Galveston Bay Prairie Preserve in Texas City, a 2,263-acre refuge that is home to the Attwater's prairie chicken, a highly endangered species (once numbering almost a million, its population had fallen to fewer than ten by the early 1990s). The conservancy has entered into an agreement with Galveston Bay Resources of Houston and Aspects Resources, LLC, of Denver to drill for oil and natural gas in the preserve. Clearly some risks attend oil drilling in the habitat of a fragile endangered species, and the conservancy has considered them, but it considers the gains sufficient to justify bearing the risks. According to Ray Johnson, East County program manager for the Nature Conservancy of Texas. "We believe this could provide a tremendous opportunity to raise funds to acquire additional habitat for the Attwater's prairie chicken, one of the most threatened birds in North America." Obviously the primary concern is to protect the endangered species, but the demand for gas and oil is helping achieve that objective. Johnson is quick to point out, "We have taken every precaution to minimize the impact of the drilling on the prairie chickens and to ensure their continued health and safety."

Back to ANWR

Without private ownership, the incentive to take a balanced and accommodating view toward competing land-use values disappears. So, it is hardly surprising that the Audubon Society and other major environmental groups categorically oppose drilling in ANWR. Because ANWR is publicly owned, the environmental groups have no incentive to take into account the benefits of drilling. The Audubon Society does not capture any of the benefits if drilling is allowed, as it does at the Rainey Sanctuary; in ANWR, it sacrifices nothing if drilling is prevented. In opposing drilling in ANWR, despite the fact that the precautions to be taken there would be greater than those required of companies operating in the Rainey Sanctuary, the Audubon Society is completely unaccountable for the sacrificed value of the recoverable petroleum.

Obviously, my recommendation to "let the environmentalists decide" whether to allow oil to be recovered from ANWR makes no sense if they are not accountable for any of the costs (sacrificed benefits) of preventing drilling. I am confident, however, that environmentalists would immediately see the advantages of drilling in ANWR if they were responsible for both the costs and the benefits of that drilling. As a thought experiment about how incentives work, imagine that a consortium of environmental organizations is given veto power over drilling, but is also given a portion (say, 10 percent) of what energy companies are willing to pay for the right to recover oil in ANWR. These organizations could capture tens of millions of dollars by giving their permission to drill. Suddenly the opportunity to realize important environmental objectives by favorably considering the benefits others gain from more energy consumption would come into sharp focus. The environmentalists might easily conclude that although ANWR is an "environmental treasure," other environmental treasures in other parts of the country (or the world) are even more valuable; moreover, with just a portion of the petroleum value of the ANWR, efforts might be made to reduce the risks to other natural habitats, more than compensating for the risks to the Arctic wilderness associated with recovering that value.

Some people who are deeply concerned with protecting the environment see the concentration on "saving" ANWR from any development as misguided even without a vested claim on the oil wealth it contains. For example, according to Craig Medred, the outdoor writer for the *Anchorage Daily News* and a self-described "development-phobic wilderness lover,"

> That people would fight to keep the scar of clearcut logging from the spectacular and productive rain-forests of Southeast Alaska is easily understandable to a shopper in Seattle or a farmer in Nebraska. That people would argue

against sinking a few holes through the surface of a frozen wasteland, however, can prove more than a little baffling even to development-phobic, wilderness lovers like me. Truth be known, I'd trade the preservation rights to any 100 acres on the [ANWR] slope for similar rights to any acre of central California wetlands. . . . It would seem of far more environmental concern that Alaska's ducks and geese have a place to winter in overcrowded, overdeveloped California than that California's ducks and geese have a place to breed each summer in uncrowded and undeveloped Alaska.

— (1996, Cl)

Even a small share of the petroleum wealth in ANWR would dramatically reverse the trade-off Medred is willing to make because it would allow environmental groups to afford easily a hundred acres of central California wetlands in exchange for what they would receive for each acre of ANWR released to drilling.

We need not agree with Medred's characterization of the ANWR as "a frozen wasteland" to suspect that environmentalists are overstating the environmental amenities that drilling would put at risk. With the incentives provided by private property, environmental groups would quickly reevaluate the costs of drilling in wilderness refuges and soften their rhetoric about how drilling would "destroy the integrity" of these places. Such hyperbolic rhetoric is to be expected when drilling is being considered on public land because environmentalists can go to the bank with it. It is easier to get contributions by depicting decisions about oil drilling on public land as righteous crusades against evil corporations out to destroy our priceless environment for short-run profit than it is to work toward minimizing drilling costs to accommodate better the interests of others. Environmentalists are concerned about protecting wildlife and wilderness areas in which they have ownership interest, but the debate over any threat from drilling and development in those areas is far more productive and less acrimonious than in the case of ANWR and other publicly owned wilderness areas.

The evidence is overwhelming that the risks of oil drilling to the arctic environment are far less than commonly claimed. The experience gained in Prudhoe Bay has both demonstrated and increased the oil companies' ability to recover oil while leaving a "light footprint" on arctic tundra and wildlife. Oil-recovery operations are now sited on gravel pads providing foundations that protect the underlying permafrost. Instead of using pits to contain the residual mud and other waste from drilling, techniques are now available for pumping the waste back into the well in ways that help maintain well pressure and reduce the risks of spills on the tundra. Improvements in arctic road construction have eliminated the need for the gravel access roads used in the development of the Prudhoe Bay oil fields. Roads are now made from ocean water pumped onto the tundra, where it freezes to form a road surface. Such roads melt without a trace during the short summers. The oversize rubber tires used on the roads further minimize any impact on the land.

Improvements in technology now permit horizontal drilling to recover oil that is far from directly below the wellhead. This technique reduces further the already small amount of land directly affected by drilling operations. Of the more than 19 million acres contained in ANWR, almost 18 million acres have been set aside by Congress—somewhat more than 8 million as wilderness and 9.5 million as wildlife refuge. Oil companies estimate that only 2,000 acres would be needed to develop the coastal plain.

This carefully conducted and closely confined activity hardly sounds like a sufficient threat to justify the rhetoric of a righteous crusade to prevent the destruction of ANWR, so the environmentalists warn of a detrimental effect on arctic wildlife that cannot be gauged by the limited acreage directly affected. Given the experience at Prudhoe Bay, however, such warnings are difficult to take seriously. The oil companies have gone to great lengths and spent tens of millions of dollars to reduce any harm to the fish, fowl, and mammals that live and breed on Alaska's North Slope. The protections they have provided for wildlife at Prudhoe Bay have been every bit as serious and effective as those the Audubon Society and the Nature Conservancy find acceptable in the Rainey Sanctuary and the Galveston Bay Prairie Preserve. As the numbers of various wildlife species show, many have thrived better since the drilling than they did before.

Before drilling began at Prudhoe Bay, a good deal of concern was expressed about its effect on caribou herds. As with many wildlife species, the population of the caribou on Alaska's North Slope fluctuates (often substantially) from year to year for completely natural reasons, so it is difficult to determine with confidence the effect of development on the caribou population. It is noteworthy, however, that the caribou population in the area around Prudhoe Bay has increased greatly since that oil field was developed, from approximately 3,000 to a high of some 23,400. . . . Some argue that the increase has occurred because the caribou's natural predators have avoided the area—some of these predators are shot, whereas the caribou are not. But even if this argument explains some or even all of the increase in the population, the increase still casts doubt on claims that the drilling threatens the caribou. Nor has it been shown that the viability of any other species has been genuinely threatened by oil drilling at Prudhoe Bay.

Caribou Versus Humans

Although consistency in government policy may be too much to hope for, it is interesting to contrast the federal government's refusal to open ANWR with some of its other oil-related policies. While opposing drilling in ANWR, ostensibly because we should not put caribou and other Alaskan wildlife at risk for the sake of getting more petroleum, we are exposing humans to far greater risks because of federal policies motivated by concern over petroleum supplies.

For example, the United States maintains a military presence in the Middle East in large part because of the petroleum reserves there. It is doubtful that the U.S. government would have mounted a large military action and sacrificed American lives to prevent Iraq from taking over the tiny sheikdom of Kuwait except to allay the threat to a major oil supplier. Nor would the United States have lost the nineteen military personnel in the barracks blown up in Saudi

Arabia in 1996 or the seventeen killed onboard the USS Cole in a Yemeni harbor in 2000. I am not arguing against maintaining a military presence in the Middle East, but if it is worthwhile to sacrifice Americans' lives to protect oil supplies in the Middle East, is it not worthwhile to take a small (perhaps nonexistent) risk of sacrificing the lives of a few caribou to recover oil in Alaska?

Domestic energy policy also entails the sacrifice of human lives for oil. To save gasoline, the federal government imposes Corporate Average Fuel Economy (CAFE) standards on automobile producers. These standards now require all new cars to average 27.5 miles per gallon and new light trucks to average 20.5 miles per gallon. The one thing that is not controversial about the CAFE standards is that they cost lives by inducing manufacturers to reduce the weight of vehicles. Even Ralph Nader has acknowledged that "larger cars are safer—there is more bulk to protect the occupant." An interesting question is, How many lives might be saved by using more (ANWR) oil and driving heavier cars rather than using less oil and driving lighter, more dangerous cars?

It has been estimated that increasing the average weight of passenger cars by 100 pounds would reduce U.S. highway fatalities by 200 a year. By determining how much additional gas would be consumed each year if all passenger cars were 100 pounds heavier, and then estimating how much gas might be recovered from ANWR oil, we can arrive at a rough estimate of how many human lives potentially might be saved by that oil. To make this estimate, I first used data for the technical specifications of fifty-four randomly selected 2001 model passenger cars to obtain a simple regression of car weight on miles per gallon. This regression equation indicates that every additional 100 pounds decreases mileage by 0.85 miles per gallon. So 200 lives a year could be saved by relaxing the CAFE standards to allow a 0.85 miles per gallon reduction in the average mileage of passenger cars. How much gasoline would be required to compensate for this decrease of average mileage? Some 135 million passenger cars are currently in use, being driven roughly 10,000 miles per year on average (1994–95 data from U.S. Bureau of the Census 1997, 843). Assuming these vehicles travel 24 miles per gallon on average, the annual consumption of gasoline by passenger cars is 56.25 billion gallons (= 135 million × 10,000/24). If instead of an average of 24 miles per gallon the average were reduced to 23.15 miles per gallon, the annual consumption of gasoline by passenger cars would be 58.32 billion gallons (= 135 million × 10,000/23.15). So, 200 lives could be saved annually by an extra 2.07 billion gallons of gas. It is estimated that ANWR contains from 3 to 16 billion barrels of recoverable petroleum. Let us take the midpoint in this estimated range, or 9.5 billion barrels. Given that on average each barrel of petroleum is refined into 19.5 gallons of gasoline, the ANWR oil could be turned into 185.25 billion additional gallons of gas, or enough to save 200 lives a year for almost ninety years (185.25/2.07 = 89.5). Hence, in total almost 18,000 lives could be saved by opening up ANWR to drilling and using the fuel made available to compensate for increasing the weight of passenger cars.

I claim no great precision for this estimate. There may be less petroleum in ANWR than the midpoint estimate indicates, and the study I have relied on may have overestimated the number of lives saved by heavier passenger cars.

Still, any reasonable estimate will lead to the conclusion that preventing the recovery of ANWR oil and its use in heavier passenger cars entails the loss of thousands of lives on the highways. Are we willing to bear such a cost in order to avoid the risks, if any, to ANWR and its caribou?

Conclusion

I am not recommending that ANWR actually be given to some consortium of environmental groups. In thinking about whether to drill for oil in ANWR, however, it is instructive to consider seriously what such a group would do if it owned ANWR and therefore bore the costs as well as enjoyed the benefits of preventing drilling. Those costs are measured by what people are willing to pay for the additional comfort, convenience, and safety that could be derived from the use of ANWR oil. Unfortunately, without the price communication that is possible only by means of private property and voluntary exchange, we cannot be sure what those costs are or how private owners would evaluate either the costs or the benefits of preventing drilling in ANWR. However, the willingness of environmental groups such as the Audubon Society and the Nature Conservancy to allow drilling for oil an environmentally sensitive land they own suggests strongly that their adamant verbal opposition to drilling in ANWR is a poor reflection of what they would do if they owned even a small fraction of the ANWR territory containing oil.

 **NO**

ANWR Minority Views

Dissenting Views of Senators Bingaman, Dorgan, Wyden, Johnson, Feinstein, Cantwell, Corzine, and Salazar

The Arctic National Wildlife Refuge has long stirred deep emotions and strong passions. To some it is the most promising place to look for oil in the [nation]. To others it is "the Last Great Wilderness," a vast and beautiful natural wonder, which deserves permanent protection for its wildlife, scenic, and recreational values. These two viewpoints are held with equal passion by roughly equal parts of the Senate and the nation as a whole. For a quarter of a century, neither side has been able to enact legislation either opening the area to oil and gas development or permanently preserving it as wilderness.

We come down squarely in favor of preserving the Arctic National Wildlife Refuge. Opening the Refuge to oil and gas development will do little to meet our energy needs and nothing to reduce our energy prices. Not one drop of oil will come from the Refuge for ten years. And even at its peak production— twenty years from now—it will reduce our reliance on imports by only 4 percent. We believe we should tap alternative sources of oil and gas and develop alternative energy technologies, rather than sacrifice the Refuge's unique wildlife and wilderness values.

Years ago, Senator Clinton P. Anderson said that our willingness to protect wilderness areas showed "that we are still a rich Nation, tending our resources as we should—not a people in despair searching every last nook and cranny of our land for . . . a barrel of oil." We believe we still are a rich Nation, rich in untapped oil and gas resources in other areas that can be developed consistent with environmental protection, and rich in the intellectual capital needed to develop new alternative energy technologies. We do not believe we need to sacrifice our wildlife refuges or other environmentally sensitive areas to fuel our cars or heat our homes.

But even if the day should come when we do need to exploit the Arctic National Wildlife Refuge for oil, we should approach the task with care. If we must open the Refuge to oil and gas development, we should do so in accordance with existing mineral leasing laws and regulations, existing environmental protections, and existing rules of administrative procedure and judicial

From ANWR Minority Views from the Senate Energy Committee, October 24, 2005.

review. We should, in short, afford the Arctic Refuge no less protection than current law affords any other refuge or public land that is open to oil and gas development. In addition, we should ensure that any oil that comes from the Arctic Refuge goes to Americans, and is not sold overseas; and that the Federal Treasury receives the full amount of royalties and bonus bids that we are promised. Regrettably, the legislation recommended by a majority of the Committee fails in every one of these respects.

1. The Mineral Leasing Act and rules. Oil and gas leasing on the public lands, including wildlife refuges, is currently conducted under the Mineral Leasing Act of 1920 and regulations adopted by the Bureau of Land Management under that Act. Among other things, they require minimum royalties, maximum lease sizes, and various performance standards and environmental protections. Oil and gas development on wildlife refuges can only take place with the concurrence of the Fish and Wildlife Service and subject to stipulations prescribed by the Fish and Wildlife Service that protect wildlife.

   It is unclear from the legislation recommended by the Committee whether any of the mineral leasing laws or regulations will apply to the leasing program for the Arctic Refuge. The legislation directs the Secretary to administer an "environmentally sound" leasing program in the refuge's Coastal Plain, but does not explicitly require it to be in accordance with the existing statutory and regulatory framework. The Secretary is directed to administer the program through "regulations, lease terms, conditions, restrictions, prohibitions, stipulations, and other provisions" of the Secretary's choosing that will ensure that the leasing program is "carried out in a manner that will ensure the receipt of fair market value by the public for the mineral resources to be leased." The legislation leaves it up to the Secretary, and ultimately the courts, to decide whether existing mineral leasing regulations and procedures will still apply or whether the new system is meant to supersede it.

2. The "compatibility" determination. Under current law, the Secretary of the Interior may permit oil and gas development in a national wildlife refuge if it is "compatible" with the purposes for which the refuge was established. If oil and gas development can take place in the Arctic Refuge without harm to the wildlife populations and habitats it was established to protect, as proponents believe, the Secretary should make the compatibility determination. Instead, the legislation recommended by the Committee "deems" the leasing program to be compatible and absolves the Secretary of any responsibility for determining whether it is or is not, in fact, compatible.

3. NEPA compliance. The National Environmental Policy Act of 1969 requires federal agencies contemplating a major federal action to prepare an environmental impact statement. The requirement is a continuing one. Agencies must supplement environmental impact statements if they make substantial changes in their proposed action or if there are significant new circumstances or information bearing on the proposed action or its impacts. The Department of the Interior last prepared an environmental impact statement on oil and gas development in ANWR in 1987, 18 years ago. In 1992, a federal court

held that significant new information was available that required the Department to supplement the 1987 environmental impact statement. If a court thought that a supplement was required 13 years ago, one must surely be needed today. Yet the legislation recommended by the Committee "deems" the 1987 statement to satisfy the requirements of NEPA "with respect to prelease activities," including the development of the regulations establishing the new leasing program.

NEPA also requires that environmental impact statements consider reasonable alternatives to a proposed action. Consideration of alternatives is said to be "the heart of the environmental impact statement." The courts have said this requirement is governed by a "rule of reason," and common sense, "bounded by some notion of feasibility." The legislation recommended by the Committee waives even this common sense requirement, and limits the Secretary to consideration of only her "preferred action for leasing and a single leasing alternative."

4. Judicial review. Under current law, a person harmed by agency action is entitled to judicial review. A person can bring suit either in the District of Columbia or where he resides. The reviewing court is empowered to "decide all relevant questions of law," and set aside agency actions found to be arbitrary, capricious, or unsupported by substantial evidence. Review is generally limited to the administrative record compiled by the agency when it made its decision, though the court may sometimes look beyond the record in NEPA cases to make sure the decision maker adequately considered environmental impacts and alternatives.

The legislation recommended by the Committee restricts the right of judicial review. It does so by requiring anyone seeking to challenge the Secretary's actions to file suit in the District of Columbia Circuit, by limiting judicial review of the Secretary's decision to conduct a lease sale and the environmental analysis of that decision solely to whether the Secretary has complied with the new legislation, and by strictly limiting review to the administrative record. It is unclear to what extent the narrow scope of review imposed by the new legislation will be read to preempt to the broad scope of review afforded under the Administrative Procedure Act. The Committee leaves that question to the courts.

5. Roads, pipelines, and other rights-of-way. Current law provides a comprehensive process for approving rights-of-ways for roads, pipelines, airstrips, and other transportation and utility systems in conservation system units in Alaska. The principal purpose of this process was to provide access to and from resource development areas, but to do so in an orderly way that would avoid or minimize harm to the environment. The legislation recommended by the Committee exempts all rights-of-way for the exploration, development, production, or transportation of oil and gas on the Coastal Plain from this process.

6. 2,000-acre limitation. The legislation recommended by the Committee restricts the surface acreage covered by oil and gas production within the Arctic Refuge to a maximum of 2,000 acres on the Coastal Plain. This limitation is cited by leasing proponents as evidence that oil

and gas development will occupy a tiny footprint on the Coastal Plain. This provision contains so many loopholes, however, that exploration and development activities could impact much of the Coastal Plain. There is no requirement that the developed surface lands be contiguous or even consolidated. The limitation only applies literally to the actual ground covered, ignoring the effect on nearby lands. As example of the hollowness of the limitation, it only counts the area occupied by the footings of a pipeline support structure towards the acreage limitation, even though the pipeline itself may run many miles across the Coastal Plain.

7. Alaska Native lands. Over 100,000 acres (over 150 square miles) that are within both the boundaries of the Refuge and the definition of the Coastal Plain are owned by Alaska Natives. The surface of over 90,000 of these acres is owned by the Kaktovik Inupiat Corporation, and their subsurface is owned by the Arctic Slope Regional Corporation [ASRC]. The remaining approximately 10,000 acres are owned by individual Alaska Natives. Under a 1983 agreement . . ., oil and gas development on the Arctic Slope Regional Corporation's lands is prohibited "until Congress authorizes such activities on Refuge lands within the coastal plain or on ASRC Lands, or both." Enactment of the legislation recommended by the Committee will plainly "authorize such activities on Refuge lands within the coastal plain," enabling the Arctic Slope Regional Corporation and the individual Alaska Native owners to develop oil and gas resources on their lands within the Refuge. However, the Alaska Native lands are within the defined Coastal Plain covered by the leasing program, and thus are subject to the overall 2,000 acre limitation. But including the Native lands within this limitation appears to abrogate the Arctic Slope Regional Corporation's right to develop all of its lands under its 1983 contract . . .

   Moreover, it is unclear what effect the Committee recommendation will have on revenues derived from oil and gas development on Native lands within the Refuge. The legislation plainly states that all receipts derived from oil and gas development on the Coastal Plain, which is defined to include the Native lands, are to be divided equally between the State of Alaska and the U.S. Treasury, though neither would have any right to revenues derived from oil and gas development on Native lands.

8. Division of receipts. The Congressional Budget Office estimates that oil and gas leasing in the Arctic Refuge will generate $5 billion in bonus, rental, and royalty receipts. Under current law, 90 percent of those receipts ($4.5 billion) are to be paid to the State of Alaska, and the remaining 10 percent ($500 million) to the U.S. Treasury. The legislation recommended by the Committee changes the allocation in current law, to permit the Federal Government to retain 50 percent ($2.5 billion) of the receipts.

   This change is necessary for the Committee to meet its reconciliation instructions. The State of Alaska contends that any such reduction of its share of receipts violates the Alaska Statehood Act and apparently intends to challenge it in court. The Committee believes that Congress has the power to reduce the State's share and that the provision will ultimately be upheld by the courts. If, however, the

State should prevail, opening the Arctic Refuge to oil and gas leasing may produce only one-fifth of the receipts the Committee expects.

9. Exports. The Mineral Leasing Act of 1920 generally prohibits the export of oil transported through pipelines over rights-of-way over public lands unless the President finds the export to be in the national interest. A 1995 amendment to the Mineral Leasing Act makes an exception for oil transported through the Trans-Alaska Pipeline. Oil moved through the Trans-Alaska Pipeline can be exported unless the President finds the export to not be in the national interest. Thus, since oil produced from the Arctic Refuge is likely to be transported through the Trans-Alaska Pipeline, it can be exported rather than used here in the [U.S.].

During consideration of the Committee recommendation, Democrats offered a series of amendments to address many of the problems in the legislation. Regrettably, all of them were rejected. Although we continue to oppose oil and gas development in the Arctic National Wildlife Refuge, adoption of our amendments would have at least ensured that development would have proceeded in accordance with the laws governing oil and gas development in other wildlife refuges, ensured that the Treasury would receive 50 percent of the receipts, and ensured that the oil would be used in the United States.

Our efforts to improve the legislation were thwarted by the fact that it is being considered under the extraordinary rules for budget reconciliation measures. All of our amendments were met with the argument that they might either reduce the amount of estimated receipts or delay their collection, or that they might not have any effect on the receipts at all, and thus would be extraneous. Such arguments only serve to remind us why important policy legislation of this sort should not be considered under the strictures of the reconciliation process.

The Senate's quarter-century debate over the future of the Arctic National Wildlife Refuge has never been about money. It has been, and continues to be, about two different sets of priorities: whether we should sacrifice a pristine wilderness to exploit the energy resources it may contain; or whether we should forego a much-needed energy resource to save a remote and frigid wilderness. Such a fundamental, deep-seated, philosophical controversy requires the deliberative process of the Senate.

Many years ago, when he introduced the bill that would become the Wilderness Act of 1964, Senator Anderson compared our wilderness areas to "our museums and our art galleries." They were "part of our cultural resource as well as our natural heritage," he said. "We should regard them as such and cherish them." We would all do well to keep Senator Anderson's words in mind as we consider the future of the Arctic National Wildlife Refuge.

POSTSCRIPT

Should the Arctic National Wildlife Refuge Be Opened to Oil Drilling?

Those who see in nature only values that can be expressed in human terms are well represented by Jonah Goldberg, who, in "Ugh, Wilderness! The Horror of 'ANWR,' the American Elite's Favorite Hellhole," *National Review* (August 6, 2001), describes the ANWR as so bleak and desolate that development can only improve it. On the other hand, Adam Kolton, testifying before the House Committee on Resources on July 11, 2001, in opposition to the National Energy Security Act of 2001 (NESA), presented the coastal plain as "the site of one of our continent's most awe-inspiring wildlife spectacles" and, thus, deserving of protection from exploitation. Kennan Ward, in *The Last Wilderness: Arctic National Wildlife Refuge* (Wildlight Press, 2001), describes a realm where human impact is still minimal and wilderness endures. John G. Mitchell, in "Oil Field or Sanctuary?" *National Geographic* (August 2001), is more balanced in his appraisal but sides with Amory B. Lovins and L. Hunter Lovins, "Fool's Gold in Alaska," *Foreign Affairs* (July/August 2001), concluding that better alternatives to developing the ANWR exist.

In "ANWR Oil: An Alternative to War Over Oil," *American Enterprise* (June 2002), Walter J. Hickle, former U.S. secretary of the interior and twice the governor of Alaska, writes, "[T]he issue is not going to go away. Given our continuing precarious dependence on overseas oil suppliers ranging from . . . the Saudis to Venezuela's Castro-clone Hugo Chavez, sensible Americans will continue to press Congress in the months and years ahead to unlock America's great Arctic energy storehouse." The recent rise in the prices of oil and gasoline renews the point: The issue is not about to go away. In fact, it is gaining urgency from growing awareness that oil production may have already passed its peak, meaning that year by year the amount of oil available to the market will decline; see the "Peak Oil Forum" essays in *World Watch* (January/ February 2006). In July 2006, backers of the failed effort to approve opening the ANWR for oil drilling introduced a new bill in the U.S. House of Representatives. This one would dedicate leasing and royalty revenues from opening the ANWR to drilling to "a trust fund to boost production of cellulosic ethanol, liquefied coal, solar and biofuel energy, among others" ("ANWR Bill Would Direct Revenue to Alternative Fuels," *CongressDaily AM,* July 27, 2006). The bill never came to a vote, but it has been called an interesting compromise and may resurface. However, in January 2007, House Democrats introduced a bill to make ANWR a permanently protected wilderness and put an end to attempts to exploit it.

Similar debate has centered on mineral exploitation in the American Southwest. President Clinton created the Grand Staircase–Escalante National Monument by executive order to protect an important part of Utah's remaining wilderness, but opposition remains. See T. H. Watkins, *The Redrock Chronicles: Saving Wild Utah* (Johns Hopkins University Press, 2000). For a survey of the wilderness system created by the 1964 Wilderness Act, see John G. Mitchell and Peter Essick, "Wilderness: America's Land Apart," *National Geographic* (November 1998).

ISSUE 8

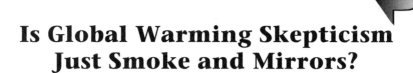

Is Global Warming Skepticism Just Smoke and Mirrors?

YES: Seth Schulman et al., from "Smoke, Mirrors & Hot Air: How ExxonMobil Uses Big Tobacco's Tactics to Manufacture Uncertainty on Climate Science," *Union of Concerned Scientists* (January 2007)

NO: Ivan Osorio, Iain Murray, and Myron Ebell, from "Liberal 'Scientists' Lead Jihad against Global-Warming Skeptics," *Human Events* (May 8, 2007)

ISSUE SUMMARY

YES: The Union of Concerned Scientists argues that opposition to the idea that global warming is real, is due to human activities, and is a threat to human well-being has been orchestrated by ExxonMobil in a disinformation campaign very similar to the tobacco industry's efforts to convince the public that tobacco was not bad for health.

NO: Ivan Osorio, Iain Murray, and Myron Ebell, all of the Competitive Enterprise Institute, argue that the Union of Concerned Scientists is a liberal-funded partisan organization that distorts facts and attempts to discredit opponents with innuendo.

Scientists have known for more than a century that carbon dioxide and other "greenhouse gases" (including water vapor, methane, and chlorofluorocarbons) help prevent heat from escaping the earth's atmosphere. In fact, it is this "greenhouse effect" that keeps the earth warm enough to support life. Yet there can be too much of a good thing. Ever since the dawn of the industrial age, humans have been burning vast quantities of fossil fuels, releasing the carbon they contain as carbon dioxide. Because of this, some estimate that by the year 2050, the amount of carbon dioxide in the air will be double what it was in 1850. By 1982 an increase was apparent. Less than a decade later, many researchers were saying that the climate had already begun to warm.

Debate over the reality of the warming trend and its significance for humanity and the environment has been vigorous. But the data are now very clear. See Richard A. Kerr, "A Worrying Trend of Less Ice, Higher Seas," *Science*

(March 24, 2006), and Jeffrey Kluger, "By Any Measure, Earth Is at the Tipping Point," *Time* (April 3, 2006). In 2007, the Intergovernmental Panel on Climate Change (IPCC; http://www.ipcc.ch/) issued its Fourth Assessment Report, saying in no uncertain terms that "Warming of the climate system is unequivocal, as is now evident from observations of increases in global average air and ocean temperatures, widespread melting of snow and ice, and rising global average sea level." The impacts on ecosystems and human well-being (especially in developing nations) will be serious. The report also outlines the steps that must be taken to prevent, ease, or cope with these impacts. The importance of early action is stressed in "Confronting Climate Change: Avoiding the Unmanageable and Managing the Unavoidable," a Scientific Expert Group Report on Climate Change and Sustainable Development, sponsored by Sigma Xi and the United Nations Foundation, prepared for the 15th Session of the UN's Commission on Sustainable Development and presented on February 27, 2007 (available at http://www.sigmaxi.org/about/news/UNSEGonline.pdf). See also William Collins, et al., "The Physical Science Behind Climate Change," *Scientific American* (August 2007).

Critics continue to stress uncertainties in the data, as well as in projections of how severe global warming will be and how severe the impact will be on human (and other) life. See Michael Shnayerson, "A Convenient Untruth," *Vanity Fair* (May 2007). Before 2007, the Bush Administration showed little interest in addressing the issue. Its plan for dealing with global warming insisted that short-term economic benefits must come first. Stuart Jordan, "The Global Warming Crisis," *The Humanist* (November/December 2005), wrote that "I see little evidence that the Bush Administration has given more than lip service to the problem. . . . There are . . . things that can be done [and] The important thing is to get started doing some of these things in earnest, and not for us to stick our heads in the sand doing business as usual until the water comes in over our eyelids. If we wait long enough, it will." After the IPCC's Fourth Assessment Report came out, the Bush Administration did admit there was a problem but it still refused to consider specific actions.

In the following selections, the Union of Concerned Scientists argues that opposition to the idea that global warming is real, is due to human activities, and is a threat to human well-being has been orchestrated by ExxonMobil in a disinformation campaign very similar to the tobacco industry's efforts to convince the public that tobacco was not bad for health. It has funded groups such as the Competitive Enterprise Institute (which was founded to fight government regulation of business), funded and orchestrated "climate contrarians" whose research lacks credibility, and lobbied government to block federal action and weaken government reports on the problem. Ivan Osorio, Iain Murray, and Myron Ebell, all of the Competitive Enterprise Institute, argue that the Union of Concerned Scientists is a liberal-funded partisan organization that distorts facts and attempts to discredit opponents with innuendo.

YES

Seth Schulman et al.

Smoke, Mirrors & Hot Air

Executive Summary

In an effort to deceive the public about the reality of global warming, Exxon-Mobil has underwritten the most sophisticated and most successful disinformation campaign since the tobacco industry misled the public about the scientific evidence linking smoking to lung cancer and heart disease. As this report documents, the two disinformation campaigns are strikingly similar. ExxonMobil has drawn upon the tactics and even some of the organizations and actors involved in the callous disinformation campaign the tobacco industry waged for 40 years. Like the tobacco industry, ExxonMobil has:

- *Manufactured uncertainty* by raising doubts about even the most indisputable scientific evidence.
- Adopted a strategy of *information laundering* by using seemingly independent front organizations to publicly further its desired message and thereby confuse the public.
- *Promoted scientific spokespeople* who misrepresent peer-reviewed scientific findings or cherry-pick facts in their attempts to persuade the media and the public that there is still serious debate among scientists that burning fossil fuels has contributed to global warming and that human-caused warming will have serious consequences.
- *Attempted to shift the focus* away from meaningful action on global warming with misleading charges about the need for "sound science."
- *Used its extraordinary access to the Bush administration* to block federal policies and shape government communications on global warming.

The report documents that, despite the scientific consensus about the fundamental understanding that global warming is caused by carbon dioxide and other heat-trapping emissions, ExxonMobil has funneled about $16 million between 1998 and 2005 to a network of ideological and advocacy organizations that manufacture uncertainty on the issue. Many of these organizations have an overlapping—sometimes identical—collection of spokespeople serving as staff, board members, and scientific advisors. By publishing and republishing the non-peer-reviewed works of a small group of scientific spokespeople, ExxonMobil-funded organizations have propped up and amplified work that has been discredited by reputable climate scientists.

From *Union of Concerned Scientists,* January 2007, pp. 1–2, 4–7, 9–12, 14–23, 24. Copyright © 2007 by Union of Concerned Scientists. Reprinted by permission.

ExxonMobil's funding of established research institutions that seek to better understand science, policies, and technologies to address global warming has given the corporation "cover," while its funding of ideological and advocacy organizations to conduct a disinformation campaign works to confuse that understanding. This seemingly inconsistent activity makes sense when looked at through a broader lens. Like the tobacco companies in previous decades, this strategy provides a positive "pro-science" public stance for ExxonMobil that masks their activity to delay meaningful action on global warming and helps keep the public debate stalled on the science rather than focused on policy options to address the problem.

In addition, like Big Tobacco before it, ExxonMobil has been enormously successful at influencing the current administration and key members of Congress. Documents highlighted in this report, coupled with subsequent events, provide evidence of ExxonMobil's cozy relationship with government officials, which enables the corporation to work behind the scenes to gain access to key decision makers. In some cases, the company's proxies have directly shaped the global warming message put forth by federal agencies. . . .

Background the Facts about ExxonMobil

ExxonMobil is a powerful player on the world stage. It is the world's largest publicly traded company: at $339 billion, its 2005 revenues exceeded the gross domestic products of most of the world's nations. It is the most profitable corporation in history. In 2005, the company netted $36 billion—nearly $100 million in profit *each day*.

As the biggest player in the world's gas and oil business, ExxonMobil is also one of the world's largest producers of global warming pollution. Company operations alone pumped the equivalent of 138 million metric tons of carbon dioxide into the atmosphere in 2004 and roughly the same level of emissions in 2005, according to company reporting. In 2005, the end use combustion of ExxonMobil's products—gasoline, heating oil, kerosene, diesel products, aviation fuels, and heavy fuels—resulted in 1,047 million metric tons of carbon dioxide–equivalent emissions. If it was a country, ExxonMobil would rank sixth in emissions.

While some oil companies like BP, Occidental Petroleum, and Shell have begun to invest in clean energy technologies and publicly committed to reduce their heat-trapping emissions, Exxon-Mobil has made no such commitment.

Lee Raymond, ExxonMobil's chief executive officer (CEO) until 2006, set a brazenly unapologetic corporate tone on global warming. During his nearly 13 years as ExxonMobil's leader, Raymond unabashedly opposed caps on carbon dioxide emissions and refused to acknowledge the scientific consensus on global warming. Under Raymond's direction, ExxonMobil positioned itself, as Paul Krugman of the *New York Times* recently put it, as "an enemy of the planet." Not only did he do nothing to curb his company's global warming emissions, during his tenure Raymond divested the company of nearly all its alternative energy holdings. During his time as CEO, ExxonMobil's board lavishly rewarded him with compensation amounting to more than $686 million. When Raymond retired at the end of 2005, he received an exorbitant retirement package worth nearly $400 million, prompting sharp criticism from shareholders. ExxonMobil is now headed by CEO Rex Tillerson, but the corporate policies Raymond forged so far remain largely intact.

ExxonMobil has played the world's most active corporate role in underwriting efforts to thwart and undermine climate change regulation. For instance, according to the Center for Responsive Politics, ExxonMobil's PAC—its political action committee—and individuals affiliated with the company made more than $4 million in political contributions throughout the 2000 to 2006 election cycles. It was consistently among the top four energy sector contributors. In the 2004 election cycle alone, ExxonMobil's PAC and individuals affiliated with the company gave $935,000 in political contributions, more than any other energy company. Much of that money went in turn to President Bush's election campaign. In addition, ExxonMobil paid lobbyists more than $61 million between 1998 and 2005 to help gain access to key decision makers.

This report does not attempt to shed light on all ExxonMobil activities related to global warming. Instead, it takes an in-depth look at how the relatively modest investment of about $16 million between 1998 and 2004 to select political organizations has been remarkably effective at manufacturing uncertainty about the scientific consensus on global warming. It offers examples to illustrate how ExxonMobil's influence over key administration officials and members of Congress has fueled the disinformation campaign and helped forestall federal action to reduce global warming emissions. And this report identifies how strategies and tactics used by ExxonMobil mirror the well-documented campaign by the tobacco industry to prevent government regulation by creating public confusion about the link between smoking and disease.

The Origins of a Strategy

In its campaign to sow uncertainty about the scientific evidence on global warming, ExxonMobil has followed a corporate strategy pioneered by the tobacco industry. Because ExxonMobil's strategy, tactics, and even some personnel draw heavily from the tobacco industry's playbook, it is useful to look briefly at this earlier campaign. The settlement of the lawsuit brought by the attorneys general of 46 states forced the major tobacco companies to place their enormous caches of internal documents online. Thanks to these archives, the details of the tobacco industry's covert strategy are now clear.

The story begins in the mid-1950s when scientific evidence began to emerge linking smoking to cancer. The tobacco industry's initial response was to fund a research consortium, initially called the Tobacco Industry Research Committee and later known as the U.S. Tobacco Institute, to "study the issue." In 1954, Big Tobacco released a seminal public document called the "Frank Statement to Cigarette Smokers," which set the industry's tone for the coming decades. This document questioned the emerging scientific evidence of the harm caused by smoking but tried to appear concerned about the issue, pledging to the public that the industry would look closely at the scientific evidence and study it themselves.

As we now know, tobacco industry lawyers advised the companies early on that they could never admit they were selling a hazardous product without opening themselves to potentially crippling liability claims. So, rather than studying the health hazards posed by their products, the tobacco industry hired Hill & Knowlton, a leading public relations firm of the day to mount a public relations campaign on their behalf. In a key memo, Hill & Knowlton framed the issue this way: "There is only one problem—confidence and how to establish it; public assurance, and how to create it." In other words, the tobacco companies should ignore the deadly health effects of smoking and focus instead on maintaining the public's confidence in their products.

As time went on, a scientific consensus emerged about a multitude of serious dangers from smoking—and the tobacco manufacturers knew it. Despite the evidence, the industry developed a sophisticated disinformation campaign—one they knew to be misleading—to deceive the public about the hazards of smoking and to forestall governmental controls on tobacco consumption.

How Big Tobacco's Campaign Worked

In executing their calculated strategy over the course of decades, tobacco industry executives employed five main tactics:

- They sought to *manufacture uncertainty* by raising doubts about even the most indisputable scientific evidence showing their products to be hazardous to human health.
- They pioneered a strategy of *"information laundering"* in which they used—and even covertly established—seemingly independent front organizations to make the industry's own case and confuse the public.
- They *promoted scientific spokespeople* and invested in scientific research in an attempt to lend legitimacy to their public relations efforts.
- They attempted to *recast the debate* by charging that the wholly legitimate health concerns raised about smoking were not based upon "sound science."
- Finally, they *cultivated close ties with government officials* and members of Congress. While many corporations and institutions seek access to government, Tobacco's size and power gave it enormous leverage.

In reviewing the tobacco industry's disinformation campaign, the first thing to note is that the tobacco companies quickly realized they did not need to prove their products were safe. Rather, as internal documents have long since revealed, they had only to "maintain doubt" on the scientific front as a calculated strategy. As one famous internal memo from the Brown & Williamson tobacco company put it: "Doubt is our product, since it is the best means of competing with the 'body of fact' that exists in the minds of the general public. It is also the means of establishing a controversy." David Michaels, professor of occupational and environmental health at George Washington University School of Public heath and former assistant secretary for the environment, safety and health at the Department of Energy during the Clinton administration, has dubbed the strategy one of "manufacturing uncertainty." As Michaels has documented, Big Tobacco pioneered the strategy and many opponents of public health and environmental regulations have emulated it. . . .

ExxonMobil's Disinformation Campaign

In the late 1980s, when the public first began to hear about global warming, scientists had already conducted more than a century of research on the impact of carbon dioxide on earth's climate. As the science matured in the late 1980s, debate, a key component of the scientific process, surfaced among reputable scientists about the scope of the problem and the extent to which human activity was responsible. Much like the status of scientific knowledge about the health effects of smoking in the early 1950s, emerging studies suggested cause for concern but many scientists justifiably argued that more research needed to be done.

Exxon (and later ExxonMobil), concerned about potential repercussions for its business, argued from the start that no global warming trend existed and that a link between human activity and climate change could not be established. Just as the tobacco companies initially responded with a coalition to address the health effects of smoking, Exxon and the American Petroleum Institute (an organization twice chaired by former Exxon CEO Lee Raymond) joined with other energy, automotive, and industrial companies in 1989 to form the Global Climate Coalition. The coalition responded aggressively to the emerging scientific studies about global warming by opposing governmental action designed to address the problem.

Drawing on a handful of scientific spokespeople during the early and mid-1990s, the Global Climate Coalition emphasized the remaining uncertainties in climate science. Exxon and other members of the coalition challenged the need for action on global warming by denying its existence as well as characterizing global warming as a natural phenomenon. As Exxon and its proxies mobilized forces to cast doubt on global warming, however, a scientific consensus was emerging that put their arguments on exceptionally shaky scientific ground.

Manufacturing Uncertainty

By 1997, scientific understanding that human-caused emissions of heat-trapping gases were causing global warming led to the Kyoto Protocol, in which the majority of the world's industrialized nations committed to begin reducing their global warming emissions on a specified timetable. In response to both the strength of the scientific evidence on global warming and the governmental action pledged to address it, leading oil companies such as British Petroleum, Shell, and Texaco changed their stance on climate science and abandoned the Global Climate Coalition.

ExxonMobil chose a different path.

In 1998, ExxonMobil helped create a small task force calling itself the "Global Climate Science Team" (GCST). Members included Randy Randol, Exxon-Mobil's senior environmental lobbyist at the time, and Joe Walker, the public relations representative of the American Petroleum Institute. One member of the GCST task force, Steven Milloy, headed a nonprofit organization called the Advancement of Sound Science Coalition, which had been covertly created by the tobacco company Philip Morris in 1993 to manufacture uncertainty about the health hazards posed by secondhand smoke.

A 1998 GCST task force memo outlined an explicit strategy to invest millions of dollars to manufacture uncertainty on the issue of global warming—a strategy that directly emulated Big Tobacco's disinformation campaign. Despite mounting scientific evidence of the changing climate, the goal the team outlined was simple and familiar. As the memo put it, "Victory will be achieved when average citizens understand (recognize) uncertainties in climate science" and when public "recognition of uncertainty becomes part of the 'conventional wisdom.'"

Regardless of the mounting scientific evidence, the 1998 GCST memo contended that "if we can show that science does not support the Kyoto treaty . . . this puts the United States in a stronger moral position and frees its negotiators from the need to make concessions as a defense against perceived selfish economic concerns."

ExxonMobil and its partners no doubt understood that, with the scientific evidence against them, they would not be able to influence reputable scientists. The 1998 memo proposed that ExxonMobil and its public relations partners "develop and implement a national media relations program to inform the media about uncertainties in climate science." In the years that followed, Exxon-Mobil executed the strategy as planned underwriting a wide array of front organizations to publish in-house articles by select scientists and other like-minded individuals to raise objections about legitimate climate science research that has withstood rigorous peer review and has been replicated in multiple independent peer-reviewed studies—in other words, to attack research findings that were well established in the scientific community. The network ExxonMobil created masqueraded as a credible scientific alternative, but it publicized discredited studies and cherry-picked information to present misleading conclusions.

Information Laundering

A close review reveals the company's effort at what some have called "information laundering": projecting the company's desired message through ostensibly independent nonprofit organizations. First, ExxonMobil underwrites well-established groups such as the American Enterprise Institute, the Competitive Enterprise Institute, and the Cato Institute that actively oppose mandatory action on global warming as well as many other environmental standards. But the funding doesn't stop there. ExxonMobil also supports a number of lesser-known organizations that help to market and distribute global warming disinformation. Few of these are household names. For instance, most people are probably not familiar with the American Council for Capital Formation Center for Policy Research, the American Legislative Exchange Council, the Committee for a Constructive Tomorrow, or the International Policy Network, to name just a few. yet these organizations—and many others like them—have received sizable donations from ExxonMobil for their climate change activities.

Between 1998 and 2005 (the most recent year for which company figures are publicly available), ExxonMobil has funneled approximately $16 million to carefully chosen organizations that promote disinformation on global warming. As the *New York Times* has reported, ExxonMobil is often the single largest corporate donor to many of these nonprofit organizations, frequently accounting for more than 10 percent of their annual budgets.

A close look at the work of these organizations exposes ExxonMobil's strategy. Virtually all of them publish and publicize the work of a nearly identical group of spokespeople, including scientists who misrepresent peer-reviewed climate findings and confuse the public's understanding of global warming. Most of these organizations also include these same individuals as board members or scientific advisers.

Why would ExxonMobil opt to fund so many groups with overlapping spokespeople and programs? By generously funding a web of organizations with redundant personnel, advisors, or spokespeople, ExxonMobil can quietly and effectively provide the appearance of a broad platform for a tight-knit group of vocal climate science contrarians. The seeming diversity of the organizations creates an "echo chamber" that amplifies and sustains scientific disinformation even though many of the assertions have been repeatedly debunked by the scientific community.

Take, for example, ExxonMobil's funding of a Washington, DC-based organization called Frontiers of Freedom. Begun in 1996 by former Senator Malcolm Wallop, Frontiers of Freedom was founded to promote property rights and critique environmental regulations like the Endangered Species Act. One of the group's staff members, an economist named Myron Ebell, later served as a member of the Global Climate Science Team, the small task force that laid out Exxon-Mobil's 1998 message strategy on global warming. Following the outline of the task force's plan in 1998, ExxonMobil began funding Frontiers of Freedom—a group that Vice President Dick Cheney recently called "an active, intelligent, and needed presence in the national debate."

Since 1998, ExxonMobil has spent $857,000 to underwrite the Frontiers of Freedom's climate change efforts. In 2002, for example, ExxonMobil made a grant to Frontiers of Freedom of $232,000 (nearly a third of the organization's annual budget) to help launch a new branch of the organization called the Center for Science and Public Policy, which would focus primarily on climate change.

A recent visit to the organization's website finds little information about the background or work of the Center for Science and Public Policy. The website offers no mention of its staff or board members other than its current executive director Robert Ferguson, for whom it offers no biographical information. As of September 2006, however, the website did prominently feature a 38-page non-peer-reviewed report by Ferguson on climate science, heavily laden with maps, graphs, and charts, entitled "Issues in the Current State of Climate Science: A Guide for Policy Makers and Opinion Leaders." The document offers a hodge-podge of distortions and distractions posing as a serious scientific review. Ferguson questions the clear data showing that the majority of the globe's glaciers are in retreat by feebly arguing that not all glaciers have been inventoried, despite the monitoring of thousands of glaciers worldwide. And, in an attempt to dispute solid scientific evidence that climate change is causing extinctions of animal species, Ferguson offers the non sequitur that several new butterfly and frog species were recently discovered in New Guinea.

Perhaps most notable are Ferguson's references, citing a familiar collection of climate science contrarians such as Willie Soon. In fact, although his title is not listed on the organization's website, Soon is the Center for Science and Public Policy's "chief science researcher," according to a biographical note accompanying a 2005 *Wall Street Journal* op-ed co-authored by Ferguson and Soon. Ferguson's report was not subject to peer review, but it is nonetheless presented under the auspices of the authoritative-sounding Center for Science and Public Policy.

Another organization used to launder information is the George C. Marshall Institute. During the 1990s, the Marshall Institute had been known primarily for its work advocating a "Star Wars" missile defense program. however, it soon became an important home for industry-financed "climate contrarians," thanks in part to ExxonMobil's financial backing. Since 1998, ExxonMobil has paid $630,000 primarily to underwrite the Marshall Institute's climate change effort. William O'Keefe, CEO of the Marshall Institute, formerly worked as executive vice president and chief operating officer of the American Petroleum Institute, served on the board of directors of the Competitive Enterprise Institute, and is chairman emeritus of the Global Climate Coalition.

Since ExxonMobil began to support its efforts, the Marshall Institute has served as a clearinghouse for global warming contrarians, conducting roundtable events and producing frequent publications. Most recently, the Marshall Institute has been touting its new book, *Shattered Consensus: The True State of Global Warming,* edited by longtime climate contrarian Patrick Michaels (a meteorologist). Michaels has, over the past several years, been affiliated with at least ten organizations funded by ExxonMobil. Contributors to the book include others with similar affiliations with Exxon-funded groups: Sallie Baliunas, Robert Balling, John Christy, Ross McKitrick, and Willie Soon.

The pattern of information laundering is repeated at virtually all the private, nonprofit climate change programs ExxonMobil funds. The website of the Chicago-based Heartland Institute, which received $119,000 from ExxonMobil in 2005, offers recent articles by the same set of scientists. A visit to the climate section of the website of the American Legislative Exchange Council, which received $241,500 from ExxonMobil in 2005, turns up yet another non-peer-reviewed paper by Patrick Michaels. The Committee for a Constructive Tomorrow, which received $215,000 from ExxonMobil over the past two funding cycles of 2004 and 2005, boasts a similar lineup of articles and a scientific advisory panel that includes Sallie Baliunas, Robert Balling, Roger Bate, Sherwood Idso, Patrick Michaels, and Frederick Seitz—all affiliated with other ExxonMobil-funded organizations.

A more prominent organization funded by ExxonMobil is the Washington, DC-based Competitive Enterprise Institute (CEI). Founded in 1984 to fight government regulation on business, CEI started to attract significant ExxonMobil funding when Myron Ebell moved there from Frontiers of Freedom in 1999.

Since then, CEI has not only produced a steady flow of vituperative articles and commentaries attacking global warming science, often using the same set of global warming contrarians; it has also sued the federal government to stop the dissemination of a National Assessment Synthesis Team report extensively documenting the region-by-region impacts of climate change in the United States. For its efforts, CEI has received more than $2 million in funding from ExxonMobil from 1998 through 2005.

The irony of all these efforts is that ExxonMobil, a company that claims it is dedicated to supporting organizations favoring "free market solutions to public policy problems," is actively propping up discredited studies and misleading information that would otherwise never thrive in the scientific marketplace of ideas. . . .

Confounding the matter further is ExxonMobil's funding of established research institutions that seek to better understand science, policies, and technologies to address global warming. . . .

The funding of academic research activity has provided the corporation legitimacy, while it actively funds ideological and advocacy organizations to conduct a disinformation campaign.

Promoting Scientific Spokespeople

Inextricably intertwined with ExxonMobil's information laundering strategy of underwriting multiple organizations with overlapping staff is the corporation's promotion of a small handful of scientific spokespeople. Scientists are trusted messengers among the American public. Scientists can and do play an important and legitimate role in educating the public and policymakers about issues that have a scientific component, including global warming. Early on, Exxon (and later ExxonMobil) sought to support groups that worked with the handful of scientists, such as Frederick Singer (a physicist), John Christy (an atmospheric scientist), and Patrick Michaels, who had persistently voiced doubt about humancaused global warming and its consequences, despite mounting evidence.

However, to pull off the disinformation campaign outlined in the 1998 GCST task force memo, ExxonMobil and its public relations partners recognized they would need to cultivate new scientific spokespeople to create a sense among the public that there was still serious debate among scientists. Toward that end, the memo suggested that the team "identify, recruit and train a team of five independent scientists to participate in media outreach. These will be individuals who do not have a long history of visibility and/or participation in the climate change debate. Rather, this team will consist of new faces who will add their voices to those recognized scientists who already are vocal."

By the late 1990s, the scientific evidence on global warming was so strong that it became difficult to find scientists who disputed the reality of human-caused climate change. But ExxonMobil and its public relations partners persevered. The case of scientists Willie Soon and Sallie Baliunas is illustrative.

Soon and Baliunas are astrophysicists affiliated with the Harvard-Smithsonian Center for Astrophysics who study solar variation (i.e., changes in the amount of energy emitted by the Sun). Solar variation is one of the many factors influencing Earth's climate, although according to the IPCC it is one of the minor influences over the last century. In the mid-1990s, ExxonMobil-funded groups had already begun to spotlight the work of Soon and Baliunas to raise doubts about the human causes of global warming. To accomplish this, Baliunas was initially commissioned to write several articles for the Marshall Institute positing that solar activity might be responsible for global warming. With the Baliunas articles, the Marshall Institute skillfully amplified an issue of minor scientific importance and implied that it was a major driver of recent warming trends.

In 2003, Baliunas and Soon were catapulted into a higher profile debate when they published a controversial review article about global warming in the peer-reviewed scientific literature. Writing in the journal *Climate Research*, the two contrarians reviewed the work of a number of previous scientists and alleged that the twentieth century was not the warmest century of the past 1,000 years and that the climate had not changed significantly over that period. The Soon-Baliunas paper was trumpeted widely by organizations and individuals funded by ExxonMobil. It was also seized upon by like-minded politicians, most notably James Inhofe (R-OK), chair (until January 2007) of the Senate Environment and Public Works Committee, who has repeatedly asserted that global warming is a hoax. Inhofe cited the Soon-Baliunas review as proof that natural variability, not human activity, was the "overwhelming factor" influencing climate change.

Less widely publicized was the fact that three of the editors of *Climate Research*—including incoming editor-in-chief Hans von Storch—resigned in protest over the Soon-Baliunas paper. Storch stated that he suspected that "some of the skeptics had identified *Climate Research* as a journal where some editors were not as rigorous in the review process as is otherwise common" and described the manuscript as "flawed." In addition, thirteen of the scientists cited in the paper published a rebuttal explaining that Soon and Baliunas had seriously misinterpreted their research.

The National Research Council recently examined the large body of published research on this topic and concluded that, "It can be said with a high level of confidence that global mean surface temperature was higher during the last few decades of the 20th century than during any comparable period during the preceding four centuries. . . . Presently available proxy evidence indicates that temperatures at many, but not all, individual locations were higher in the past 25 years than during any period of comparable length since A.D. 900." The brouhaha in the scientific community had little public impact. The echo chamber had already been set in motion reverberating among the mainstream media, while the correction became merely a footnote buried in the science sections of a few media outlets.

This controversy did not stop Soon and Baliunas from becoming central "new voices" in ExxonMobil's effort to manufacture uncertainty about global warming. Both scientists quickly established relationships with a network of organizations underwritten by the corporation. Over the past several years, for example, Baliunas has been formally affiliated with no fewer than nine organizations receiving funding from ExxonMobil. Among her other affiliations, she is now a board member and senior scientist at the Marshall Institute, a scientific advisor to the Annapolis Center for Science-Based Public Policy, an advisory board member of the Committee for a Constructive Tomorrow, and a contributing scientist to the online forum Tech Central Station, all of which are underwritten by ExxonMobil.

Another notable case is that of Frederick Seitz, who has ties to both Big Tobacco and ExxonMobil. Seitz is the emeritus chair of the Marshall Institute. He is also a prominent solid state physicist who was president of the National Academy of Sciences (NAS) from 1962 to 1969.

In an example of the tobacco industry's efforts to buy legitimacy, the cigarette company R.J. Reynolds hired Seitz in 1979. His role was to oversee a tobacco industry–sponsored medical research program in the 1970s and 1980s. "They didn't want us looking at the health effects of cigarette smoking," Seitz, who is now 95, admitted recently in an article in *Vanity Fair*, but he said he felt no compunction about dispensing the tobacco company's money.

While working for R.J. Reynolds, Seitz oversaw the funding of tens of millions of dollars worth of research. Most of this research was legitimate. For instance, his team looked at the way stress, genetics, and lifestyle issues can contribute to disease. But the program Seitz oversaw served an important dual purpose for R.J. Reynolds. It allowed the company to tout the fact that it was funding health research (even if it specifically proscribed research on the health effects of smoking) and it helped generate a steady collection of ideas and hypotheses that provided "red herrings" the company could use to disingenuously suggest that factors other than tobacco might be causing smokers' cancers and heart disease.

Aside from giving the tobacco companies' disinformation campaign an aura of scientific credibility, Seitz is also notable because he has returned from retirement to play a prominent role as a global warming contrarian involved in organizations funded by ExxonMobil. Consider, for instance, one of Seitz's most controversial efforts. In 1998, he wrote and circulated a letter asking scientists to sign a petition from a virtually unheard-of group called the Oregon Institute of Science and Medicine calling upon the U.S. government to reject the Kyoto Protocol. Seitz signed the letter identifying himself as a former NAS president. He also enclosed with his letter a report co-authored by a team including Soon and Baliunas asserting that carbon dioxide emissions pose no warming threat. The report was not peer reviewed. But it was formatted to look like an article from *The Proceedings of the National Academy of Sciences* (PNAS), a leading scientific journal.

The petition's organizers publicly claimed that the effort had attracted the signatures of some 17,000 scientists. But it was soon discovered that the list contained few credentialed climate scientists. For example, the list was riddled with the names of numerous fictional characters. Likewise, after investigating a random sample of the small number of signers who claimed to have a Ph.D. in a climate-related field, *Scientific American* estimated that approximately one percent of the petition signatories might actually have a Ph.D. in a field related to climate science. In a highly unusual response, NAS issued a statement disavowing Seitz's petition and disassociating the academy from the PNAS-formatted paper. None of these facts, however, have stopped organizations, including those funded by

ExxonMobil, from touting the petition as evidence of widespread disagreement over the issue of global warming. For instance, in the spring of 2006, the discredited petition surfaced again when it was cited in a letter to California legislators by a group calling itself "Doctors for Disaster Preparedness," a project of the Oregon Institute of Science and Medicine.

Shifting the Focus of the Debate

One prominent component of ExxonMobil's disinformation campaign on global warming is the almost unanimous call for "sound science" by the organizations it funds. Like the Bush administration's "Healthy Forests" program, which masks a plan to augment logging, the rallying call for "sound science" by ExxonMobil-funded organizations is a clever and manipulative cover. It shifts the focus of the debate away from ExxonMobil's irresponsible behavior regarding global warming toward a positive concept of "sound science." By keeping the discussion focused on refining scientific understanding, ExxonMobil helps delay action to reduce heat-trapping emissions from its company and products indefinitely. For example, like the company itself, ExxonMobil-funded organizations routinely contend, despite all the solid evidence to the contrary, that scientists don't know enough about global warming to justify substantial reductions in heat-trapping emissions. As ExxonMobil explains prominently on the company's website:

> While assessments such as those of the IPCC [Intergovernmental Panel on Climate Change] have expressed growing confidence that recent warming can be attributed to increases in greenhouse gases, these conclusions rely on expert judgment rather than objective, reproducible statistical methods. Taken together, gaps in the scientific basis for theoretical climate models and the interplay of significant natural variability make it very difficult to determine objectively the extent to which recent climate changes might be the result of human actions.

In contrast, 11 of the world's major national scientific academies issued a joint statement in 2005 that declared, "The scientific understanding of climate change is now sufficiently clear to justify nations taking prompt action. It is vital that all nations identify cost-effective steps that they can take now to contribute to substantial and long-term reduction in net global greenhouse gas emissions."

There is no denying that the tactic of demanding "certainty" in every aspect of our scientific understanding of global warming is a rhetorically effective one. If manufactured uncertainty and governmental inaction is the goal, science will arguably never be "sound enough," or 100 percent certain, to justify action to protect public health or the environment.

Again, the tobacco industry paved the way. The calculated call for "sound science" was successfully used by tobacco firms as an integral part of a tobacco company's pioneering "information laundering" scheme. As we now know from internal tobacco industry documents, a campaign to demand "sound science" was a key part of a strategy by the cigarette manufacturer Philip Morris to create uncertainty about the scientific evidence linking disease to "second-hand" tobacco smoke, known in the industry as "environmental tobacco smoke" or ETS. Toward this end, in 1993, Philip Morris covertly created a front organization called "The Advancement of Sound Science Coalition" or TASSC.

In setting up the organization, Philip Morris took every precaution. The company opted not to use its regular public relations firm, Burson-Marsteller, choosing instead APCO Associates, a subsidiary of the international advertising and PR firm of GCI/Grey Associates. For a sizable retainer, APCO agreed to handle every aspect of the front organization. . . .

The public relations firm introduced TASSC to the public through a decentralized launch outside the large markets of Washington, DC, and New York in order to "avoid cynical reporters from major media" who might discover the truth that the organization was nothing more than a front group created by Philip Morris. Top Philip Morris media managers compiled lists of reporters they deemed most sympathetic to TASSC's message. But they left all press relations to APCO so as to, in the words of one internal memo, "remove any possible link to PM."

The TASSC campaign was a particularly obvious example of information laundering. But it also represented an important messaging strategy by using the concept of "sound science" to attach Philip Morris's disinformation about second-hand smoke to a host of other antiregulation battles. Philip Morris sought to foil any effort by the Environmental Protection Agency (EPA) to promulgate regulations to protect the public from the dangers of ETS. But the company realized that it could build more support for its discredited position that ETS was safe by raising the broader "sound science" banner. As a result, it took stands against government efforts to set safety regulations on everything from asbestos to radon. "The credibility of EPA is defeatable," one Philip Morris strategy document explained, "but not on the basis of ETS alone. It must be part of a large mosaic that concentrates all of the EPA's enemies against it at one time."

The important point in reviewing this history is that it is not a coincidence that ExxonMobil and its surrogates have adopted the mantle of "sound science." In so doing, the company is simply emulating a proven corporate strategy for successfully deflecting attention when one's cause lacks credible scientific evidence. From the start in 1993, in TASSC's search for other antiregulation efforts to provide political cover, the organization actively welcomed global warming contrarians like Frederick Seitz, Fred Singer, and Patrick Michaels to its scientific board of advisors. Thanks to the online archive of tobacco documents, we know that in 1994, when Philip Morris developed plans with APCO to launch a TASSC-like group in Europe, "global warming" was listed first among suggested topics with which the tobacco firm's cynical "sound science" campaign could profitably ally itself.

Given these historical connections, it is disturbing that ExxonMobil would continue to associate with some of the very same TASSC personnel who had overseen such a blatant and shameful disinformation campaign for Big Tobacco. The most glaring of ExxonMobil's associations in this regard is with Steven Milloy, the former executive director of TASSC. Milloy's involvement with ExxonMobil is more than casual. He served as a member of the small 1998

Global Climate Science Team task force that mapped out ExxonMobil's disinformation strategy on global warming.

Milloy officially closed TASSC's offices in 1998 as evidence of its role as a front organization began to surface in the discovery process of litigation against Big Tobacco. Thanks in part to ExxonMobil, however, the "sound science" disinformation campaign continued unabated. Resuscitating TASSC under the slightly altered name The Advancement of Sound Science *Center* (rather than Coalition), Milloy continues to operate out of his home in Maryland. Between 2000 and 2004, ExxonMobil gave $50,000 to Milloy's Advancement of Sound Science Center, and another $60,000 to an organization called the Free Enterprise Education Institute (a.k.a. Free Enterprise Action Institute), which is also registered to Milloy's home address. According to its 2004 tax return, this group was founded to "educate the public about the American system of free enterprise," employed no staff, and incurred approximately $48,000 in expenses categorized as "professional services."

In addition to serving as a columnist on *FoxNews.com,* Milloy is also a contributor to Tech Central Station and an adjunct scholar at the Competitive Enterprise Institute, both funded by ExxonMobil.

The irony of the involvement of tobacco disinformation veterans like Milloy in the current campaign against global warming science is not lost on close watchers. Representative Henry Waxman (D-CA), for instance, chaired the 1994 hearings where tobacco executives unanimously declared under oath that cigarettes were not addictive. As Waxman marveled recently about the vocal contrarians like Milloy on global warming science: "Not only are we seeing the same tactics the tobacco industry used, we're seeing some of the same groups." Of course, unlike the tobacco companies, ExxonMobil has yet to receive a court order to force to light internal documents pertaining to its climate change activities. Nonetheless, even absent this information, the case could hardly be clearer: ExxonMobil is waging a calculated and familiar disinformation campaign to mislead the public and forestall government action on global warming.

Buying Government Access

Tobacco companies have historically been very successful at cultivating close ties in government and hiring former government officials to lobby on their behalf. This list includes, among others, Craig Fuller, who served in the Reagan and Bush administrations, and former GOP chair Haley Barbour as well as former Senate majority leader George Mitchell, who was recruited in 1997 by the tobacco industry firm Verner, Liipfert, Bernhard, McPherson, and Hand to help negotiate a settlement.

When it comes to exerting influence over government policy, however, ExxonMobil, in its global warming disinformation campaign, may have even surpassed the tobacco industry it so clearly emulates. During the 2000 to 2006 election cycles, ExxonMobil's PAC and individuals affiliated with the company gave more than $4 million to federal candidates and parties. Shortly after President Bush's inauguration, ExxonMobil, like other large corporate backers in the energy sector, participated in Vice President Dick Cheney's "Energy Task Force" to set the administration's goals for a national energy plan. ExxonMobil successfully urged the Bush administration to renege on the commitments to the Kyoto Protocol made by previous administrations. Paula Dobriansky, who currently serves as undersecretary for global affairs in the State Department and who has headed U.S. delegations negotiating follow-ons to the Kyoto Protocol in Buenos Aires and Montreal, explicitly said as much in 2001. Just months after she had been confirmed by the U.S. Senate, Dobriansky met with ExxonMobil lobbyist Randy Randol and other members of the Global Climate Coalition. Her prepared talking points, uncovered through a Freedom of Information Act request, reveal that Dobriansky thanked the group for their input on global warming policy. One of her notes reads: "POTUS [the President of the United States] rejected Kyoto, in part, based on input from you."

A Freedom of Information Act request also revealed that in February 2001, immediately following the release of the authoritative 2001 report on global warming from the Intergovernmental Panel on Climate Change (IPCC), ExxonMobil successfully lobbied the Bush administration to try to oust the chair of the IPCC. In a memo sent to the White House, Randol complained that Robert Watson, who had chaired the IPCC since 1996, had been "hand-picked by Al Gore." Watson is an internationally respected scientist who has served as the director of the science division at NASA and as chief scientist at the World Bank. His work at the IPCC had met with widespread international approval and acclaim. Nonetheless, the ExxonMobil memo urged: "Can Watson be replaced now at the request of the U.S.?" At its next opportunity, the Bush administration's State Department refused to re-nominate Dr. Watson for a second five-year term as head of the IPCC, instead backing an Indian engineer-economist for the post. In April 2002, lacking U.S. support, Dr. Watson lost his position as chair. The Bush administration's move outraged many in the scientific community who saw it as a blatantly political attempt to undermine an international scientific effort. At the time, however, ExxonMobil's behind-the-scenes role in the incident remained secret.

Meanwhile, in an equally consequential recommendation, the 2001 ExxonMobil memo suggested that President Bush's climate team hire Harlan Watson (no relation), a staff member on the House Science Committee who had served as a climate negotiator at the 1992 Rio Earth Summit for the administration of George Bush Senior and had worked closely with members of Congress who opposed action on global warming. Shortly thereafter, the Bush administration announced Harlan Watson's appointment as its chief climate negotiator. He has steadfastly opposed any U.S. engagement in the Kyoto process.

As successful as ExxonMobil's efforts to lobby the Bush administration have been, perhaps even more striking is the way the company's disinformation campaign on global warming science has managed to permeate the highest echelons of the federal government. Between 2001 and 2005, the nerve center for much of this censorship and control resided in the office of Philip Cooney, who served during this time as chief of staff in the White House Council on Environmental Quality. Thanks to a whistle-blowing researcher named Rick Piltz in the U.S. government's interagency Climate Change Science Program who resigned in protest over the practice, we now know that Cooney spent a significant amount of time censoring and distorting government reports so as to exaggerate scientific uncertainty about global warming.

Cooney, a lawyer with an undergraduate degree in economics, had no scientific credentials that might qualify him to rewrite the findings of top government scientists. Rather, before coming to the Bush administration in 2001, Cooney had spent roughly a decade as a lawyer for the American Petroleum Institute, the oil industry lobby that worked with ExxonMobil in 1998 to develop a global warming disinformation campaign. In that capacity, Cooney served as a "climate team leader" seeking to prevent the U.S. government from entering into any kind of international agreement or enacting any domestic legislation that might lead to mandatory limits on global warming emissions. After joining the White House staff in 2001, Cooney furthered much the same work agenda from the top ranks of the Bush administration.

During his tenure, Cooney altered and compromised the accuracy of numerous official scientific reports on climate change issued by agencies of the federal government. For instance, in 2002, as U.S. government scientists struggled to finalize the Climate Change Science Program's strategic plan, Cooney dramatically altered the document, editing it heavily and repeatedly inserting qualifying words to create an unwarranted aura of scientific uncertainty about global warming and its implications.

As Rick Piltz explained in his resignation letter when he exposed Cooney's efforts, the government agencies had adapted to the environment created within the Bush administration by "engaging in a kind of anticipatory self-censorship on this and various other matters seen as politically sensitive under this administration." Even beyond the outright suppression and distortion by Cooney and others, according to Piltz, this self-censorship on the part of career professionals marked one of the most insidious and "deleterious influences of the administration" on climate research efforts within the government.

On June 10, 2005, Cooney resigned, two days after the *New York Times* first reported Piltz's revelations. Despite the suspicious timing, the White House claimed that Cooney's resignation was unrelated to Piltz's disclosures. But it was not surprising when Cooney announced, one week after he left the White House, that he was accepting a high-ranking public relations position at ExxonMobil.

One of the most damning incidents involving Cooney also illustrates the extent of ExxonMobil's influence over the Bush administration policy on global warming. In May 2002, the administration issued the "U.S. Climate Action Report," which the U.S. State Department was obligated by treaty to file with the United Nations. Major elements of the report were based on an in-depth, peer-reviewed government research report analyzing the potential effects of global warming in the United States. That report, titled "U.S. National Assessment of the Potential Consequences of Climate Variability and Change," predates the Bush administration and had already been attacked by ExxonMobil. The report generated widespread headlines such as one in the *New York Times* proclaiming: "Climate Changing, US Says in Report."

Not surprisingly, ExxonMobil vociferously objected to the conclusion of the multiagency "Climate Action Report" that climate change posed a significant risk and was caused by humanmade emissions. Concerned about the matter,

Cooney contacted Myron Ebell at the ExxonMobil-funded Competitive Enterprise Institute. "Thanks for calling and asking for our help," Ebell responded in a June 3, 2002, email to Cooney that surfaced as a result of a Freedom of Information Act request. Ebell urged that the President distance himself from the report. Within days, President Bush did exactly that, denigrating the report in question as having been "put out by the bureaucracy."

In the June 3 email, Ebell explicitly suggests the ouster of then-EPA head Christine Todd Whitman. "It seems to me that the folks at the EPA are the obvious fall guys and we would only hope that the fall guy (or gal) should be as high up as possible," Ebell wrote. "Perhaps tomorrow we will call for Whitman to be fired." Sure enough, Whitman would last for less than a year in her post, resigning in May 2003. Finally, Ebell pledged he would do what he could to respond to the White House's request to "clean up this mess."

A major piece of Ebell's "clean-up" effort presumably came on August 6, 2003, when the Competitive Enterprise Institute filed the second of two lawsuits calling for the Bush administration to invalidate the National Assessment (a peer-reviewed synthesis report upon which the U.S. Climate Action Report was based). The CEI lawsuit called for it to be withdrawn because it was not based upon "sound science."

Given the close, conspiratorial communication between Ebell and Cooney that had come to light, the lawsuit prompted the attorneys general of Maine and Connecticut to call upon the U.S. Justice Department to investigate the matter. However, the Bush administration Justice Department, then led by John Ashcroft, refused to launch such an investigation, despite the fact that the Maine and Connecticut attorneys general stated forcefully that the evidence suggested that Cooney had conspired with Ebell to cause the Competitive Enterprise Institute to sue the federal government. As Maine Attorney General Steven Rowe noted: "The idea that the Bush administration may have invited a lawsuit from a special interest group in order to undermine the federal government's own work under an international treaty is very troubling."

A key piece of evidence, unnoticed at the time, strongly suggests just how the scheme fit together. In 2002, in a move virtually unprecedented in its corporate giving program, ExxonMobil offered an additional $60,000 in support for the Competitive Enterprise Institute—specifically earmarked to cover the organization's unspecified "legal activities."

In addition to a high level of administration access, ExxonMobil has cultivated close relationships with members of Congress. In July 2005, ExxonMobil's generous campaign contributions paid off when Congress passed the Energy Policy Act of 2005. This bill, modeled on the President's 2001 energy plan, provides more than $7.4 billion in tax breaks and subsidies to the oil and gas industry over 10 years and excludes any provisions that would mandate reductions in U.S. global warming emissions.

Joe Barton (R-TX), chair of the house Energy and Commerce Committee from 2004 through 2006 and the lead author of the 2005 energy bill, has received more than $1 million from the oil and gas industry over the course of his career, including $22,000 in PAC contributions from ExxonMobil between 2000 and 2006. In addition to shepherding through the massive oil and gas subsidies in that bill, Representative Barton has played a key role in elevating misleading information and delaying congressional action on global warming. Before he became chair of the full committee in 2004, Barton chaired the Energy and Air Quality Subcommittee. In that capacity, he stated at a March 2001 hearing that as long as he was the subcommittee chair, regulation of global warming emissions would be "off the table indefinitely." As Barton put it: "I don't want there to be any uncertainty about that." In his capacity as chair of the full committee, Barton has held true to his word, holding only two climate-related hearings, both aimed at attacking reputable climate scientists.

Meanwhile, the most vocal opponent to climate action in the Senate is James Inhofe (R-OK), chair—until January 2007—of the Environment and Public Works Committee. He adamantly denies the reality of global warming and has prevented consideration of climate bills by his committee during his tenure as chair from 2003 to 2006. In September 2005, he went so far as to invite Michael Crichton, a science fiction writer, to testify at a hearing on climate science and policy. Despite Crichton's lack of expertise, he attempted to undermine peer-reviewed climate science in his testimony. Inhofe was also a coplaintiff in the first Competitive Enterprise Institute lawsuit, filed in 2000, which attempted to bar the distribution or use of the National Assessment. Senator Inhofe has received a total of $847,123 from ExxonMobil and others in the oil and gas industry over the course of his career.

Like Big Tobacco before it, ExxonMobil has been enormously successful at influencing the current administration and key members of Congress. From successfully recommending the appointment of key personnel in the Bush administration, to coordinating its disinformation tactics on global warming with high-ranking Bush administration personnel, to funding climate change contrarians in Congress, ExxonMobil and its proxies have exerted extraordinary influence over the policies of the U.S. government during the Bush administration. The cozy relationship ExxonMobil enjoys with government officials has enabled the corporation to work effectively behind the scenes to block federal policies and shape government communications on global warming.

Ivan Osorio, Iain Murray,
and Myron Ebell

 **NO**

Liberal "Scientists" Lead Jihad against Global-Warming Skeptics

On Oct. 30, 2006, Senators John D. Rockefeller (D.-W.Va.) and Olympia Snowe (R.-Maine) wrote an extraordinary open letter to ExxonMobil CEO Rex Tillerson urging him to end his company's support of "climate-change-denial front groups." The only organization mentioned by name is the one that the authors of this article work for—the Competitive Enterprise Institute (CEI). The senators' letter then goes on to announce: "A study to be released in November by an American scientific group will expose ExxonMobil as the primary funder of no fewer than 29 climate-change-denial front groups in 2004 alone."

The letter brought a strong reaction from the *Wall Street Journal* editorial page: "[I]f the senators are so afraid that a handful of policy wonks at a single small think tank are in danger of winning this debate, they must not have much confidence in their own case." Home state newspapers also chided the senators. In West Virginia, the *Charleston Daily Mail* called the letter "an intemperate attempt to squelch debate," while in Maine, an editor at the *Portland Press Herald* wrote that a spokeswoman for Snowe told him that "the senator is greatly worried that the average moke on the street can't figure all this out on his own. So she and her colleague were just trying to clarify the issue, that is, by telling someone they disagreed with to shut up." The accompanying news release from Snowe's office was headlined: "Rockefeller and Snowe demand that ExxonMobil end funding of campaign that denies global climate change."

Be a Concerned Scientist for $35

It's astonishing that elected officials would use their taxpayer-funded offices to bully a company's president into changing his corporation's philanthropic giving practices. And it's deplorable that in trying to discredit critics of catastrophic global warming, Rockefeller and Snowe would stoop to using smear rhetoric that alludes to "Holocaust denial"—and is inaccurate to boot. No one seriously denies that the Earth is warming. The debate is over the extent and consequences of such warming. Remarkable, too, is the senators' cryptic reference to a "scientific group" that would soon issue its "findings," words that

As seen in *Human Events,* May 8, 2007, by Ivan Osorio, Iain Murray, and Myron Ebell, original to *Organization Trends,* March 2007. Copyright © 2007 by Capital Research Center. Reprinted by permission. www.capitalresearch.org.

bestow authority on what, as we expected, would turn out to be no more than a political attack.

It didn't take us long to figure out what this "scientific group" might be and who is behind it: For almost four decades, the Union of Concerned Scientists (UCS) has manipulated the high reputation of "science" to serve the low ends of politics. It has done a good job of cherry-picking scientific facts to stir up public fears to advance its agenda. This time it is promoting alarmist claims about global warming by leveraging the prestige of the "concerned scientist."

The credulous media usually fall all over themselves to defer to UCS every time the group takes a political position. For instance, when it issued a report in 2004 criticizing President George W. Bush's handling of science policy, the union was described as "a scientific advocacy group" (*New York Times*), "a group of scientists" (Reuters), "an independent Cambridge-based organization" (*Boston Globe*), and a "nonprofit . . . advocacy group in Cambridge, Mass." (*Newsday*). After all, who but concerned scientists would pass judgment on President Bush and conclude that he was a scientific ignoramus manipulating science in order to advance a partisan agenda?

That UCS is a highly partisan operation—well funded by left-leaning foundations and Hollywood celebrities and happy to ignore established scientific methodologies for its own purposes—is apparently not newsworthy. The group has a long history of being just plain wrong on many scientific issues, and its current agenda conforms to the extremes of environmentalist ideology. Moreover, UCS is neither representative of the scientific community at large nor is it a gathering of top scientists. Instead, a cadre of senior staff whose credentials are steeped more in Washington policy-making than in scientific research rides herd over a grassroots membership from all walks of life. You too can be a Concerned Scientist for a new member fee of $35!

In 2006, UCS decided to attack ExxonMobil, the world's largest private energy company, over the issue of global warming. It also decided on its tactics: It would demonize the oil company by comparing it to cigarette companies. ExxonMobil, said UCS, was "adopt[ing] the tobacco industry's disinform-ation tactics . . . to cloud the scientific understanding of climate change and delay action on the issue."

In a paper issued Jan. 3, 2007, UCS accuses ExxonMobil of funding "front groups" opposed to the climate-alarmist agenda of groups such as UCS and of former Vice President Al Gore. The company, said the UCS report, had distributed $16 million to 43 advocacy groups from 1998 to 2005 "to confuse the public on global-warming science."

Let's leave aside the fact that $16 million over eight years can't match the $2 billion that the federally funded Climate Change Science Program spends each year on global warming, or even the $4 million annual budget of just one of the many well-funded global-warming advocacy groups, Strategies for the Global Environment (the umbrella organization for the Pew Center on Global Climate Change). Moreover, the UCS document is hardly an investigative breakthrough. ExxonMobil itself publishes its philanthropic contributions to nonprofit organizations online.

Most interesting, however, about the UCS report are its distortions of fact and what they reveal about UCS political tactics. These should have undermined the group's credibility long ago were it not for that high-minded name: Union of Concerned Scientists.

Conspiracy? No.

UCS plays the game of Washington politics using hardball tactics, including innuendo, and its report on ExxonMobil epitomizes this approach. The UCS document cites what it terms a "conspiratorial communication" between one of the authors of this article, Myron Ebell of the Competitive Enterprise Institute, and Phil Cooney, then-chief of staff to the chairman of the White House Council on Environmental Quality (CEQ). In 2002, Ebell sent Cooney an e-mail expressing his distress over the Bush Administration's handling of the global-warming issue.

A front-page *New York Times* story had reported that the Bush Administration was conceding that global warming was a big problem. According to *Times* reporter Andrew Revkin, the administration had quietly sent a document titled "Climate Action Report 2002" (CAR) to the United Nations Framework Convention on Climate Change. CAR contained extremely inaccurate materials from the "National Assessment," an overview of the climate-change issue produced by the Clinton Administration. The Bush Administration had disavowed the "National Assessment" as a result of a lawsuit filed by CEI. But Revkin reported that CAR used a big chunk of the "National Assessment." As soon as this story broke, CEI sent out a press release sharply criticizing the Environmental Protection Agency (EPA) and then-Administrator Christine Todd Whitman for sending the report to the UN.

At that point, Phil Cooney left a message on Ebell's answering machine asking Ebell to call him, saying that he needed his help. Soon after, Ebell left a message on Cooney's answering machine. After not hearing back from him for several hours, Ebell sent him an e-mail saying that CEI would be glad to help. This e-mail is described in the UCS report as a "conspiratorial communication." If anyone from the Union of Concerned Scientists had bothered to ask Ebell, they could have gotten the facts straight.

When Phil Cooney called Ebell back, he explained that he wanted CEI to stop attacking EPA and refrain from calling on President Bush to fire Whitman, because she had nothing to do with the report. Cooney said EPA was not ultimately responsible for what was an interagency document on an environmental issue. Cooney further told Ebell that CEQ was in charge of conducting the interagency review and producing the final version of the report. As CEQ chief of staff, Cooney had directed the review and made the final edits. Cooney said that if Ebell wanted anyone fired, it should be him. Ebell replied that CEI would stop attacking Whitman but would not attack Cooney because he was not an appointee nominated by the President and confirmed by the Senate. So much for this "conspiracy."

The story doesn't end there. CEI unsuccessfully petitioned President Bush to rescind his submission of the flawed CAR to the UN and subsequently filed a second lawsuit in federal court against the "National Assessment" on the grounds that it did not meet the minimal requirements of the Federal Data Quality Act. CEI dropped the suit after the White House Office of Science and Technology Policy agreed to put a disclaimer on the "National Assessment" website that states that the document had not been subjected to the Federal Data Quality guidelines. That caused some of the report's authors to claim that the Bush Administration was suppressing scientific research. But if administration officials are burying research, they're not doing a very good job. The "National Assessment" is still available on more than one federal website.

A subsequent *New York Times* front-page story further fueled the controversy. While one *Times* story claimed that "Climate Action Report 2002" constituted an admission by the Bush Administration that global warming is real and serious, another story claimed that, in producing CAR, the White House had doctored the science because Phil Cooney had edited the text. Yet, since CAR is not a scientific report but an official U.S. government policy document, editing the text to reflect accurately the administration's official policies should be obligatory. Rather than doing anything wrong, Cooney was doing his job.

And rather than suppress science, Cooney was trying to get the science right in the document he was editing. What Cooney was trying to do was correct the "National Assessment" text by replacing the most obvious junk science claims with information and conclusions taken from the UN Intergovernmental Panel on Climate Change Third Assessment Report.

UCS doesn't focus its attacks on the actual work produced by the organizations it targets. Instead, it tries to discredit its opponents by using *ad hominem* innuendo. And that's what gets the attention of the media. For instance, when astrophysicist Sallie Baliunas determined that the Earth's temperature had actually been warmer at earlier times in history—a premise endorsed by a National Academy of Sciences (NAS) panel—UCS ignored the research but attacked the researcher personally, noting that Baliunas was affiliated with the George C. Marshall Institute, which UCS said had received $630,000 in ExxonMobil grants for its climate-science program.

Politics before Science

The UCS track record is bursting with examples of how it puts politics ahead of science. Founded in 1969 by a group of Massachusetts Institute of Technology (MIT) scientists concerned about the threat of nuclear war, the group currently claims a membership base of more than 100,000 "citizens and scientists" and an annual budget of more than $10 million.

True to its peacenik roots, UCS organized opposition to President Ronald Reagan's Strategic Defense Initiative (SDI) in the 1980s, fearing that it would push the world to war. But history showed otherwise. British Prime Minister Margaret Thatcher would call Reagan's decision to go ahead with SDI the "one vital factor in the ending of the Cold War." UCS continues its anti-nuclear activism today.

The 1980s were not a good decade for the union's predictive powers in other ways. In 1980, UCS claimed that "it is now abundantly clear that the world has entered a period of chronic energy shortages." As is now abundantly clear, known energy reserves are higher than ever. Middle East oil reserves alone are estimated to have increased from 431 billion barrels in 1985 to 742 billion in 2005. Of course, if UCS achieved its stated aim of capping energy production from fossil fuel sources and closing down nuclear plants, then the world most certainly would face a major energy shortage today.

More recently, UCS has been consistently wrong in its stated concerns about genetically modified crops. In 1999, it publicized reports that corn modified with the natural pesticide Bacillus thuringiensis (Bt) is harmful to the monarch butterfly, findings that were subsequently rejected by NAS.

Another NAS report found that increasing CAFE (corporate average fuel economy) standards contributed to between 1,300 and 2,600 additional traffic deaths per year because manufacturers downsize cars to increase their fuel economy and comply with the regulation. Yet the UCS website still says, "To reduce fuel consumption and address global warming, CAFE standards must increase."

UCS helped coordinate a campaign to attempt to discredit Danish statistician Bjørn Lomborg, whose 2001 best-selling book, *The Skeptical Environmentalist,* presents compelling statistical evidence refuting many of the modern environmental movement's alarmist claims.

In other matters—abortion, suburban "sprawl" and the war in Iraq—UCS stakes out policy positions that are predictably those of a far-left pressure group.

Funding Sources

The database of campaign contributions assembled by the Center for Responsive Politics contains abundant evidence of the partisan political leanings of UCS officials. For instance, UCS Chairman Cornell physicist Kurt Gottfried has donated more than $10,000 to Democratic Party organizations since 1990, mostly to the Democratic National Committee.

Signatories to a 2004 statement attacking President Bush over alleged manipulation of science donated more than $300,000 to Democratic candidates and liberal organizations since 1990—long before the supposed Bush "assault on science." In contrast, they donated only $5,050 to Republicans—the majority of that to liberal Sen. Arlen Specter (Pa.). The signatories donated $28,000 to the presidential campaign funds of Democratic Senators John Kerry (Mass.) and John Edwards (N.C.). Nobel laureates affiliated with UCS have contributed $97,000 to Democrats.

All of this undermines the credibility of UCS President Kevin Knobloch, who claimed in the 2003 UCS annual report: "Several key principles and beliefs will guide my leadership. Nonpartisanship is one." Knobloch, an environmental activist, spent six years on Capitol Hill, where he worked for Sen. Tim Wirth (D.-Colo.) and Rep. Ted Weiss (D.-N.Y.).

UCS Director of Strategy and Policy Alden Meyer is also a longtime environmental activist. Prior to joining the UCS staff in 1989, Meyer worked as executive director at a series of green groups: League of Conservation Voters, Americans for the Environment, and Environmental Action. Meyer's academic background isn't in the natural sciences. His biography on the UCS website notes that he received an undergraduate degree from Yale in 1975 "concentrating in political science and economics" and that "he received a Master of Science degree in human resource and organization development from American University in 1990."

UCS likes to attack free-market groups for accepting corporate donations, but much of its own funding comes from foundations established by conservative businessmen but subsequently hijacked by left-wing partisans. Unlike the leftists on many foundation boards, companies like ExxonMobil make grants from money that they actually earned.

The John D. and Catherine T. MacArthur Foundation has given the Union of Concerned Scientists $3.09 million since 2000. Long a major funder of leftist peace and environmental causes, the foundation owes its independence and generosity to its careless founder, John D. MacArthur, who neither formulated a mission for his foundation nor provided clear instructions as to how its money should be spent. In 1987, the foundation's then-president admitted to *USA Today* that if MacArthur were alive to see how his money was spent, "I think a lot of it would just make him furious." And how! In a 1974 interview, MacArthur, an insurance entrepreneur, denounced environmentalists as "bearded jerks and little old ladies" who "are obstructionists and just throw rocks in your path."

Other prominent businessmen whose name-bearing foundations fund UCS include Henry Ford ($950,000 from the Ford Foundation since 2000), *Time* magazine founder Henry Luce ($400,000 from the Henry Luce Foundation during 2001-2002), and J. Howard Pew ($1 million from the Pew Memorial Trust during 2002-2003). In a 1957 deed establishing the J. Howard Pew Freedom Trust, he wrote that the trust's mission was "to acquaint the American public" with "the evils of bureaucracy" and "the values of a free market."

UCS also receives funding from Hollywood celebrities and explicitly activist leftist foundations such as the Barbra Streisand Foundation ($10,000 for "general use" in 2004), the [Ted] Turner Foundation (nearly $500,000 since 2000), and the Energy Foundation, which states on its website that its "mission is to advance energy efficiency and renewable energy" ($5.08 million since 1999).

Indeed, UCS took in more money from 1998 to 2005 than ExxonMobil contributed to global-warming skeptics during the same period. In that seven-year span, ExxonMobil contributed $16 million in grants to groups that combat climate-change alarmism, while UCS alone received nearly $24 million in foundation grants.

UCS Political Circus

The Union of Concerned Scientists has powerful allies in Congress. Recently, UCS took advantage of a congressional hearing to publicize yet another report smearing its political opponents. On Jan. 30, 2007, House Oversight and Government Reform Chairman Henry Waxman (D.-Calif.) held a hearing on "Political Interference on Government Climate Change-Scientists." Waxman, a member of the radical Congressional Progressive Caucus [See "Fringe-Left Democrats Wield New Influence," by Cheryl Chumley, *Human Events,* February 26.] and other Democrats took the opportunity to once again pillory former CEQ chief of staff Phil Cooney over his editing of "Climate Action Report 2002" and his allegedly conspiratorial e-mail exchange with CEI's Myron Ebell. Rep. Darrell Issa (R.-Calif.) commented that the committee has "been trashing a lawyer we've never met."

One of the witnesses, UCS spokeswoman Francesca Grifo, announced a new UCS survey that allegedly showed political interference by the Bush Administration into climate science. But, as Issa noted, this survey was hardly representative, since only 19% of the 1,600 scientists polled responded—a response rate so low that it suggests bias in favor of a self-selected minority with a political axe to grind. Grifo had no response to this criticism.

Media Allies

UCS also has powerful allies in the media. On Jan. 31, 2007, CNN's "Larry King Live" hosted a debate on global warming featuring Bill Nye, best known for his television appearances as "The Science Guy." On the show, Nye boasted about being "a member of the advisory board of the Union of Concerned Scientists." He also warned that fresh water from melting ice caps flowing into the sea would upset "the salt-heat driven ocean currents," which are "what makes the Gulf Stream go . . . and if the Gulf Stream stops. . . ."

MIT professor of atmospheric science Richard Lindzen, a highly respected scientist, responded on air that there is no danger of the Gulf Stream's stopping, since it would require one of two physical impossibilities. "The Gulf Stream is driven by wind," he said. "To shut it down, you'd have to stop the rotation of the Earth or shut off the wind." After further debate, Lindzen noted, "I was saying textbook material. And if the textbooks are out-voiced by environmental advocacy groups like the Union of Concerned Scientists by 100,000 to one, that would be bizarre. We should close down our schools."

The Union of Concerned Scientists is not about to relent in its green climate crusade. Yet UCS does not speak for the scientific community. Instead, it is a well-funded, left-wing pressure group, which politicizes science while claiming to be its true guardian. A partisan is no less a partisan because he has won the Nobel Prize, but a scientist is less of a scientist if he allows ideology to color his research.

POSTSCRIPT

Is Global Warming Skepticism Just Smoke and Mirrors?

The United Nations Conference on Environment and Development in Rio de Janeiro, Brazil, took place in 1992. High on the agenda was the problem of global warming, but despite widespread concern and calls for reductions in carbon dioxide releases, the United States refused to consider rigid deadlines or set quotas. The uncertainties seemed too great, and some thought the economic costs of cutting back on carbon dioxide might be greater than the costs of letting the climate warm.

The nations that signed the UN Framework Convention on Climate Change in Rio de Janeiro in 1992 met again in Kyoto, Japan, in December 1997 to set carbon emissions limits for the industrial nations. The United States agreed to reduce its annual greenhouse gas emissions 7 percent below the 1990 level between 2008 and 2012 "but still has not ratified the Kyoto Treaty. In November 2005, they met in Montreal, Canada, and decided to begin formal talks on mandatory post-2012 reductions in greenhouse gases. The U.S. agreed to talk but ruled out any future commitments." Ross Gelbspan, in "Rx for a Planetary Fever," *American Prospect* (May 8, 2000), blames much of that opposition on "big oil and big coal [which] have relentlessly obstructed the best-faith efforts of government negotiators." See Fred Pearce, "State of Denial," *New Scientist* (November 4, 2006). At the most recent meeting of the nations that signed the Kyoto Protocol, held in Nairobi, Kenya, in November 2006, Kofi Annan, Secretary-General of the UN, said that the facts are so clear that skeptics about global warming are "out of step, out of arguments, and out of time." James Lovelock, creator of the Gaia metaphor for the living Earth beloved by many environmental activists, warns in *The Revenge of Gaia: Earth's Climate in Crisis and the Fate of Humanity* (Basic Books, 2006) that global warming may prove catastrophic. Tim Appenzeller, "The Big Thaw," *National Geographic* (June 2007), discusses the rapid loss of ice from ice caps and glaciers, with serious impacts expected on economies dependent on winter sports and—much more ominously—on the summertime flow of water from melting snow and ice in nearby mountains. According to Richard A. Kerr, "Pushing the Scary Side of Global Warming," *Science* (June 8, 2007), some climate researchers are concerned that we are seriously underestimating how disastrous global warming will be.

In June 2006, the National Academy of Sciences reported that the Earth is now warmer than it has been in the last 400 years, and perhaps in the last 1,000 (*Surface Temperature Reconstructions for the Last 2,000 Years*, National Academies Press, 2006). Concerns have been raised about the risks to coastal populations from rising seas and changes in storm patterns; see John Young, "Black Water Rising," *World Watch* (September/October 2006). States and environmental groups have brought before the U.S. Supreme Court a case demanding that the U.S. Environmental Protection Agency regulate carbon dioxide as a threat to public health. In April 2007 the Court ruled against the EPA, saying that carbon dioxide is an air pollutant that the EPA can regulate under the Clean Air Act. No one yet knows what form regulations (if any) will take, but requirements for better automobile fuel efficiency seem likely. See Charles Q. Choi, "Warming to Law: After the U.S. Supreme Court Ruling, How Stiff Will Greenhouse Gas Regulations Be?" *Scientific American* (July 2007).

Potential solutions include capturing carbon dioxide and storing it deep underground; see Robert H. Socolow, "Can We Bury Global Warming?" *Scientific American* (July 2005), and Eli Kintisch, "Making Dirty Coal Plants Cleaner," *Science* (July 13, 2007). "It's Not Too Late," a special report published by *Technology Review* (July/August 2006), discusses energy technologies that could be put into use today to forestall the worst; fossil fuels will continue to be used, but alternatives include nuclear power, ethanol, and solar energy. On a more grandiose scale, Paul Crutzen (who won the Nobel Prize for his work on ozone depletion), wrote in "Albedo Enhancement by Stratospheric Sulfur Injections: A Contribution to Resolve a Policy Dilemma?" *Climatic Change* (August 2006), that it may prove possible to reflect solar heat with clouds of what is in essence industrial pollution. Jerome Pearson, et al., suggest putting "An artificial planetary ring about the Earth, composed of passive particles or controlled spacecraft with parasols," at a cost of $125–500 billion, for a similar purpose in "Earth Rings for Planetary Environmental Control," *Acta Astronautica* (January 2006).

In January 2006, the U.K.'s Department for Environment, Food and Rural Affairs (DEFRA) published "Avoiding Dangerous Climate Change" (http://www.defra.gov.uk/environment/climatechange/research/dangerous-cc/index.htm), warning that immediate action is essential if catastrophe is to be avoided. The price tag for immediate action to prevent or—more likely—reduce the negative impacts of global warming will not be small. When the National Commission on Energy Policy published its report, *Ending the Energy Stalemate: A Bipartisan Strategy to Meet America's Energy Challenges* in December 2004 (http:/ /bcsia.ksg.harvard.edu/BCSIA_content/documents/NCEP_KSG_4-14-05.pdf), the Energy Information Administration promptly analyzed its economic impact in *Impacts of Modeled Recommendations of the National Commission on Energy Policy* (April 2005) (http://www.eia.doe.gov/oiaf/servicerpt/bingaman/index.html) and concluded that increasing automobile fuel efficiency, encouraging alternate energy sources, and increasing oil production and clean coal technology would cost the "U.S. economy . . . no more than 0.15% of GDP or about $78 per household per year, while overall GDP is projected to grow by 87%." Sir Nicholas Stern, Head of the British Government Economics Service, sees a larger bill in "Stern Review: The Economics of Climate Change" (October 30, 2006) (http:// www.hm-treasury.gov.uk/independent_reviews/stern_review_economics_climate_change/ sternreview_index.cfm). He says that "if we don't act, the overall costs and risks of climate change will be equivalent to losing at least 5% of global GDP each year, now and forever. If a wider range of risks and impacts is taken into account, the estimates of damage could rise to 20% of GDP or more. In contrast, the costs of action—reducing greenhouse gas emissions to avoid the worst impacts of climate change—can be limited to around 1% of global GDP each year . . . the benefits of strong, early action considerably outweigh the costs. . . . Ignoring climate change will eventually damage economic growth. . . . Tackling climate change is the pro-growth strategy for the longer term, and it can be done in a way that does not cap the aspirations for growth of rich or poor countries. The earlier effective action is taken, the less costly it will be." Gregg Easterbrook, "Global Warming: Who Loses—and Who Wins?" *The Atlantic* (April 2007), concludes that "Keeping the world economic system and the global balance of power the way they are seems very strongly in the U.S. national interest—and keeping things the way they are requires prevention of significant climate change. That, in the end, is what's in it for us."

ISSUE 9

Is Wind Power Green?

YES: **Charles Komanoff**, from "Whither Wind?" *Orion* (September/October 2006)

NO: **Jon Boone**, from "The Wayward Wind," speech given in the Township of Perry, near Silver Lake, Wyoming County, New York (June 19, 2006)

ISSUE SUMMARY

YES: Charles Komanoff argues that the energy needs of civilization can be met without adding to global warming if we both conserve energy and deploy large numbers of wind turbines.

NO: Jon Boone argues that wind power is better for corporate tax avoidance than for providing environmentally friendly energy. It is at best a placebo for our energy dilemma.

For centuries, windmills have exploited the pressure exerted by blowing wind to grind grain and pump water. On U.S. farms, before rural electrification programs brought utility power, windmills provided small amounts of electricity. These windmills fell out of favor when the powerlines arrived, bringing larger amounts of electricity more reliably (even when the wind was not blowing). The source of electricity became hydroelectric dams, coal and oil-fired power plants, and even nuclear power plants.

The oil crisis of the 1970s brought new attention to wind and other renewable energy sources. The federal government poured money into research and development, but when the price of oil dropped, most of the funding went away. Now we are concerned about global warming caused by the carbon dioxide emitted by the burning of fossil fuels. Researchers are looking for ways to reduce carbon emissions (see Jay Apt, David W. Keith, and M. Granger Morgan, "Promoting Low-Carbon Electricity Production," *Issues in Science and Technology*, Spring 2007). Low-carbon techniques include capturing carbon dioxide instead of releasing it, biofuels, nuclear power, hydroelectric power, wind power, and geothermal power, all of which play minor roles at present. At present, "the electric power industry is the single largest emitter of carbon dioxide in the United States, accounting for 40% of emissions of CO_2 in 2006, up from 36% in 1990 and 25% in 1970."

Wind power is growing rapidly. According to Ryan H. Wiser of Lawrence Berkeley National Laboratory, testifying before the Senate Finance Committee on March 29, 2007, "recent growth in the U.S. renewable electricity sector has been led by wind power. In fact, the year 2006 was the largest on record in the U.S. for wind power capacity additions, with over 2,400 MW of capacity added to the U.S. grid. And, for the second consecutive year, this made wind power the second largest new resource added to the U.S. electrical grid in capacity terms, well behind new natural gas plants, but ahead of coal. New wind plants contributed roughly 19% of the new capacity added to the U.S. grid in 2006, compared to 12% in 2005. On a worldwide basis, 15,200 MW of wind capacity was added in 2006, up from 11,500 MW in 2005, for a cumulative total of 74,200 MW." Yet these advances are not without resistance. Wind turbines take up space and affect views. They provide fatal obstacles to birds and bats. They make noise, and some people find that noise objectionable (see Johann Tasker, "Wind Farm Noise Is Driving Us Out of Our House," *Farmers Weekly*, January 12, 2007). Yet they do generate electricity without burning fossil fuels and emitting CO_2 (or other pollutants). The question is whether they offer enough advantages to be worth their price.

In the following selections, Charles Komanoff argues that the energy needs of civilization can be met without adding to global warming if we both conserve energy and deploy large numbers of wind turbines. Acceptance of wind farms, he says, could be "our generation's way of avowing our love for the next." Jon Boone, speaking in Wyoming County, New York, about a local windfarm proposal, argues that wind power is better for corporate tax avoidance than for providing environmentally friendly energy. It is at best a placebo for our energy dilemma. "The only environmentally responsible short-range solution to the problem of our dependence upon fossil fuels must combine effective conservation with much higher efficiency standards." For a longer version of his argument, see "Less for More, the Rube Goldberg Nature of Industrial Wind Development" (http://www.stopillwind.org/downloads/LessForMore.pdf).

YES

Charles Komanoff

Whither Wind?

It was a place I had often visited in memory but feared might no longer exist. Orange slabs of calcified sandstone teetered overhead, while before me, purple buttes and burnt mesas stretched over the desert floor. In the distance I could make out southeast Utah's three snowcapped ranges—the Henrys, the Abajos, and, eighty miles to the east, the La Sals, shimmering into the blue horizon.

No cars, no roads, no buildings. Two crows floating on the late-winter thermals. Otherwise, stillness . . . Now, perched on a precipice above Goblin Valley, stoked on endorphins and elated by the beauty before me, I had what might seem a strange, irrelevant thought: I didn't want windmills here.

Not that any windmills are planned for this Connecticut-sized expanse— the winds are too fickle. But wind energy is never far from my mind these days. As Earth's climate begins to warp under the accumulating effluent from fossil fuels, the increasing viability of commercial-scale wind power is one of the few encouraging developments.

Encouraging to me, at least. As it turns out, there is much disagreement over where big windmills belong, and whether they belong at all.

Fighting fossil fuels, and machines powered by them, has been my life's work. In 1971, shortly after getting my first taste of canyon country, I took a job crunching numbers for what was then a landmark exposé of U.S. power plant pollution, *The Price of Power*. The subject matter was drier than dust— emissions data, reams of it, printed out on endless strips of paper by a main-frame computer. Dull stuff, but nightmarish visions of coal-fired smokestacks smudging the crystal skies of the Four Corners kept me working 'round the clock, month after month.

A decade later, as a New York City bicycle commuter fed up with the oil-fueled mayhem on the streets, I began working with the local bicycle advocacy group, Transportation Alternatives, and we soon made our city a hotbed of urban American anti-car activism. The '90s and now the '00s have brought other battles—"greening" Manhattan tenement buildings through energy efficiency and documenting the infernal "noise costs" of Jet Skis, to name two— but I'm still fighting the same fight.

Why? Partly it's knowing the damage caused by the mining and burning of fossil fuels. And there's also the sheer awfulness of machines gone wild, their groaning, stinking combustion engines invading every corner of life.

But now the stakes are immeasurably higher. As an energy analyst, I can tell you that the science on global warming is terrifyingly clear: to have even a shot at fending off climate catastrophe, the world must reduce carbon dioxide emissions from fuel burning by at least 50 percent within the next few decades. If poor countries are to have any room to develop, the United States, the biggest emitter by far, needs to cut back by 75 percent.

Although automobiles, with their appetite for petroleum, may seem like the main culprit, the number one climate change agent in the U.S. is actually electricity. The most recent inventory of U.S. greenhouse gases found that power generation was responsible for a whopping 38 percent of carbon dioxide emissions. Yet the electricity sector may also be the least complicated to make carbon free. Approximately three-fourths of U.S. electricity is generated by burning coal, oil, or natural gas. Accordingly, switching that same portion of U.S. electricity generation to nonpolluting sources such as wind turbines, while simultaneously ensuring that our ever-expanding arrays of lights, computers, and appliances are increasingly energy efficient, would eliminate 38 percent of the country's CO_2 emissions and bring us halfway to the goal of cutting emissions by 75 percent.

To achieve that power switch entirely through wind power, I calculate, would require 400,000 windmills rated at 2.5 megawatts each. To be sure, this is a hypothetical figure, since it ignores such real-world issues as limits on power transmission and the intermittency of wind, but it's a useful benchmark just the same.

What Would That Entail?

To begin, I want to be clear that the turbines I'm talking about are huge, with blades up to 165 feet long mounted on towers rising several hundred feet. Household wind machines like the 100-foot-high Bergey 10-kilowatt BWC Excel with 11-foot blades, the mainstay of the residential and small business wind turbine market, may embody democratic self-reliance and other "small is beautiful" virtues, but we can't look to them to make a real dent in the big energy picture. What dictates the supersizing of windmills are two basic laws of wind physics: a wind turbine's energy potential is proportional to the square of the length of the blades, and to the cube of the speed at which the blades spin. I'll spare you the math, but the difference in blade lengths, the greater wind speeds higher off the ground, and the sophisticated controls available on industrial-scale turbines all add up to a market-clinching five-hundred-fold advantage in electricity output for a giant General Electric or Vestas wind machine.

How much land do the industrial turbines require? The answer turns on what "require" means. An industry rule of thumb is that to maintain adequate exposure to the wind, each big turbine needs space around it of about 60 acres. Since 640 acres make a square mile, those 400,000 turbines would need 37,500 square miles, or roughly all the land in Indiana or Maine.

On the other hand, the land actually occupied by the turbines—their "footprint"—would be far, far smaller. For example, each 3.6-megawatt Cape Wind turbine proposed for Nantucket Sound will rest on a platform roughly

22 feet in diameter, implying a surface area of 380 square feet—the size of a typical one-bedroom apartment in New York City. Scaling that up by 400,000 suggests that just six square miles of land—less than the area of a single big Wyoming strip mine—could house the bases for all of the windmills needed to banish coal, oil, and gas from the U.S. electricity sector.

Of course, erecting and maintaining wind turbines can also necessitate clearing land: ridgeline installations often require a fair amount of deforestation, and then there's the associated clearing for access roads, maintenance facilities, and the like. But there are also now a great many turbines situated on farmland, where the fields around their bases are still actively farmed.

Depending, then, on both the particular terrain and how the question is understood, the land area said to be needed for wind power can vary across almost four orders of magnitude. Similar divergences of opinion are heard about every other aspect of wind power, too. Big wind farms kill thousands of birds and bats . . . or hardly any, in comparison to avian mortality from other tall structures such as skyscrapers. Industrial wind machines are soft as a whisper from a thousand feet away, and even up close their sound level would rate as "quiet" on standard noise charts . . . or they can sound like "a grinding noise" or "the shrieking sound of a wild animal," according to one unhappy neighbor of an upstate New York wind farm. Wind power developers are skimming millions via subsidies, state-mandated quotas, and "green power" scams . . . or are boldly risking their own capital to strike a blow for clean energy against the fossil fuel Goliath.

Some of the bad press is warranted. The first giant wind farm, comprising six thousand small, fast-spinning turbines placed directly in northern California's principal raptor flyway, Altamont Pass, in the early 1980s rightly inspired the epithet "Cuisinarts for birds." The longer blades on newer turbines rotate more slowly and thus kill far fewer birds, but bat kills are being reported at wind farms in the Appalachian Mountains; as many as two thousand bats were hacked to death at one forty-four-turbine installation in West Virginia. And as with any machine, some of the nearly ten thousand industrial-grade windmills now operating in the U.S. may groan or shriek when something goes wrong. Moreover, wind power does benefit from a handsome federal subsidy; indeed, uncertainty over renewal of the "production tax credit" worth 1.9 cents per kilowatt-hour nearly brought wind power development to a standstill a few years ago.

At the same time, however, there is an apocalyptic quality to much anti-wind advocacy that seems wildly disproportionate to the actual harm, particularly in the overall context of not just other sources of energy but modern industry in general. New York State opponents of wind farms call their website "Save Upstate New York," as if ecological or other damage from wind turbines might administer the coup de grace to the state's rural provinces that decades of industrialization and pollution, followed by outsourcing, have not. In neighboring Massachusetts, a group called Green Berkshires argues that wind turbines "are enormously destructive to the environment," but does not perform the obvious comparison to the destructiveness of fossil fuel-based power. Although the intensely controversial Cape Wind project "poses an imminent threat to navigation and raises many serious maritime safety issues," according to the anti-wind Alliance to Protect Nantucket Sound, the alliance was strangely silent when an oil barge bound for the region's electric power plant spilled ninety-eight thousand gallons of its deadly, gluey cargo into Buzzards Bay three years ago.

Of course rhetoric is standard fare in advocacy, particularly the environmental variety with its salvationist mentality—environmentalists always like to feel they are "saving" this valley or that species. It all comes down to a question of what we're saving, and for whom. You can spend hours sifting through the anti-wind websites and find no mention at all of the climate crisis, let alone wind power's potential to help avert it.

In fact, many wind power opponents deny that wind power displaces much, if any, fossil fuel burning. Green Berkshires insists, for example, that "global warming [and] dependence on fossil fuels . . . will not be ameliorated one whit by the construction of these turbines on our mountains."

This notion is mistaken. It is true that since wind is variable, individual wind turbines can't be counted on to produce on demand, so the power grid can't necessarily retire fossil fuel generators at the same rate as it takes on windmills. The coal- and oil-fired generators will still need to be there, waiting for a windless day. But when the wind blows, those generators can spin down. That's how the grid works: it allocates electrons. Supply more electrons from one source, and other sources can supply fewer. And since system operators program the grid to draw from the lowest-cost generators first, and wind power's "fuel," moving air, is free, wind-generated electrons are given priority. It follows that more electrons from wind power mean proportionately fewer from fossil fuel burning.

What about the need to keep a few power stations burning fuel so they can instantaneously ramp up and counterbalance fluctuations in wind energy output? The grid requires this ballast, known as spinning reserve, in any event both because demand is always changing and because power plants of any type are subject to unforeseen breakdowns. The additional variability due to wind generation is slight—wind speeds don't suddenly drop from strong to calm, at least not for every turbine in a wind farm and certainly not for every wind farm on the grid. The clear verdict of the engineers responsible for grid reliability—a most conservative lot—is that the current level of wind power development will not require additional spinning reserve, while even much larger supplies of wind-generated electricity could be accommodated through a combination of energy storage technologies and improved models for predicting wind speeds.

With very few exceptions, then, wind output can be counted on to displace fossil fuel burning one for one. No less than other nonpolluting technologies like bicycles or photovoltaic solar cells, wind power is truly an anti-fossil fuel.

I made my first wind farm visit in the fall of 2005. I had seen big windmills up close in Denmark, and I had driven through the big San Gorgonio wind farm that straddles Highway I-10 near Palm Springs, California. But this trip last November had a mission. After years of hearing industrial wind turbines in the northeastern United States characterized as either monstrosities or crowns of creation, I wanted to see for myself how they sat on the land. I also wanted to measure the noise from the turning blades, so I brought the professional noise meter I had used in my campaign against Jet Skis.

Madison County occupies the broad middle of New York State, with the Catskill Mountains to the south, Lake Ontario to the northwest, and the Adirondacks to the northeast. Its rolling farms sustain seventy thousand residents and, since 2001, two wind farms, the 20-windmill Fenner Windpower Project in the western part of the county and the 7-windmill Madison Windpower Project twenty miles east.

At the time of my visit Fenner was the state's largest wind farm, although that distinction has since passed to the 120-windmill Maple Ridge installation in the Tug Hill region farther north. It was windy that day, though not unusually so, according to the locals. All twenty-seven turbines were spinning, presumably at their full 1.5-megawatt ratings. For me the sight of the turning blades was deeply pleasing. The windmills, sleek, white structures more than three hundred feet tall sprinkled across farmland, struck me as graceful and marvelously useful. I thought of a story in the *New York Times* about a proposed wind farm near Cooperstown, New York, in which a retiree said that seeing giant windmills near your house "would be like driving through oil derricks to get to your front door." To my eye, the Fenner turbines were anti-derricks, oil rigs running in reverse.

For every hour it was in full use, each windmill was keeping a couple of barrels of oil, or an entire half-ton of coal, in the ground. Of course wind turbines don't generate full power all the time because the wind doesn't blow at a constant speed. The Madison County turbines have an average "capacity factor," or annual output rate, of 34 percent, meaning that over the course of a year they generate about a third of the electricity they would produce if they always ran at full capacity. But that still means an average three thousand hours a year of full output for each turbine. Multiply those hours by the twenty-seven turbines at Fenner and Madison, and a good 200,000 barrels of oil or 50,000 tons of coal were being kept underground by the two wind farms each year—enough to cover an entire sixty-acre farm with a six-inch-thick oil slick or pile of coal.

The windmills, spinning easily at fifteen revolutions per minute—that's one leisurely revolution every four seconds—were clean and elegant in a way that no oil derrick or coal dragline could ever be. The nonlinear arrangement of the Fenner turbines situated them comfortably among the traditional farmhouses, paths, and roads, while at Madison, a grassy hillside site, the windmills were more prominent but still unaggressive. Unlike a ski run, say, or a power line cutting through the countryside, the windmills didn't seem like a violation of the landscape. The turning vanes called to mind a natural force—the wind—in a way that a cell phone or microwave tower, for example, most certainly does not.

They were also relatively quiet. My sound readings, taken at distances ranging from one hundred to two thousand feet from the tower base, topped out at 64 decibels and went as low as 45—the approximate noise range given for a small-town residential cul-de-sac on standard noise charts. It's fair to say that the wind turbines in Madison County aren't terribly noisy even from up close and are barely audible from a thousand feet or more away. The predominant sound was a low, not unpleasant hum, or hvoohmm, like a distant seashore, but perhaps a bit thicker.

Thinking back on that November day, I've come to realize that a windmill, like any large structure, is a signifier. Cell-phone towers signify the intrusion of quotidian life—the reminder to stop off at the 7-Eleven, the unfinished business at the office. The windmills I saw in upstate New York signified, for me, not just displacement of destructive fossil fuels, but acceptance of the conditions of inhabiting the Earth. They signified, in the words of environmental lawyer and MIT research affiliate William Shutkin, "the capacity of environmentalists—of citizens—to match their public positions with the private choices necessary to move toward a more environmentally and economically sustainable way of life."

The notion of choices points to another criticism of wind turbines: the argument that the energy they might make could be saved instead through energy-efficiency measures. The Adirondack Council, for example, in a statement opposing the 10-windmill Barton Mines project on a former mountaintop mine site writes, "If the Barton project is approved, we will gain 27 to 30 megawatts of new, clean power generation. Ironically, we could save more than 30 megawatts of power in the Adirondack Park through simple, proven conservation methods in homes and businesses."

The council's statement is correct, of course. Kilowatts galore could be conserved in any American city or town by swapping out incandescent light bulbs in favor of compact fluorescents, replacing inefficient kitchen appliances, and extinguishing "vampire" loads by plugging watt-sucking electronic devices into on-off power strips. If this notion sounds familiar, it's because it has been raised in virtually every power plant dispute since the 1970s. But the ground has shifted, now that we have such overwhelming proof that we're standing on the threshold of catastrophic climate change.

Those power plant debates of yore weren't about fuels and certainly not about global warming, but about whether to top off the grid with new megawatts of supply or with "negawatts"—watts that could be saved through conservation. It took decades of struggle by legions of citizen advocates and hundreds of experts (I was one) to embed the negawatt paradigm in U.S. utility planning. But while we were accomplishing that, inexorably rising fossil fuel use here and around the world was overwhelming Earth's "carbon sinks," causing carbon dioxide to accumulate in the atmosphere at an accelerating rate, contributing to disasters such as Hurricane Katrina and Europe's 2003 heat wave, and promising biblical-scale horrors such as a waning Gulf Stream and disappearing polar icepacks.

The energy arena of old was local and incremental. The new one is global and all-out. With Earth's climate, and the world as we know and love it, now imperiled, topping off the regional grid pales in comparison to the task at hand. In the new, ineluctable struggle to rescue the climate from fossil fuels, efficiency and "renewables" (solar and biomass as well as wind) must all be pushed to the max. Those thirty negawatts that lie untapped in the kitchens and TV rooms of Adirondack houses are no longer an alternative to the Barton wind farm—they're another necessity.

In this new, desperate, last-chance world—and it is that, make no mistake—pleas like the Adirondack Council's, which once would have seemed reasonable, now sound a lot like fiddling while the Earth burns. The same goes for the urgings by opponents of Cape Wind and other pending wind farms to "find a more suitable site"; those other suitable wind farm sites (wherever they exist) need to be developed in addition to, not instead of, Nantucket Sound, or Barton Mines, or the Berkshires.

There was a time when the idea of placing immense turbines in any of these places would have filled me with horror. But now, what horrifies me more is the thought of keeping them windmill free. . . .

Intruding the unmistakable human hand on any landscape for wind power is, of course, a loss in local terms, and no small one, particularly if the site is a verdant ridgeline. Uplands are not just visible markers of place but fragile environments, and the inevitable access roads for erecting and serving the turbines can be damaging ecologically as well as symbolically. In contrast, few if any benefits of the wind farm will be felt by you in a tangible way. If the thousands of tons of coal a year that your wind farm will replace were being mined now, a mile from your house, it might be a little easier to take. Unfortunately, our society rarely works that way. The bread you cast upon the waters with your vote will not come back to you in any obvious way—it will be eaten in Wyoming, or Appalachia. . . .

But what if the big push for wind power simply "provides more energy for people to waste?" as Carl Safina, an oceanographer who objects to the Cape Wind project, asked me recently. Safina is unusual among Cape Wind opponents, not just because he is a MacArthur Fellow and prize-winning author (*Song for the Blue Ocean*, *Voyage of the Turtle*), but because he is completely honest about the fact that his objections are essentially aesthetic.

"I believe the aesthetics of having a national seashore with a natural view of the blue curve of the planet are very important," he wrote in an e-mail from coastal Long Island, where he lives. "I think turbines and other structures should be sited in places not famed for natural beauty"—a statement that echoed my feelings about Utah's Goblin Valley. . . .

Yet for all his fierce attachment to that view, Safina says he might give it up if doing so made a difference. "If there was a national energy strategy that would make the U.S. carbon neutral in fifty years," he wrote, "and if Cape Wind was integral and significant, that might be a worthwhile sacrifice." But the reality, as Safina described in words that could well have been mine, is that "Americans insist on wasting energy and needing more. We will affect the natural view of a famously beautiful piece of America's ocean and still not develop a plan to conserve energy."

Safina represents my position and, I imagine, that of others on both sides of the wind controversy when he pleads for federal action that could justify local sacrifice for the greater good. If Congress enacted an energy policy that harnessed the spectrum of cost-effective energy efficiency together with renewable energy, thereby ensuring that fossil fuel use shrank starting today, a windmill's contribution to climate protection might actually register, providing psychic reparation for an altered viewshed. And if carbon fuels were taxed for their damage to the climate, wind power's profit margins would widen, and surrounding communities could extract bigger tax revenues from wind farms. Then some of that bread upon the waters would indeed come back—in the form of a new high school, or land acquired for a nature preserve.

It's very human to ask, "Why me? Why my ridgeline, my seascape, my viewshed?" These questions have been difficult to answer; there has been no framework—local or national—to guide wind farm siting by ranking potential wind power locales for their ecological and community suitability. That's a gap that the Appalachian Mountain Club is trying to bridge, using its home state of Massachusetts as a model.

According to AMC research director Kenneth Kimball, who heads the project, Massachusetts has ninety-six linear miles of "Class 4" ridgelines, where wind speeds average fourteen miles per hour or more, the threshold for profitability with current technology. Assuming each mile can support seven to nine large turbines of roughly two megawatts each, the state's uplands could theoretically host 1,500 megawatts of wind power. (Coastal areas such as Nantucket Sound weren't included in the survey.)

Kimball's team sorted all ninety-six miles into four classes of governance— Appalachian Trail corridor or similar lands where development is prohibited; other federal or state conservation lands; Massachusetts open space lands; and private holdings. They then overlaid these with ratings denoting conflicts with recreational, scenic, and ecological values. The resulting matrix suggests the following rankings of wind power suitability:

1. *Unsuitable*—lands where development is prohibited (Appalachian Trail corridors, for example) or "high conflict" areas: 24 miles (25 percent).
2. *Less than ideal*—federal or state conservation lands rated "medium conflict": 21 miles (22 percent).
3. *Conditionally favorable*—Conservation or open space lands rated "low conflict," or open space or private lands rated "medium conflict": 27 miles (28 percent).
4. *Most favorable*—Unrestricted private land and "low conflict" areas: 24 miles (25 percent).

Category 4 lands are obvious places to look to for wind farm development. Category 3 lands could also be considered, says the AMC, if wind farms were found to improve regional air quality, were developed under a state plan rather than piecemeal, and were bonded to assure eventual decommissioning. If these conditions were met, then categories 3 and 4, comprising approximately fifty miles of Massachusetts ridgelines, could host four hundred wind turbines capable of supplying nearly 4 percent of the state's annual electricity— without grossly endangering wildlife or threatening scenic, recreational, or ecological values (e.g., critical habitat, roadless areas, rare species, old growth, steep slopes).

Whether that 4 percent is a little or a lot depends on where you stand and, equally, on where we stand as a society. You could call the four hundred turbines mere tokenism against our fuel-besotted way of life, and considering them in isolation, you'd be right. But you could also say this: Go ahead and halve the state's power usage, as could be done even with present-day technology, and "nearly 4 percent" doubles to 7-8 percent. Add the Cape Wind project and other offshore wind farms that might follow, and wind power's statewide share might reach 20 percent, the level in Denmark.

Moreover, the windier and emptier Great Plains states could reach 100 percent wind power or higher, even with a suitability framework like the AMC's, thereby becoming net exporters of clean energy. But even at 20 percent, Massachusetts would be doing its part to displace that 75 percent of U.S. electricity generated by fossil fuels. If you spread the turbines needed to achieve that goal across all fifty states, you'd be looking to produce roughly eight hundred megawatt-hours of wind output per square mile—just about what Massachusetts would be generating in the above scenario. And the rest of New England and New York could do the same, affording these "blue" states a voice in nudging the rest of the country greenward.

So goes my notion, anyway. You could call it wind farms as signifiers, with their value transcending energy-share percentages to reach the realm of symbols and images. That is where we who love nature and obsess about the environment have lost the high ground, and where *Homo americanus* has been acting out his (and her) disastrous desires—opting for the "manly" SUV over the prim Prius, the macho powerboat over the meandering canoe, the stylish halogen lamp over the dorky compact fluorescent.

Throughout his illustrious career, wilderness champion David Brower called upon Americans "to determine that an untrammeled wildness shall remain here to testify that this generation had love for the next." Now that all wild things and all places are threatened by global warming, that task is more complex.

Could a windmill's ability to "derive maximum benefit out of the site-specific gift nature is providing—wind and open space," in the words of aesthetician Yuriko Saito, help Americans bridge the divide between pristine landscapes and sustainable ones? Could windmills help Americans subscribe to the "higher order of beauty" that environmental educator David Orr defines as something that "causes no ugliness somewhere else or at some later time"? Could acceptance of wind farms be our generation's way of avowing our love for the next?

I believe so. Or want to.

Jon Boone  **NO**

The Wayward Wind

Introduction

The United States east of the Mississippi River has about 5% of the nation's onshore wind potential. If industrial wind developers achieve their goals, the region will be saturated with approximately 300,000 wind turbines spread over millions of acres. New York, with .3 of 1% of the nation's onshore wind, could be saturated with about 20,000 400 foot tall machines, covering more than a thousand miles of terrain. Wyoming County alone could absorb around 400, of which 105 have already been proposed. Although most of the country's wind-rich lands are in the upper Midwest, relatively inexpensive access to existing transmission lines makes eastern states such as New York attractive for wind development, spurred by the state's renewable portfolio law. Despite a desire by many to reduce our reliance on fossil fuels in order to shrink the rate of carbon dioxide emissions that are a by-product of burning fossil fuels, demand for electricity has increased 2% each year since 1975; at this rate, we will double current usage in about thirty years. Nationally, fossil fuels produce nearly 75% of our electricity. Because it is perceived as non-polluting and renewable, wind energy has become popular with the public and with politicians. If all these wind turbines are constructed, what will it mean to ordinary people, to energy policy and to an improved environment?

Nearly four years ago, I set out to investigate these questions. In a number of forums, I examined the claims of wind industry developers, their trade organization, the American Wind Energy Association, and the National Renewable Energy Lab, an agency of the US Department of Energy. I'll share with you how those claims withstand the evidence of real world experience. At first, I'd hoped to support wind energy because, as an environmentalist, I have long been concerned about our society's addiction to fossil fuels and such malignant coal mining practices as mountaintop removal. I'm alarmed at such statistics as the number of asthma cases in the nation doubling every five years. High levels of mercury contamination in our rivers and oceans are also by-products of fossil fuel consumption. However, I seek effective solutions for these and other environmental problems. Although I support efforts to reduce demand by living off the grid with small-scale wind and solar power, I'm mindful of the initial costs of doing so, making this kind of enterprise difficult to apply at industrial scale.

As a student of history and science, I wanted to understand the nature of "renewable" energy, and to provide some context for wind energy. The quest for renewable energy has a long history. A few hundred years ago, timber seemed inexhaustible, but our demand made short work of the supply. Coal, too, is renewable, but again, our demand will at some time overrun supply—and our meager lifespan won't extend the tens of millions of years necessary to replenish it. A few generations ago, hydroelectric dams were all the rage. Although these do produce a lot of electricity from a renewable source, they are so environmentally damaging that many are now being dismantled around the country, at taxpayer expense. Because time seems to be running out on fossil fuels and the lure of non-polluting wind energy is so seductive, some people are now promoting wind energy initiatives at any cost, without investigating potential negative consequences—and with no apparent knowledge of even recent environmental history. . . .

Wind developers and their supporters make a number of claims for wind facilities, stating they lessen dependence on foreign oil; improve air quality; reduce global warming by replacing fossil-fueled power plants; and improve public health; while providing electrical power for many thousands of homes and adding significant revenue and jobs to local economies. They also promise their technology will not pose much risk to wildlife, nor will it alter the landscape in perceptible ways, nor decrease the value for surrounding properties, nor introduce disturbances that might jeopardize the right of neighbors to quietly enjoy their property. Conversely, they never mention the extraordinary subsidies that taxpayers and ratepayers provide, *subsidies not indexed to reductions in CO_2 and various toxic emissions.*

Throughout my experience, I could not substantiate a single claim developers made for industrial wind energy, including the one justifying its existence: that massive wind installations would meaningfully reduce our reliance on fossil fuels. When you understand this, you realize the wind business is not really that complex. But there are a lot of complicated issues swirling around it that obscure and distract from this main point, issues such as global warming, property values, the nature of wind leases, local revenues and taxes, wildlife, natural views, and a host of others. So how does one know the truth of it all? How does one go about separating the reality from spin?

Perhaps you'll agree that if something seems too good to be true, it almost always is. You should ask good questions and demand solid proof, not relying upon unsecured promises—and realize that the responsibility for substantiation lies with those making the claim. I could address all the complications surrounding this issue, but that would take far too much time. Consequently, I'll touch on a few prominent ones, then focus on wind energy effectiveness and economics, showing what we might get, and what we would have to give, for industrial wind energy, examining the two windplant operations presently targeting Wyoming County.

Property Values

Do you believe industrial facilities stretching many miles across your landscape, with 105 spinning skyscraper sized structures creating a cascade of noise, are not going to negatively affect property values for those in the neighborhood, as the wind industry maintains a government study proves? One of the most validated real estate precepts is that prominent natural views and historic scenery have premium value, and intrusions restricting those views erode value.

Independent inquiry in Britain, Denmark, and New England suggests the likelihood of significant property devaluations. In his June 10, 2005 direct testimony before the Wisconsin Public Service Commission, Kevin Zarem, an appraiser, estimated that residential property near a proposed windplant "will likely be in the 17%–20% loss range." And this based solely upon visual impact. He did not assess potential loss due to wind turbine noise, motion, or shadows. Russell Bounds, one of western Maryland's leading realtors in large property transactions, has already lost sales in the area of proposed windplants. Mr. Bounds testified in a PSC hearing that, over the last several years, he has had at least 25 people who expressed interest in buying land in the area targeted by wind developers. However, when he advised them about the plans for wind facilities, not one of those people expressed further interest.

Wind Turbine Noise

Developers claim modern wind turbines make little noise beyond the sound of "wind rustling through the leaves," pointing to their own studies showing low decibel levels and the experience of observers who've been taken directly under a wind turbine. However, the reality for people living near a windplant does not substantiate this claim. Turbine noise is so irritating and disconcerting that it can cause people to seek medical attention, as Rodger Hutzell of Meyersdale, Pennsylvania, had to do. The problem is so acute in Europe that the world's first International Wind Turbine Noise conference was convened in Berlin last fall. A Malone, New York, physician, Nina Pierpont, who has studied the symptoms of several families, especially the d'Entremont family of Nova Scotia, has called for systematic medical study of what she has termed Wind Turbine Noise Syndrome.

An explanation for turbine noise was published earlier this year by a Dutch researcher, G.P. van den Berg of the University of Groningen in the Netherlands. Van den Berg demonstrated that loud aerodynamic sounds are generated when the moving propeller blade passes the turbine's tower mast, creating sound pressure fluctuations. Such fluctuations may not be great from an individual turbine, but when several turbines operate together, the pulses amplify each other, greatly magnifying the sound. Van den Berg also noted a "distinct audible difference between the night and daytime sound at some distance [more than one mile] from the turbines." . . .

There are windplant-generated nuisances that have been verified across three continents. The failure of many local governments to provide appropriate leadership on this issue is appalling. After-the-fact lawsuits brought because of predictable nuisances are difficult, expensive, and time consuming. These massive windplants precipitate incivility, pitting neighbor against neighbor. A major reason for government to exist is to anticipate and mitigate this incivility. Recent changes to the Perry Township Wind Ordinance strongly suggest they were prepared under the guidance, if not the dictation, of the wind industry, without consulting any comprehensive plan that honors the rural integrity of this region. This is a prescription for discord.

Safety for Birds, Bats and Other Wildlife

The wind industry asserts its technology is safe for migrating wildlife, using avian experts and industry-sponsored studies to bolster its claims. None of these have withstood the scrutiny of independent inquiry. Tall structures—buildings, cell and communication towers—are responsible for annually killing millions of migrating birds. Wind turbines, each more than 400 feet tall—with propeller blades moving at nearly 160 miles per hour at their tips and placed atop prominent ridges where large numbers of birds concentrate in migration—kill birds of prey, songbirds, and especially bats. Despite industry insistence this won't happen, it already has. When confronted with this reality, the industry argument morphs into a ten wrongs make a right scenario: "Cats and communication towers kill millions of birds annually, and we won't kill that many." When challenged about the appropriateness of this defense, the industry shifts gears once more: "The need for clean energy justifies the loss of wildlife," implying that wind energy will displace significant fossil fuel production. Some might recall this same ends justifies the means rationale promoted use of DDT.

Recent radar studies—one in Vermont, one in Virginia—documented significant potential problems for migratory wildlife. Last week, I also talked with graduate students from Frostburg University in Maryland who recently used radar to chart numbers of birds and bats flying overhead along the mountains of Western Maryland. The preliminary data shows that wind turbines may pose a high risk to bird and bat populations. On many nights during the season as many as 300,000 birds and bats flew low enough to collide with huge wind turbines. Last year, Ed Arnett, a biologist with Bat Conservation International, released his study of two Florida Power and Light windplants in Pennsylvania and West Virginia. His research reaffirmed earlier studies showing major bat mortality. Faced with the news that its wind turbines were killing thousands of bats, Florida Power and Light reacted quickly. It barred scientists from pursuing follow-up work, removed its $75,000 contribution from the research cooperative studying bat mortality and ended the doctoral work of a graduate student who had produced two years of data showing unusually high rates of bat death at the Pennsylvania and West Virginia sites. Although Florida Power and Light has pulled the plug on further research into avian and bat mortality on its properties, the company plans to construct hundreds more turbines in the mountainous areas of the region.

Braddock Bay near Rochester along Lake Ontario is a major destination point for thousands of birds of prey, many of which use the ridges along Silver Lake to help guide their spring migration. The July issue of *Wildbird* (2006) contains an excellent article about Braddock Bay and raptor migration. When I told Donald Heintzelman, one of this country's leading bird of prey specialists, about plans to construct several large windplants in this part of New York, he expressed concern. Avian migration is an extremely complex phenomenon, with many influencing factors, including the changing conditions of weather and climate over many years. Adequate preconstruction study for wind projects does not mean that, because such study is made, therefore windplants can be built. Rather, risk studies should be made to determine whether or not they should be built at all. Wind developers plan thousands of turbines along major avian migration flyways from Georgia through New Hampshire, creating a gauntlet of risk for birds and bats, some species of which have extremely vulnerable populations. We should take great care to avoid the unintended adverse consequence wrought by uninformed decisions of the well intentioned.

Not Here There

Just how clean, green, functional and effective is wind energy? Is it worth the high level of public subsidy sustaining it? On a per kilowatt-hour basis, wind is the most heavily subsidized source of industrial power in the nation, according to John Sherwell, spokesman for Maryland's Power Plant Research Program. Conventional power producers enjoy a high level of public subsidy, but they also provide significant, reliable electricity.

Presently, Horizon Wind seeks to place 65 turbines around the county. Orion Wind has stated it wants to put 40 turbines here—for a total of 105. The rated capacity of each Horizon turbine is 2MW; Orion's turbines may be rated at 1.5 MW. These 105 turbines would have a combined rated capacity of 190 MW. *Rated capacity* means the amount of electricity a wind turbine could put into the electricity grid over a year's time *if* it were working at full strength all the time.

Engineers use the term *capacity factor* to assess what percentage of its rated capacity a power plant will actually deliver over a specified time. Wind turbines don't begin generating electricity until wind speeds hit around 8 mph and they shut off at wind speeds exceeding 55 mph to avoid damage. They achieve rated capacity typically at wind speeds of 32–37 mph. Because of wind's unpredictable intermittency and wide fluctuations, along with the downtime for maintenance, no windplants located in the eastern United States have achieved a capacity factor of more than 30%. Consequently, Horizon/Orion's combined windplants might contribute on average 57 MW of sporadic energy annually to the electricity grid, which is 30% of their rated capacity. Although no conventional power plants work at peak capacity all the time, this is due to operator choice, not because of the limitations of their power source.

Conventional units must pass stringent tests for reliability and effectiveness. Generators that satisfy basic levels of demand, such as nuclear, large coal plants, and hydro, have capacity factors approaching 100%. Smaller, more flexible load balancing units, which may be used only a few hours a year, may have capacity factors of less than 5%. But the random, desultory nature of the wind, which rapidly changes the energy level contributed to the electricity grid at frequent intervals, limits what wind machinery can do, condemning wind turbines to *intrinsically* low capacity factors. Moreover, the wind typically blows hardest at night, at times of least demand, and much less during the afternoon, at times of peak demand. And in summer months, when demand for electricity is greatest, there may be no wind at all. The *capacity credit* for wind technology, a measure of the percentage of time wind technology can be relied upon to have energy available when needed to meet demand, is in the low single digits. Conversely, the capacity credit for conventional units approaches certainty.

An electricity *grid* is a complex network of regional power sources to supply demand for a variety of customers—residential, commercial, industrial, along with public and quasi-public operations like police, schools, hospitals, traffic control. The grid management creates and maintains a dynamic equilibrium between demand and supply, fine-tuning it on a second-by-second basis to within +/- one percent. The New York grid is known as the New York Independent System Operator, and it serves over 19 million people with a required annual installed capacity of 37, 534 MW. Coal generates around 15% of its power; nuclear—27%; hydro—19 %; natural gas—20%; and petroleum—15%. All of these conventional fuels provide steady, reliable power, and are carefully orchestrated to achieve effective responses at reasonable cost; their predictable performance provides *capacity*, a measure of firm generation and transmission capability.

No one knows what hours in a day the wind will blow sufficiently to produce electricity; or when the wind will blow at speeds providing only a fraction of a plant's rated capacity; or when wind will stop suddenly or change direction. One minute a wind facility may be producing 80% of its rated capacity and in the next produce nothing. A few minutes later, it may hit the grid with 10% of its rated capacity and, a bit later, jump up to 40%, and a few minutes after that, come back down to zero. Since electricity produced at industrial scales cannot be stored and must be used immediately, the grid, to assure reliability and stability, must enlist compensatory companion generation to follow and balance wind's skittering activity. This typically consists of smaller, very flexible conventional generators such as hydro or natural gas, which, unlike large coal and nuclear plants, can be throttled forward and back quickly, giving grid operators the kind of rapid response they must have to "smooth" the ongoing flux.

As wind developers are quick to say, grid operators already are engaged in the process of balancing the fluctuations of demand, which occurs continually as people turn on and off their appliances. The number and variety of power sources on the grid and the fluctuations of demand do indeed provide a real challenge for grid operators to integrate and balance all of this activity, even without the addition of highly fluctuating wind energy, which is many times more unpredictable and volatile than the time-tested patterns of demand flux. Paradoxically, fluctuating wind energy is treated as grid operators treat sudden shifts caused by people turning on and off their lights unexpectedly.

It is treated as if it were fluctuating demand, and not as a source of supply that can be relied upon to satisfy demand.

According to energy expert Tom Hewson, wind technology generates energy, not capacity—that is, reliable energy on demand. It is therefore inimical to the process by which power grids work to insure reliability and system security at reasonable cost. Moreover, since it has virtually no capacity credit, it cannot obviate the need for new reliable conventional generating plants, especially in light of increasing demand. Most of this new generation would likely be in the form of coal, since there is so little discretionary hydro (most of it is used to supply either basic demand or to follow and balance demand flux), and natural gas is so expensive, much of it diverted for heating purposes, while new nuclear plants do not seem to be an option anywhere in this country.

Because wind energy must be followed and balanced second-by-second by conventional generation, typically flexible fossil-fueled units, to achieve functional integration within the grid, this creates higher costs in the form of increased rates (someone must pay for the accompanying generation) and in the form of increased CO_2 emissions, since those wind-following fossil-fueled units are operating extremely inefficiently. For example, the Judith Gap windplant in Montana, producing in one hour 80 MW of electricity, and falling to 20 MW the next, is causing major headaches for the grid, even though the amount of wind energy relative to the grid's total supply is minute. The Montana utility was forced to buy more short-term power than expected from other energy sources to balance the grid's supply, driving costs upward.

If wind energy's electricity approaches, say, 10% of the grid's online supply, it would threaten the grid's ability to function. What would happen if the grid had insufficient supply to compensate for a sudden loss of wind power at this level of production? The same kind of blackout that occurred recently in Spain. What would happen if a sudden burst of high energy hit the grid because the wind increased unexpectedly? Such a power surge at 10% of the grid's total would shut the grid down. These situations demonstrate why E.On Netz, Germany's largest utility, has concluded that if wind penetration approaches 5%–10% of the grid's actual generation, *additional* conventional power must be added to the system at 90% of the wind installations' rated capacity—to protect the grid from wind's higher levels of fluctuating volatility.

Several sobering observations emerge from all this. First, any CO_2 emission saving that wind production creates is offset to some degree by the need for accompanying and backup generation, since much of that would come from "dirty" power sources working overtime. It is not clear whether wind energy creates even more CO_2 emissions throughout the grid because of this factor, given the need to continuously ramp the various companion power sources up and down, using more fuel much in the way a car engine does in stop-and-go traffic. Generally, fossil-fueled generators operating even 2% less efficiently than they were designed to do for following demand flux produce 16% more carbon emissions. No independent, peer-reviewed analyses measuring the operation of existing wind facilities have been conducted to show how—or whether—wind-produced electricity actually reduces CO_2 emissions

throughout a grid system. If hydro units, which emit no hydrocarbons, balance the variability of wind turbines, that wind energy cannot be considered
green, given how destructive hydro plants are to sensitive ecosystems. If fossil-
fueled generators follow and balance wind flux, these wind facilities can
hardly be considered clean or green.

Second, no unpredictably intermittent, highly variable power source can,
of itself, provide power for anyone, given modern expectations of reliability
and performance—despite all the media puffery implying it can. Wyoming
County's two proposed wind facilities might contribute, on a hit-or-miss basis,
an annual average of 57 MW to the NYISO's annual installed capacity of more
than 37,000 MW, which would be little more than one tenth of 1% of the
grid's current total. Since we increase our demand for electricity at 2% a year
(and New York is presently doing so at more than 3%), these 57 MW would be
swallowed up within the first three weeks of the new demand year. And this is
if everything works the way the wind industry hopes it will. If this amount of
power were generated by coal or nuclear, with capacity factors approaching
90% and with a predictable and constant stream of energy, it would service
about 50,000 homes. However, because of the intermittent, unpredictable
nature of wind, no homes would be powered by this source. How will such
energy stabilize natural gas prices or serve to "diversify" the mix of power
sources, as wind developers claim?

An electricity grid generally accepts wind energy not because it works
well, but because it has to, the result of political decisions to legislate Renewable Portfolio Standards that now exist in over 20 states, requiring utilities to
purchase a certain percentage of renewable energy. The only two renewables
capable of achieving industrial levels of electricity are hydro and wind—and
no one outside China and some third world countries is constructing new
hydro facilities.

Industrial Wind and Climate Change

I'd also like to dispel an untruth that wind developers spread to generate support
among environmental groups: that wind energy will contravene the forces
contributing to global warming and reduce dependence on foreign oil. Wind
only generates electricity. Electricity generation is only part of our energy
production. More than 60% of the nation's energy use does not involve the
making of electricity. Coal and gas-fired electricity power plants do pollute
the air with hydrocarbons. But the sheer volume of automobile exhausts and
an insatiable demand for heat are the major contributors to the problem. It is
folly to suggest that thousands of wind turbines blanketing the eastern US
would do anything to mitigate these other energy uses. Nationally, oil contributes only 3% of our electricity production. Even if industrial wind generated 10% of the nation's electricity, it would not staunch the fossil fuel
emissions thought to be involved in accelerating global warming, given our
nation's increasing energy consumption and given that wind can only inter-
mittently address the electricity portion of our energy production problem—
the minor portion. Since wind energy only produces electricity and we use so

little oil for the production of electricity, even if large numbers of wind turbines displaced the 3% of our electricity now powered by oil, we would still be heavily dependent on coal and gas—and we would still be mightily dependent on foreign oil.

These realities doom wind technology as anything other than a Rube Goldberg operation. To reduce fossil-fueled companion and back-up power generation for wind energy, we would have to build three windplants for every one, each in a separate geographical area to provide statistical redundancy for the intermittent flux of wind—and, even then, we would have to spend billions for new grid equipment to provide the lowest level capacity value. By this reckoning, about 6000-2.5 MW turbines would be needed to equal the output of one 1600 MW coalplant—and perhaps 8000 or more to compensate for the low wind capacity credit during the summer months, when, of course, demand is greatest. This far-fetched scenario might minimize windplant-induced CO_2 emissions at the operational level. However, energy expert Tom Tanton has calculated that, to compensate for the CO_2 emissions required for the concrete pads necessary to anchor each wind turbine, a large windplant would have to operate CO_2 free for at least seven years. There is no free lunch with the laws of thermodynamics.

To me, this is the smoking gun that challenges notions that wind energy may be a splendid idea. For wind-rich areas in the East, industrial wind is a non-starter anywhere in the region because, although the industry will have reasonably easy access to transmission lines, it's not going to produce enough reliable power to dent a grape in the scheme of things. Contrary to a wind developer's claim that our choice is between the belching smokestacks of coal and the twirling blades of huge wind turbines, we'll have both. Even if these turbines were fully deployed in the uplands of the East, coal plants would still be puffing away, their numbers actually increasing, while many thousands of gigantic wind machines would glut the landscape—killing wildlife, destroying culturally significant viewsheds, devaluing property, and creating major disturbances for those who live nearby. And, because the air would be getting dirtier, people everywhere would be getting sicker while paying more in rates and taxes.

In the Midwest, however, and certainly in the deep oceans, there is enough wind resource to make a big difference. European nations such as Denmark, Germany, Spain, and the Netherlands, each of which has encountered a political uproar about onshore windplants, are now seriously examining offshore wind potential. The problem is that, because of the need for redundancy and our ever-increasing demand for electricity, millions of turbines must be constructed and linked to new transmission systems, as well as retooled grid systems—at a cost of trillions of dollars. The ecological implications would be enormous.

Wind Economics

A discussion of the wind industry's vast array of public subsidies would fill a book. In a nutshell, Kenneth Lay's Enron, the "energy" company that, before its demise, owned and operated the nation's largest collection of wind facilities, cultivated them; Enron pioneered the tax shelter as a commodity. Publicly funded tax avoidance schemes reimburse wind energy developers as much as two-thirds of the capital cost of each $2 million turbine, with many states creating incentives to cover on average an additional 10% of these costs. A recent Beacon Hill Institute study showed that such incentive programs would allow the Cape Wind project to be reimbursed up to 78% of its capital costs over the life of the facility. Windplant owners can use these tax shelters themselves, or sell them, or enter into "equity partnerships" with other companies—all to reduce their corporate tax obligations by tens of millions each year, as the Marriott Corporation did a few years ago with a similar clean energy scheme, within a year reducing its corporate tax obligations from 36% to 6%—generating tax credits worth $159 million and a return of 246% on its investment in just one year.

The Florida Power and Light Group, the parent of FPL Energy, paid no federal income tax in 2002 and 2003, according to Citizens for Tax Justice, despite having revenues of $2.2 billion during those years. The FPL Group made large investments in wind energy during that time and now claims to be one of the nation's leading wind energy producers. It is also the parent company of Meyersdale Wind in Pennsylvania and Mountaineer Wind in West Virginia, both of which have delivered only a fraction of promised local taxes to date.

Wind is the perfect vehicle for tax shelter generation. Its unearned environmental credit brings a public relations cachet while trading in wind's renewable energy credits allows outfits like Florida Power and Light, GE, AES, and BP, which collectively own most of the nation's wind facilities, to avoid cleaning up their dirtiest burning plants. And the politicians who support all this give the appearance of challenging the status quo when in reality they're reinforcing it, especially since more wind facilities very likely will result in more coal plants. These costs to the federal treasury, which are actually transfers of wealth from average tax and ratepayers to a few corporate investors, aren't worth the benefits accruing to a handful of full time employees and to undisclosed annual lease payments for a very few property owners. Many new reliable conventional plants will be needed in the next 30 years. The subsidies for each will encourage efficient, dependable, cost effective electricity. The subsidies for dysfunctional industrial wind energy, which can provide virtually no capacity to the system and can deliver only sporadic energy, will be used to make ineffective and uneconomical technology falsely appear to be effective and economical. . . .

Carl Sagan once said, "Extraordinary claims require extraordinary proof." I have never seen an independent peer-reviewed analysis in a reputable scientific publication substantiating any of the claims made for wind energy. Reports from wind developers, their trade association, and from the Renewable Energy Lab, all of which stand to gain from the "enhancement" of wind technology, do not stem from the rigorous science enabling good public policy.

I have worked to understand this issue from a citizen's perspective, reasoning from basic principles, knowing when answers I've been given are unresponsive, and saying, like the boy in the story about the emperor's new clothes, that I don't see the evidence when it was not provided. As citizens interested in more effective energy solutions, I trust you will do the same. One can be concerned about how our fossil fuel combustion practices may help accelerate the process of global warming and injure public health without agreeing that the intrusive and ineffectual nature of wind energy technology is even a partial solution to the problem. For me, the harsh reality is that massive wind turbines are much more functional (and lucrative) as corporate tax avoidance generators than they are as environmentally friendly producers of energy, symbolic not of a more enlightened energy future but rather of our continuing attraction to the forces of ignorance and greed. They represent at best a placebo for our energy dilemma, distracting from the level of discourse—and political action—necessary for achieving genuinely effective responses. There are no magic bullets, unfortunately. The environmentally responsible short-range solution to the problem of our dependence upon fossil fuels must combine effective conservation with much higher efficiency standards—heavy lifting indeed for the most wasteful culture in the history of the planet.

POSTSCRIPT

Is Wind Power Green?

Despite the rapid growth of wind power, the debate over its benefits is far from over. Yet the debate centers more on industrial-scale wind farms, which deploy large numbers of large wind turbines, than on small home-scale systems. According to Jennifer Alsever, "Wind Power, The Home Edition," *Business 2.0* (January/February 2007), a dozen U.S. companies are offering wind turbines to be installed next to private homes. The Skystream costs about $10,000. Glen Salas, "A Great Wind Is Blowing," *Professional Builder* (January 2007), says that home builders can cater to "a lot of people [who] are willing to spend the extra money to generate their own electricity" and notes that "Over its life, a small residential wind turbine can offset about 1.2 tons of air pollutants and 200 tons of greenhouse gases, according to the American Wind Energy Association" (which offers fact sheets and more at http://www.awea.org/).

Wind farms require more space and faster, steadier wind than home systems. This is one major reason why many wind farms in North America and Europe have been proposed for off-shore sites. Objections have included damage to the view and interference with navigation and fishing. A more novel proposal is to use kites that could lift wind turbines into the jet stream, where the wind is very fast; see "Getting Wind Farms off the Ground," *Economist* (June 9, 2007). Another proposal, described by Todd Woody in "Tower of Power," *Business 2.0* (August 2006), calls for a 1,600-foot tall tower that will exploit the temperature difference between the bottom and the top to set a powerful wind flowing up the tower. EnviroMission, the Australian company behind the idea, says the tower will yield 50 megawatts of electricity, and larger versions are in the planning stages.

Like standard wind farms, both kites and towers will surely provoke protests. It is worth bearing in mind that no matter how humanity generates energy in the future, there will be environmental impacts. The only way to avoid this is to stop using energy, which is not a practical proposal. On the other hand, reducing energy use through conservation and efficiency is practical, and as Komanoff writes, it is an essential part of the solution. One of the primary exponents of efficiency is Amory B. Lovins, whose "More Profit with Less Carbon," *Scientific American* (September 2005), urges that "Using energy more efficiently offers an economic bonanza—not because of the benefits of stopping global warming but because saving fossil fuel is a lot cheaper than buying it." Lovins is profiled by Elizabeth Kolbert in "Mr. Green," *New Yorker* (January 22, 2007).

In May 2007, the National Research Council released *Environmental Impacts of Wind-Energy Projects.* The report says that by 2020 wind farms could offset about 4.5 percent of the carbon dioxide released by electricity generation but may harm bird and bat populations. It also discusses aesthetic and economic impacts on human communities, provides a guide for evaluating wind-energy projects, and recommends more government guidance for planners of such projects.

ISSUE 10

Should Cars Be More Efficient?

YES: David Friedman, from "CAFE Standards," Testimony before Committee on Senate Commerce, Science and Transportation, March 6, 2007

NO: Charli E. Coon, from "Why the Government's CAFE Standards for Fuel Efficiency Should Be Repealed, Not Increased," The Heritage Foundation Backgrounder #1458, July 11, 2001

ISSUE SUMMARY

YES: David Friedman, Research Director at the Union of Concerned Scientists, argues that the technology exists to improve the fuel efficiency standards for new cars and trucks and requiring improved efficiency can cut oil imports, save money, create jobs, and help with global warming.

NO: Charli E. Coon, Senior Policy Analyst with the Heritage Foundation, argues that the 1975 Corporate Average Fuel Economy (CAFE) program failed to meet its goals of reducing oil imports and gasoline consumption and has endangered human lives. It needs to be abolished and replaced with market-based solutions.

Automobiles have been a much-beloved feature of modern technology for over a century. Their advent increased mobility, made it possible for city workers to live outside the city, enabled goods to reach stores near almost everyone, and solved a growing environmental problem. Few realize how dirty the streets can be, how smelly cities can be, or how many flies can fill the air when transportation relies on horses! Ralph Turvey, "Horse Traction in Victorian London," *Journal of Transport History* (September 2005), says that these problems drew little mention at the time, but neither did most people mention the noisiness and smokiness (and poor gas mileage) of cars and trucks through the first half of the twentieth century. Still, most people thought cars and trucks were a vast improvement over horses. Those who had made millions in the oil business beginning in Texas, Oklahoma, California, and some other states, and later in the Middle East and elsewhere were quite happy with the change. See Leonardo Maugeri, *The Age of Oil: The Mythology, History, and Future of the World's Most Controversial Resource* (Praeger, 2006), and Lisa Margonelli, *Oil on the Brain: Adventures from the Pump to the Pipeline* (Nan A. Talese, 2007).

By the 1970s, there were a great many cars and trucks on the road. Local railways were extinct, and horses were no longer bred in the large numbers of the past. There really was no substitute for gasoline-powered transportation, even though it consumed large amounts of oil imported from distant parts of the world. American dependence on these imports was highlighted when the Organization of Petroleum-Exporting Countries (OPEC) cut supplies and raised prices in the Oil Crisis of 1973. In the wake of the crisis, foreign cars with better gas mileage increased their sales at the expense of American-made cars, highway speed limits were reduced, and the U.S. Congress passed the Energy Policy and Conservation Act of 1975, which included the Corporate Average Fuel Economy (CAFE) program. The goal was to double average fuel efficiency by 1985. Fuel efficiency standards for passenger cars started at 18 mpg in 1978 and rose to 27.5 mpg for 1985; these standards were lowered in the late 1980s, but rose again to 27.5 mpg in 1990, where they have remained ever since. Light trucks had lower standards; in 2007, the standard was 22.2 mpg for light trucks.

Prompted by these requirements, as well as by foreign competition for the U.S. market, manufacturers successfully improved the performance of drive trains and engines and developed lighter materials for bodies. But as oil supplies became ample, even though the price of gasoline continued to rise, they also converted light truck designs to passenger versions, now known as Sports Utility Vehicles or SUVs, and sold them as roomier, more powerful, and even safer (largely because of size and weight) automobiles. SUVs are infamous for poor gasoline mileage, compared to passenger cars, but they are still classified as light trucks and held only to that standard. Improving their performance offers "the greatest potential to reduce fuel consumption on a total-gallons-saved basis" (*Effectiveness and Impact of Corporate Average Fuel Economy [CAFE] Standards*, National Academy Press, 2002).

In the following selections, David Friedman, Research Director at the Union of Concerned Scientists, argues that the technology exists to improve the fuel efficiency standards for new cars and trucks and requiring improved efficiency can cut oil imports, save money, create jobs, and help with global warming. In addition, more fuel-efficient vehicles are actually safer to drive. Charli E. Coon, Senior Policy Analyst with the Heritage Foundation, argues that the 1975 Corporate Average Fuel Economy (CAFE) program failed to meet its goals of reducing oil imports and gasoline consumption and has endangered human lives. It needs to be abolished and replaced with market-based solutions.

YES

David Friedman

CAFE Standards

. . . I think we have reached an important milestone on fuel economy. It would appear that some leaders in Congress, including members of this committee, and the president are basically in agreement on how far we should increase fuel economy standards in about a ten year period.

In the president's state of the union speech, he set a goal for America to conserve up to 8.5 billion gallons of gasoline by 2017. To do so, we would need to increase fuel economy standards for cars and trucks to about 34 miles per gallon by 2017, or about 4 percent per year. At the same time, the bill recently introduced by the chairman and many members of this committee establishes a fuel economy target of 35 mpg by 2019. . . . The oil savings benefits of S. 357 are almost the same as the president's goal. Other members of the Senate and House have put forth bills with similar requirements in this and recent years.

In addition, Senator Stevens has introduced a bill to raise fuel economy standards for passenger cars to 40 mpg by 2017, or about a 39 percent increase compared to the average fuel economy of cars today. If Senator Stevens applied the same improvement to the rest of the fleet, it would average just over 34 mpg by 2017. As it stands, the oil savings from S. 183 are half of the others since only half the fleet is included.

I consider this a milestone because this significant agreement on fuel economy goals means that we can focus now on how best to reach them. By reforming and strengthening fuel economy standards for cars and trucks, this committee has a significant opportunity to help cut our oil addiction, save consumers money, create new jobs, and tackle the largest long term environmental threat facing the country and the world today, global warming.

Global Warming

Carbon dioxide, the main heat trapping gas blanketing our planet and warming the earth, has reached a concentration of about 380 parts per million. That is higher than the globe has experienced in the past 650,000 years. We are already seeing the impacts of these elevated concentrations as eleven of the last 12 years rank among the 12 hottest on record.

The worldwide costs of global warming could reach at least five percent of global GDP each year if we fail to take steps to cut emissions. These costs would come in lives and resources as tropical diseases and agricultural pests move

From FDCH Congressional Testimony, March 6, 2007.

north due to our warming continent. These costs could also come from losing 60–80 percent of the snow cover in the Sierras by the end of the century and the resulting impacts on agriculture in California and similar states that rely on snow melt for water. We will also see increased asthma and lung disease because higher temperatures will make urban smog worse than it is today.

Global warming is a worldwide problem and our cars and trucks have impacts that are worldwide in scale. Only the entire economies of the United States, China, and Russia exceed the global warming pollution resulting from our cars and trucks alone. It is clear that the scope of pollution from our cars and trucks requires special attention as we begin to address climate change.

Oil Addiction

In addition to the costs created by the pollution from our cars and trucks, our vehicles also contribute to 40 percent of our oil addiction. Overall, data from the Energy Information Administration indicates that we imported about sixty percent of our oil and other petroleum products in 2006. Last year alone, our net imports were more than 12 million barrels per day. When oil is at $60 per barrel, every minute that passes means over $500,000 that could have been spent creating U.S. jobs and strengthening our economy instead leaves this country. At the end of the day, high oil and gasoline prices and continued increases in our oil addiction represent one of the single biggest threats to U.S. auto jobs today.

Fuel Economy Background

One of the main reasons our vehicles contribute so much to U.S. oil dependence and global warming is that the average fuel economy of the fleet of new cars and trucks sold in the U.S. in 2006 was lower than it was in 1986. And while automakers note the number of models on the market that get more than 30 miles per gallon on the highway, a look at EPA's 2007 fuel economy guide shows that there are more than 300 car and truck configurations that get 15 mpg or less in the city. Even if you exclude pickups and work vans, automakers still flood the market with nearly 200 car, minivan, and SUV configurations that get 15 mpg or less in the city. Consumers simply do not have enough high fuel economy choices when it comes to cars, minivans, SUVs and pickups.

Fuel economy standards were created to solve this exact problem. Just as we see today, automakers were not ready for the problems created in part by our gas guzzling in the early 1970s. As a result consumers jumped on the only option they had at the time, relatively poorly designed smaller cars. However, as fuel economy standards were fully phased in automakers switched from giving consumers poor choices to putting technology in all cars and trucks so car buyers could have options in the showroom with 70% higher fuel economy than they had in 1975 (*2006 EPA Fuel Economy Trends Report*). If the fuel economy of today's cars and trucks was at the level the fleet experienced in 1975 instead of today's 25 miles per gallon, we would be using an additional 80 billion gallons of gasoline on top of the 140 billion gallons we will use this year. That would represent an increase in oil demand by 5.2 million barrels of oil per day, or a 25 percent increase in our oil addiction. At today's average price for regular gasoline, about $2.50 per gallon, that represents $200 billion dollars saved. That number could have been much better, however, if fuel economy standards had not remained essentially unchanged for the past two decades.

Technology to Create Consumer Choice

Driving in America has become a necessity. Because of this and a lack of options, even the spikes in gasoline prices over the past five years have not been enough to push consumers to significantly reduce their gasoline consumption. Better fuels and more alternatives to driving are important to helping consumers and cutting pollution, but the quickest route to reduced gasoline consumption and saving consumers money is put to more high fuel economy choices in the showroom.

The automobile industry has been developing technologies that can safely and economically allow consumers to get more miles to the gallon in cars, minivans, pickups and SUVs of all shapes and sizes. . . . These technologies [can] dramatically increase the fuel economy of an SUV with the size and acceleration of a Ford Explorer. This could be achieved using direct injection gasoline engines, high efficiency automatic manual transmissions, engines that shut off instead of wasting fuel while idling, improved aerodynamics, better tires and other existing efficiency technologies. These technologies have no influence on the safety of the vehicle. Others, such as high-strength steel and aluminum and unibody construction could actually help make highways safer.

For just over $2,500 a consumer could have the choice of an SUV that gets more than 35 mpg. This is an SUV that alone could meet the fuel economy targets laid out by members of this committee and the president. At $2.50 per gallon, this SUV would save consumers over $7,800 on fuel costs during the vehicle's lifetime. The technologies needed for this better SUV would even pay for themselves in about three years. Automakers do already have vehicles on the road that can match this fuel economy, but most are compact cars. That leaves a mother with three children in car-seats or a farmer who needs a work truck with few vehicle choices until these technologies are packaged into higher fuel economy minivans, SUVs, pickups and other vehicles.

The technologies in this better SUV could be used across the fleet to reach more than 40 miles per gallon over the next ten years. The 2002 study by the National Academies on CAFE showed similar results. Data in the report indicate that the technology exists to reach 37 mpg in a fleet of the same make-up as the NAS analyzed, even ignoring hybrids and cleaner diesels.

The question now is whether automakers will use these tools to increase fuel economy. Automakers have spent the past twenty years using similar technologies to nearly double power and increase weight by twenty-five percent instead of increasing fuel economy (*EPA Fuel Economy Trends Report, 2006*). As a result, consumers today have cars and trucks with race-car like acceleration and plenty of room for children, pets and weekend projects. What consumers need now is to keep the size and performance they have today, while getting higher fuel economy. Without increased fuel economy standards, however, this future is unlikely. We are already seeing automakers market muscle hybrids, vehicles that use hybrid technology for increased power instead of increased fuel economy. And technologies such as cylinder cut-off, which increases fuel economy by shutting off engine cylinders when drivers need less power, are being used to offset increased engine power rather than increased fleetwide fuel economy.

This committee is in a position to ensure that consumers can keep the power, size and safety they have in their vehicles today, and save thousands of dollars while cutting both global warming pollution and our oil addiction through deployment of technology aimed at better fuel economy across the vehicle fleet.

Economic and Employment Impacts of Setting Fuel Economy Targets

Contrary to claims by the auto industry, investments in fuel economy technology, just like other investments in the economy, will lead to prosperity. No automaker would simply shut down a plant if it was making gas guzzlers that don't meet national fuel economy targets. Instead, they would make investments to upgrade their tooling to build more fuel efficient vehicles. A 2006 study from Walter McManus at the University of Michigan shows that automakers that invest in fuel economy, even as early as 2010, will improve their competitive position (*Can Proactive Fuel Economy Strategies Help Automakers Mitigate Fuel-Price Risks?*). According to the study, Detroit's Big Three could increase profits by $1.3 billion if they invest in fuel economy, even if gasoline costs only $2 per gallon. However, if they follow a business-as-usual approach their lost profits could be as large as $3.6 billion if gasoline costs $3.10 a gallon.

UCS has also sought to quantify the benefits of increased fuel economy (Friedman, 2004, *Creating Jobs, Saving Energy and Protecting the Environment*). We estimated the effect of moving existing technologies into cars and trucks over 10 years to reach an average of 40 miles per gallon (mpg). We found that:

In 10 years, the benefits resulting from investments in fuel economy would lead to 161,000 more jobs throughout the country, with California, Michigan, New York, Florida, Ohio, and Illinois topping the list.

In the automotive sector, projected jobs would grow by 40,800 in 10 years. A similar analysis done by the economic-research firm Management Information Services (MIS) evaluated the potential job impacts of increasing fuel economy to about 35–36 mpg by 2015 and found even greater growth at more than 350,000 new jobs in 2015 (Bezdek, 2005, *Fuel Efficiency and the Economy*). This job growth included all of the major auto industry states.

In both the UCS and the MIS studies these new jobs would be created both because of investments in new technologies by the automakers and because consumers would shift spending away from gasoline to more productive products and services.

Requiring all automakers to improve fuel economy will increase the health of the industry. Companies like Ford, General Motors and the Chrysler division are currently in bad financial condition due to poor management decisions and elevated gas prices, not fuel economy standards, which have been stagnant for the past two decades. Those poor decisions have put them in a place where, just as in the 1970s, they do not have the products consumers need at a time of increased gasoline prices, and they are continuing the slide in market share that began the first time they made this mistake.

By requiring Ford, GM, Chrysler and all automakers [to] give consumers the choices they need, Congress can ensure automaker jobs stay in the U.S. and models like the Ford Explorer and Chevrolet Tahoe are still on the market ten years from now though they will go farther on a gallon of gas.

Safety Impacts of Setting Fuel Economy Targets

While the NAS study clearly states that fuel economy can be increased with no impact on the safety of our cars and trucks, critics of fuel economy standards often point to the chapter, which takes a retrospective look at safety. Despite the fact that this chapter did not represent a consensus of the committee (a dissenting opinion from two panel members was included in the appendices) and the fact that three major analyses have since shown that fuel economy and safety are not inherently linked, claims are still made to the contrary.

First, David Greene (one of the NAS panel members) produced a report with Sanjana Ahmad in 2004 (*The Effect of Fuel Economy on Automobile Safety: A Reexamination*), which demonstrates that fuel economy is not linked with increased fatalities. In fact, the report notes that, "higher mpg is significantly correlated with fewer fatalities." In other words, a thorough analysis of data from 1966 to 2002 indicates that Congress can likely increase fuel economy without harming safety if the past is precept.

Second, Marc Ross and Tom Wenzel produced a report in 2002 (*An Analysis of Traffic Deaths by Vehicle Type and Model*), which demonstrates that large vehicles do not have lower fatality rates when compared to smaller vehicles. Ross and Wenzel analyzed federal accident data between 1995 and 1999 and showed that, for example, the Honda Civic and VW Jetta both had lower fatality rates for the driver than the Ford Explorer, the Dodge Ram, or the Toyota 4Runner. Even the largest vehicles, the Chevrolet Tahoe and Suburban had fatality rates that were no better than the VW Jetta or the Nissan Maxima. In other words, a well-designed compact car can be safer than an SUV or a pickup. Design, rather than weight, is the key to safe vehicles.

Finally, a study by Van Auken and Zellner in 2003 (*A Further Assessment of the Effects of Vehicle Weight and Size Parameters on Fatality Risk in Model Year 1985–98 Passenger Cars and 1985–97 Light Trucks*) indicates that increased weight is associated with increased fatalities, while increased size is associated with decreased fatalities. While this study was not able to bring in the impacts of design as well as size, it helped inform NHTSA as they rejected weight-based standards in favor of size-based standards based on the vehicle footprint.

These studies further back up Congress' ability to set fuel economy targets as high as 40 mpg for the fleet in the next ten years without impacting highway safety.

Getting Fuel Economy Policy Right

Given broader agreement on how far fuel economy must increase, we now need policies to lay out how to get there. Congress should follow four key steps to ensure that the country gets the benefits of existing fuel economy technology:

> Establish a concrete fuel economy goal
>
> Provide NHTSA with additional flexibility to establish size based standards
>
> Institute a backstop to ensure that the fuel savings benefits are realized
>
> Provide consumers and/or automakers with economic incentives to invest in technology for increased fuel economy, Set a target of 34–35 mpg

Congress can set a standard either meeting the president's goals of 34 mpg by 2017 or 35 mpg by 2019 as in S. 357. Both of these fuel economy levels are supported by the guidance requested and received from the NAS and UCS analysis. By adopting S. 357, Congress would cut global warming pollution by more than 230 million metric tons 2020, the equivalent of taking more than 30 million of today`s automobiles off the road. The bill would also cut oil dependency by 2.3 million barrels of oil per day in 2027, as much oil as we currently import from the Persian Gulf.

The key to reaching these goals, however, is that Congress must set these targets and not leave it up to NHTSA. NHTSA has proven to have a poor track record when setting fuel economy standards so far. Their recent rulemaking on light trucks will save less than two weeks of gasoline each year for the next two decades. This happened in part because they did not value the important benefits of cutting oil dependence and reducing global warming pollution from cars and trucks. By setting specific standards based on where technology can take us, Congress can make clear the importance of tackling these important problems which are hard to quantify analytically, but easy to qualify based on consumer discontent with gasoline prices last summer, political instability from dependence on oil from the Persian Gulf, and the surge in concern over global warming.

Congress should not defer its regulatory authority to the Administration and it need not as it can base such standards on the scientific research it requested. Congress can be confident that the goals are technically feasible, cost effective, and safe.

Provide NHTSA Authority to Establish Size-Based Standards

The bills in the Senate and the president's plans include the ability for NHTSA to set car and light truck standards based on vehicle attributes such as vehicle size. These size-based standards give manufacturers who make everything from compact cars to minivans to large pickups the flexibility they have been asking for and eliminate any arguments automakers have made about CAFE standards treating them inequitably.

Size-based standards designed to increase fleet fuel economy to 35 mpg might require a family car to reach 40 mpg, but a pickup would only have to reach about 28 mpg because it is larger. This is good news for farmers and contractors who rely on these vehicles. With existing technology, pickups could readily reach 28 mpg and would save their owners over $6,000 on gasoline during the life of the vehicle. The pickup would have the same power, performance, size and safety it has today, and would cost an additional $1,500. However, the added fuel economy technology would pay for itself in less than two years with gasoline at $2.50 per gallon. Higher fuel economy standards will help farmers and small businesses who rely on trucks as much or even more than the average consumer.

Ensure No Backsliding

The one challenge with size-based standards is that automakers can game the system and drive down fuel economy. Much as automakers switched to marketing SUVs because of the lower standards required of light trucks to date, automakers may also upsize their vehicles to classes with lower fuel economy targets when they redesign their vehicles every four to seven years. Our analysis of NHTSA's most recent light truck rule shows that we could lose as much as half of the promised fuel economy gains, as small as they are, if the fleet of light trucks increased in size by just 10 percent over ten years. Congress must require a backstop to ensure that fuel savings that would be generated from a 10 mpg fuel economy increase are not lost due to automakers who game the system.

Provide Incentives

Because increased fuel economy will provide a wide variety of benefits for the nation, it is in the nation's interest to help automakers and suppliers who make cars and trucks in the U.S. that go farther on a gallon of gasoline. One way to help the auto industry is to provide tax credits, loan guarantees, or grants to companies that guarantee fuel economy improvements by investing in the equipment and people who will be needed to make these more efficient vehicles. This policy could be further supported by a set of charges and rebates applied to vehicles based on their fuel economy. These "feebates" will send market signals to producers and consumers in support of higher fuel economy standards and can even be made revenue neutral.

Conclusions

Climate change represents the largest long term environmental threat facing our country and the world today and the costs of our oil addiction continue to grow. Setting a fleet-wide target sufficient to meet the president's goal and guarantee fuel economy improvements of at least 10 mpg over the next decade while giving the president the authority to reach that target through size-based standards will save consumers money, stimulate the economy, create and protect jobs and preserve the safety of our vehicles. All of these benefits will come in addition to cutting our oil dependence and emissions of global warming pollutants from our cars and trucks.

Consumers are clearly happy with the size and acceleration of their vehicles today. We don't have to change that. But consumers are clearly unhappy with the growing impacts of global warming and the high cost of gasoline and the pumps and on our economy and security.

Congress has the opportunity to ensure that automakers spend the next decade or more using technology to curb our oil addiction. This is not a surprising role for Congress, the Federal government has helped drive every major transportation revolution this country has seen, whether it was trains, planes, or automobiles. The next transition will be no different.

Charli E. Coon

 **NO**

Why the Government's CAFE Standards for Fuel Efficiency Should Be Repealed, Not Increased

Congress may soon decide to increase the standards for fuel economy imposed on manufacturers of vehicles sold in the United States. This would be a mistake.

In 1975, Congress reacted to the 1973 oil embargo imposed by the Organization of Petroleum Exporting Countries (OPEC) by establishing the Corporate Average Fuel Economy (CAFE) Program as part of the Energy Policy and Conservation Act. The goal of the program was to reduce U.S. dependence on imported oil and consumption of gasoline. Advocates also hoped it would improve air quality. But the evidence shows that it has failed to meet its goals; worse, it has had unintended consequences that increase the risk of injury to Americans. Instead of perpetuating such a program, Congress should consider repealing the CAFE standards and finding new market-based solutions to reduce high gasoline consumption and rising prices.

There is significant pressure on Members of Congress, however, not only to continue this failed program, but also to raise fuel efficiency standards even higher. The current CAFE standards require auto manufacturers selling in the United States to meet certain fuel economy levels for their fleets of new cars and light trucks (pickups, minivans, and sport utility vehicles, or SUVs). The standard for passenger cars is currently 27.5 miles per gallon; for light trucks, it is 20.7 mpg.

Manufacturers face stiff fines for failing to meet these standards based on the total number of vehicles in each class sold, but compliance is taken out of their hands. The government measures compliance by calculating a sales-weighted mean of the fuel economies for the fleets of new cars and light trucks a manufacturer sells each year, and it measures domestically produced and imported vehicles separately.

Clearly, the CAFE program has failed to accomplish its purposes. Oil imports have not decreased. In fact, they have increased from about 35 percent of supply in the mid-1970s to 52 percent today. Likewise, consumption has not decreased. As fuel efficiency improves, consumers have generally increased their driving, offsetting nearly all the gains in fuel efficiency. Not only has the CAFE program failed to meet its goals; it has had tragic even if unintended consequences. As vehicles were being made lighter to achieve more miles per gallon and meet the standards, the number of fatalities from crashes rose.

Politicians should stop distorting the marketplace with unwise policies and convoluted regulations and allow the market to respond to consumer demand for passenger vehicles. In addition to free-market considerations, there are other compelling reasons to reject the CAFE standards. For example:

- CAFE standards endanger human lives;
- CAFE standards fail to reduce consumption; and
- CAFE standards do not improve the environment.

How Cafe Increases Risks to Motorists

The evidence is overwhelming that CAFE standards result in more highway deaths. A 1999 USA TODAY analysis of crash data and estimates from the National Highway Traffic Safety Administration and the Insurance Institute for Highway Safety found that, in the years since CAFE standards were mandated under the Energy Policy and Conservation Act of 1975, about 46,000 people have died in crashes that they would have survived if they had been traveling in bigger, heavier cars. This translates into 7,700 deaths for every mile per gallon gained by the standards.

While CAFE standards do not mandate that manufacturers make small cars, they have had a significant effect on the designs manufacturers adopt—generally, the weights of passenger vehicles have been falling. Producing smaller, lightweight vehicles that can perform satisfactorily using low-power, fuel-efficient engines is the most affordable way for automakers to meet the CAFE standards.

More than 25 years ago, research established that drivers of larger, heavier cars have lower risks in crashes than do drivers of smaller, lighter cars. A 2000 study by Leonard Evans, now the president of the Science Serving Society in Michigan, found that adding a passenger to one of two identical cars involved in a two-car frontal crash reduces the driver fatality risk by 7.5 percent. If the cars differ in mass by more than a passenger's weight, adding a passenger to the lighter car will reduce total risk.

The Evans findings reinforce a 1989 study by economists Robert Crandall of the Brookings Institution and John Graham of the Harvard School of Public Health, who found that the weight of the average American automobile has been reduced 23 percent since 1974, much of this reduction a result of CAFE regulations. Crandall and Graham stated that "the negative relationship between weight and occupant fatality risk is one of the most secure findings in the safety literature."

Harvard University's John Graham reiterated the safety risks of weight reduction in correspondence with then-U.S. Senator John Ashcroft (R-MO) in June 2000. Graham was responding to a May 2000 letter distributed to Members of the House from the American Council for an Energy-Efficient Economy

(ACEEE) and the Center for Auto Safety. Graham sought to correct its misleading statements, such as its discussion of weight reduction as a compliance strategy without reference to the safety risks associated with the use of lighter steel. For example, an SUV may be more likely to roll over if it is constructed with lighter materials, and drivers of vehicles that crash into guardrails are generally safer when their vehicle contains more mass rather than less. Further, according to Graham, government studies have found that making small cars heavier has seven times the safety benefit than making light trucks lighter.

The evidence clearly shows that smaller cars have significant disadvantages in crashes. They have less space to absorb crash forces. The less the car absorbs, the more the people inside the vehicle must absorb. Consequently, the weight and size reductions resulting from the CAFE standards are linked with the 46,000 deaths through 1998 mentioned above, as well as thousands of injuries. It is time that policymakers stop defending the failed CAFE program and start valuing human lives by repealing the standards.

Why Cafe Fails to Reduce Consumption

Advocates of higher CAFE standards argue that increasing miles per gallon will reduce gas consumption. What they fail to mention is the well-known "rebound effect"—greater energy efficiency leads to greater energy consumption. A recent article in *The Wall Street Journal* noted that in the 19th century, British economist Stanley Jevons found that coal consumption initially decreased by one-third after James Watt's new, efficient steam engine began replacing older, more energy-hungry engines. But in the ensuing years (1830 to 1863), consumption increased tenfold—the engines were cheaper to run and thus were used more often than the older, less efficient models. In short, greater efficiency produced more energy use, not less.

The same principle applies to CAFE standards. A more fuel-efficient vehicle costs less to drive per mile, so vehicle mileage increases. As the author of *The Wall Street Journal* article notes, "[s]ince 1970, the United States has made cars almost 50% more efficient; in that period of time, the average number of miles a person drives has doubled." This increase certainly offsets a portion of the gains made in fuel efficiency from government mandated standards.

Why Cafe Standards Do Not Improve the Environment

Proponents of higher CAFE standards contend that increasing fuel economy requirements for new cars and trucks will improve the environment by causing less pollution. This is incorrect.

Federal regulations impose emissions standards for cars and light trucks, respectively. These standards are identical for every car or light truck in those two classes regardless of their fuel economy. These limits are stated in grams per mile of acceptable pollution, not in grams per gallon of fuel burned. Accordingly, a Lincoln Town Car with a V-8 engine may not by law emit more emissions in a mile, or 10 miles, or 1,000 miles, than a Chevrolet Metro with a three-cylinder engine.

As noted by the National Research Council (NRC) in a 1992 report on automobile fuel economy, "Fuel economy improvements will not directly affect vehicle emissions." In fact, the NRC found that higher fuel economy standards could actually have a negative effect on the environment:

Improvements in vehicle fuel economy will have indirect environmental impacts. For example, replacing the cast iron and steel components of vehicles with lighter weight materials (e.g., aluminum, plastics, or composites) may reduce fuel consumption but would generate a different set of environmental impacts, as well as result in different kinds of indirect energy consumption.

Nor will increasing CAFE standards halt the alleged problem of "global warming." Cars and light trucks subject to fuel economy standards make up only 1.5 percent of all global man-made greenhouse gas emissions. According to data published in 1991 by the Office of Technology Assessment, a 40 percent increase in fuel economy standards would reduce greenhouse emissions by only about 0.5 percent, even under the most optimistic assumptions.

The NRC additionally noted that "greenhouse gas emissions from the production of substitute materials, such as aluminum, could substantially offset decreases of those emissions achieved through improved fuel economy."

Conclusion

The CAFE program has failed to achieve its goals. Since its inception, both oil imports and vehicle miles driven have increased while the standards have led to reduced consumer choice and lives lost that could have survived car crashes in heavier vehicles.

The CAFE standards should not be increased. They should be repealed and replaced with free market strategies. Consumers respond to market signals. As past experience shows, competition can lead to a market that makes gas guzzlers less attractive than safer and more fuel-efficient vehicles. That is the right way to foster energy conservation.

POSTSCRIPT

Should Cars Be More Efficient?

In June 2007, the U.S. Senate passed the Energy Act with an amendment that increased the average fuel economy standard for cars, trucks, and SUVs by 10 miles per gallon over ten years, reaching 35 mpg for passenger cars by 2020. After that date, further increases in the standard will be at the discretion of the Secretary of Transportation. A second amendment requires the Secretary to establish a plan to make half of all automobiles sold in 2015 ones that run on alternative fuels such as biodiesel, alcohol, or hydrogen.

The Energy Act is not yet law, for it requires action from the House of Representatives and a signature from the President, but it clearly reinforces the title of Mark Clayton's and Mark Trumbull's "Fuel Economy Back on U.S. Agenda," *Christian Science Monitor* (May 10, 2007). Whether its goals can be achieved depends on a number of factors. *Effectiveness and Impact of Corporate Average Fuel Economy (CAFE) Standards* (National Academy Press, 2002) says while the technology exists to improve gasoline mileage by 20–40 percent or more, consumers may have to pay more, and if gasoline prices drop they may refuse. In addition, CAFE standards apply to vehicles sold in a particular year, not to all the vehicles on the road in that year. It can take a decade or more (depending on the economy) to replace all the vehicles on the road. Thus efficiency of fuel use cannot possibly rise as rapidly as the standards for new cars.

Some of the technologies that can improve fuel efficiency are still in the research and development stage. Some, however, are well established. In Europe, the Smart Car (developed by Daimler-Benz engineers; see http://www.smartusa.com/) is very common on the streets; using a diesel engine, it gets about 70 mpg. The gasoline-powered version introduced in California in June 2007 (see Robert Salonga, "Car Small in Size, Big on Fuel Mileage," *Salinas Californian*, June 25, 2007) gets about 50 mpg. The secret is in part size: The Smart Car is a small car designed for commuters and others who do not have to haul much cargo. It carries two, plus a bag or two of groceries.

Mark Clayton, "Safe Cars Versus Fuel Efficiency? Not So Fast," *Christian Science Monitor* (June 12, 2007), notes that some automakers are insisting that raising fuel efficiency requires reducing vehicle size, which then reduces safety (see also Moira Herbst, "Fighting for the Right to Make Big Cars," *Business Week Online*, May 30, 2007). But good design can change that. The Smart Car, as just one example, has an egg-shaped frame that holds up very well in crashes despite its size.

ISSUE 11

Do Biofuels Enhance Energy Security?

YES: Bob Dinneen, from Testimony before Committee on Senate Energy and Natural Resources, April 12, 2007

NO: Mark Anslow, from "Biofuels Facts and Fiction," *The Ecologist* (March 2007)

ISSUE SUMMARY

YES: Bob Dinneen, president and CEO of the Renewable Fuels Association, the national trade association representing the U.S. ethanol industry, argues that government support of the renewable fuels industry has created jobs, saved consumers money, and reduced oil imports. The industry's potential is great, and continued support will contribute to ensuring America's future energy security.

NO: Journalist Mark Anslow argues that producing biofuels consumes more energy than it makes available for use, will take too long to make a contribution large enough to help fight global warming, generates dangerous quantities of waste, and because of government subsidies fails to make economic sense.

The threat of global warming has spurred a great deal of interest in finding new sources of energy that do not add to the amount of carbon dioxide in the air. Among other things, this has meant a search for alternatives to fossil fuels, which modern civilization uses to generate electricity, heat homes, and power transportation. Finding alternatives for electricity generation (which relies much more on coal than on oil or natural gas) or home heating (which relies more on oil and natural gas) is easier than finding alternatives for transportation (which relies on oil, refined into gasoline and diesel oil). In addition, the transportation infrastructure, consisting of refineries, pipelines, tank trucks, gas stations, and an immense number of cars and trucks that will be on the road for many years, is well designed for handling liquid fuels. It is not surprising that industry and government would like to find non-fossil liquid fuels for cars and trucks (as well as ships and airplanes).

There are many suitably flammable liquids. Among them are the so-called biofuels or renewable fuels, plant oils and alcohols that can be distilled from plant sugars. According to Daniel M. Kammen, "The Rise of Renewable Energy," *Scientific American* (September 2006), the chief biofuel in the United States so far is ethanol, distilled from corn and blended with gasoline. Production is subsidized with $2 billion of federal funds, and "when all the inputs and outputs were correctly factored in, we found that ethanol" contains about 25 percent more energy (to be used when it is burned as fuel) than was used to produce it. At least one study says the "net energy" is actually less than the energy used to produce ethanol from corn. If other sources, such as cellulose-rich switchgrass or cornstalks, can be used, the "net energy" is much better. However, generating ethanol requires first converting cellulose to fermentable sugars, which is so far an expensive process. A significant additional concern is the amount of land needed for growing crops to be turned into biofuels; in a world where hunger is widespread, this means land is taken out of food production. If additional land is cleared to grow biofuel crops, this must mean loss of forests and wildlife habitat, increased erosion, and other environmental problems. See "Ethanol: Energy Well Spent, A Survey of Studies Published Since 1990," Natural Resources Defense Council and Climate Solutions (February 2006) (http://www.nrdc.org/air/ transportation/ethanol/ethanol.pdf).

Under the Energy Policy Act of 2005, the U.S. Environmental Protection Agency (EPA) requires that gasoline sold in the United States contain a minimum volume of renewable fuel. Under the Renewable Fuel Program (also known as the Renewable Fuel Standard Program, or RFS Program), that volume will increase over the years, reaching 7.5 billion gallons by 2012. According to the EPA (http://www.epa.gov/otaq/renewablefuels/), "the RFS program was developed in collaboration with refiners, renewable fuel producers, and many other stakeholders." However, some think it is premature to put so much emphasis on biofuels. Pat Thomas, introducing *The Ecologist*'s special report on biofuels in the March 2007 issue (http://www.theecologist.org/archive_detail. asp?content_id=838), wrote that "the science is far from complete, the energy savings far from convincing and, although many see biofuels as a way to avoid the kind of resource wars currently raging in the Middle East and elsewhere, going down that road may in the end provoke a wider series of resource wars this time over food, water and habitable land."

In the following selections, Bob Dinneen, president and CEO of the Renewable Fuels Association, the national trade association representing the U.S. ethanol industry, argues that government support of the renewable fuels industry has created jobs, saved consumers money, and reduced oil imports. The industry's potential is great, and continued support will contribute to ensuring America's future energy security. Journalist Mark Anslow, a contributor to *The Ecologist*'s special report on biofuels, argues that producing biofuels consumes more energy than it makes available for use, will take too long to make a contribution large enough to help fight global warming, generates dangerous quantities of waste, and because of government subsidies fails to make economic sense.

YES

Bob Dinneen

Biofuels

. . . I am pleased to be here to discuss the future of our nation's ethanol industry and how the bipartisan Biofuels for Energy Security and Transportation Act of 2007 (S. 987) can help our country achieve its energy security goals.

Due to the visionary and invaluable work of this Committee in the 109th Congress, the Energy Policy Act of 2005 (EPAct 2005) put our nation on a new path toward greater energy diversity and national security through the RFS [Renewable Fuel Standard Program]. EPAct 2005 has stimulated unprecedented investment in the U.S. ethanol industry. Since January of 2006, when the RFS went into effect, no fewer than 15 new ethanol biorefineries have begun operation, representing some 1.2 billion gallons of new production capacity. These new gallons represent a direct investment of more than $1.8 billion and the creation of more than 22,000 new jobs in small communities across rural America.

The RFS has done exactly what Congress intended. It provided our industry with the opportunity to grow with confidence. It convinced the petroleum industry that ethanol would be a significant part of future motor fuel markets and moved them toward incorporating renewable fuels into their future plans. It persuaded the financial community that biofuels companies are growth market opportunities, encouraging significant new investment from Wall Street and other institutional investors. If a farmer in Des Moines doesn't want to invest in the local co-op, he can choose to invest in a publicly traded ethanol company through the stock market. As can a schoolteacher in Boston, or a receptionist in Seattle. Americans coast-to-coast have the opportunity to invest in our domestic energy industry, and not just in ethanol, but biodiesel and bio-products.

In addition to the RFS, many of the other programs authorized by EPAct 2005, such as the loan guarantee and grant programs, will accelerate the commercialization of cellulosic ethanol and make the new goals set forth in S. 987 absolutely achievable. Many of the provisions included in S. 987 build upon the programs designed by this Committee and included in EPAct 2005 to further expand the domestic renewable fuels industry. The Senate Energy and Natural Resources Committee will have an invaluable role to play in making sure our nation successfully moves toward increasing the use of domestic, renewable energy sources.

From Testimony before the Committee on Senate Energy and Natural Resources, April 12, 2007.

Background

Today's ethanol industry consists of 115 biorefineries located in 19 different states with the capacity to process almost 2 billion bushels of grain into 5.7 billion gallons of high octane, clean burning motor fuel, and more than 12 million metric tons of livestock and poultry feed. It is a dynamic and growing industry that is revitalizing rural America, reducing emissions in our nation's cities, and lowering our dependence on imported petroleum.

Ethanol has become an essential component of the U.S. motor fuel market. Today, ethanol is blended in more than 46% of the nation's fuel, and is sold virtually from coast to coast and border to border. The almost 5 billion gallons of ethanol produced and sold in the U.S. last year contributed significantly to the nation's economic, environmental and energy security.

According to an analysis completed for the RFA1, the approximately 5 billion gallons of ethanol produced in 2006 resulted in the following impacts:

- Added $41.1 billion to gross output;
- Created 160,231 jobs in all sectors of the economy;
- Increased economic activity and new jobs from ethanol increased household income by $6.7 billion, money that flows directly into consumers' pockets;
- Contributed $2.7 billion of tax revenue for the Federal government and $2.3 billion for State and Local governments; and,
- Reduced oil imports by 170 million barrels of oil, valued at $11.2 billion.

In addition to providing a growing and reliable domestic market for American farmers, the ethanol industry also provides the opportunity for farmers to enjoy some of the value added to their commodity by further processing. Farmer-owned ethanol plants account for 43 percent of the U.S. fuel ethanol plants and almost 34 percent of industry capacity.

There are currently 79 biorefineries under construction. With seven existing biorefineries expanding, the industry expects more than 6 billion gallons of new production capacity to be in operation by the end of 2009. The following is our best estimate of when this new production will come online.

To date, the U.S. ethanol industry has grown almost exclusively from grain processing. As a result of steadily increasing yields and improving technology, the National Corn Growers Association (NCGA) projects that by 2015, corn growers will produce 15 billion bushels of grain. According to the NCGA analysis, this will allow a portion of that crop to be processed into 15 billion gallons of ethanol without significantly disrupting other markets for corn. Ethanol also represents a growing market for other grains, such as grain sorghum. Ethanol production consumed approximately 26 percent of the nation's sorghum crop in 2006 (domestic use).

Research is also underway on the use of sweet and forage sorghum for ethanol production. In fact, the National Sorghum Producers believe that as new generation ethanol processes are studied and improved, sorghum's role will continue to expand.

In the future, however, ethanol will be produced from other feedstocks, such as cellulose. Ethanol from cellulose will dramatically expand the types and amount of available material for ethanol production, and ultimately dramatically expand ethanol supplies. Many companies are working to commercialize cellulosic ethanol production. Indeed, there is not an ethanol biorefinery in production today that does not have a very aggressive cellulose ethanol research program. The reason for this is that today's ethanol producers all have cellulose already coming into the plant in the form of corn fiber. Producers are making good use of all parts of the corn kernel—beyond just the starch. Several ethanol producers are working on technology to turn the fiber in a corn kernel into ethanol through fermentation. Since fiber represents 11 percent of the kernel, this could lead to dramatic increases in ethanol production efficiency. If today's producers can process these cellulosic materials into ethanol, they will have a significant marketplace advantage. The RFA believes cellulose ethanol will be commercialized first by current producers who have these cellulosic feedstocks at their grain-based facilities. It is essential to the advancement of the ethanol industry that these "bridge technology" cellulosic feedstocks be included in the definition of advanced biofuels.

Further, biotechnology will play a significant role in meeting our nation's future ethanol needs.

Average yield per acre is not static and will increase incrementally, especially with the introduction of new biotech hybrid varieties. According to NCGA, corn yields have consistently increased an average of about 3.5 bushels per year over the last decade. Based on the 10-year historical trend, corn yield per acre could reach 180 bushels by 2015. For comparison, the average yield in 1970 was about 72 bushels per acre. Agricultural companies like Monsanto believe we can achieve corn yields of up to 300 bushels per acre by 2030. It is not necessary to limit the potential of any feedstock—existing or prospective. Ultimately, the marketplace will determine which feedstocks are the most economically and environmentally feasible.

While there are indeed limits to what we will be able to produce from grain, cellulose ethanol production will augment, not replace, grain-based ethanol. The conversion of feedstocks like corn stover, corn fiber and corn cobs will be the "bridge technology" that leads the industry to the conversion of other cellulosic feedstocks and energy crops such as wheat straw, switchgrass, and fast-growing trees. Even the garbage, or municipal solid waste, Americans throw away today will be a future source of ethanol.

Research & Development, Deployment and Commercialization of New Technologies

The ethanol industry today is on the cutting edge of technology, pursuing new processes, new energy sources and new feedstocks that will make tomorrow's ethanol industry unrecognizable from today's. Ethanol companies are already utilizing cold starch fermentation, corn fractionation, and corn oil extraction. Companies are pursuing more sustainable energy sources, including biomass gasification and methane digesters. And, as stated, there is not an ethanol company represented by the RFA that does not have a cellulose-to-ethanol research program.

These cutting edge technologies are reducing energy consumption and production costs, increasing biorefinery efficiency, improving the protein content of feed co-products, utilizing new feedstocks such as cellulose, and reducing emissions by employing best available control technologies.

The technology exists to process ethanol from cellulose feedstocks; however, commercialization of cellulosic ethanol remains a question of economics. The capital investment necessary to build cellulosic ethanol facilities remains about five times that of grain-based facilities. Those costs will, of course, come down once the first handful of cellulosic facilities are built, the bugs in those "first mover" facilities are worked out, and the technology continues to advance. The enzymes involved in the cellulosic ethanol process remain a significant cost, as well. While there has been a tremendous amount of progress over the past few years to bring the cost of those enzymes down, it is still a significant cost relative to processing grain-based ethanol.

To continue this technological revolution, however, continued government support will be critically important. The biomass, bioresearch, and biorefinery development programs included in S. 987 will be essential to developing these new technologies and bringing them to commercialization. Competitively awarded grants and loan guarantees that build upon the existing programs authorized in EPAct 2005 and enhanced in S. 987 will allow technologically promising cellulosic ethanol projects [that] move the industry forward [to] become a reality.

Infrastructure

Ethanol today is largely a blend component with gasoline, adding octane, displacing toxics and helping refiners meet Clean Air Act specifications. But the time when ethanol will saturate the blend market is on the horizon, and the industry is looking forward to new market opportunities.

As rapidly as ethanol production is expanding, it is possible the industry will saturate the existing blend market before a meaningful E-85 market develops. In such a case, it would be most beneficial to allow refiners to blend ethanol in greater volumes, e.g., 15 or 20 percent. The ethanol industry today is engaged in testing on higher blend levels of ethanol, beyond E-10.

There is evidence to suggest that today's vehicle fleet could use higher blends. An initial round of testing is underway, and more test programs will be needed. A study of increased blend levels of ethanol, included in S. 987, will be an essential and necessary step to moving to higher blend levels with our current vehicle fleet. Higher blend levels would have a significant positive impact on the U.S. ethanol market, without needing to install new fuel pumps and wait for a vehicle fleet to turn over in the next few decades. It would also allow for a smoother transition to E-85 by growing the infrastructure more steadily.

Enhancing incentives to gasoline marketers to install E-85 refueling pumps will continue to be essential. There are now more than 1,000 E-85 refueling stations across the country, more than doubling in number since the passage of EPAct 2005. The RFA also supports the concept of regional "corridors" that concentrate the E-85 markets first where the infrastructure already exists, which is reflected in S. 987 in the infrastructure pilot program for renewable fuels.

Over the past several years, the ethanol industry has worked to expand a "Virtual Pipeline" through aggressive use of the rail system, barge and truck traffic. As a result, we can move product quickly to those areas where it is needed. Many ethanol plants have the capability to load unit trains of ethanol for shipment to ethanol terminals in key markets. Unit trains are quickly becoming the norm, not the exception, which was not the case just a few years ago.

Railroad companies are working with our industry to develop infrastructure to meet future demand for ethanol. We are also working closely with terminal operators and refiners to identify ethanol storage facilities and install blending equipment. We will continue to grow the necessary infrastructure to make sure that in any market we need to ship ethanol there is rail access at gasoline terminals, and that those terminals are able to take unit trains. Looking to the future, studying the feasibility of transporting ethanol by pipeline from the Midwest to the East and West coasts, as proposed in S. 987, will be critical.

As flexible fuel vehicle (FFV) production is ramped up, it is important to encourage the use of the most efficient technologies. Some FFVs today experience a reduction in mileage when ethanol is used because of the differences in BTU content compared to gasoline. But the debit can be easily addressed through continued research and development. For example, General Motors has introduced a turbo-charged SAAB that experiences no reduction in fuel efficiency when E-85 is used. There is also technology being developed that utilizes "variable compression ratio engines" that would adjust the compression ratio depending on the fuel used.

Thus, if the car's computer system recognized E-85 was being used, it would adjust the compression ratio to take full advantage of ethanol's properties. RFA supports the further study of how best to optimize technologies of alternative fueled vehicles to use E-85 fuel as included in S. 987. The study of new technologies could dramatically improve E-85 economics by eliminating or substantially reducing the mileage penalty associated with existing FFV technology.

Conclusion

The continued commitment of the 110th Congress, this Committee, and the introduction of legislation such as S. 987 will all contribute to ensuring America's future energy security. Chairman Bingaman and Ranking Member Domenici, you have made clear your commitment to the hardworking men and women across America who are today's newest energy producers.

There have been numerous bills introduced in the first few months of the 110th Congress to further expand the rapidly growing domestic biofuels industry that will soon eclipse the current RFS. Many of the sound provisions included in those bills to move the industry forward and create new market opportunities for biofuels are incorporated in S. 987. With minimal modifications, S. 987 strikes the right balance between incentivizing cellulosic ethanol technologies, developing the necessary infrastructure, moving beyond existing blend markets for ethanol, and capitalizing on the momentum created by EPAct 2005. The RFA looks forward to working with you to further develop this important legislation.

Mark Anslow

 **NO**

Biofuels—Facts and Fiction

Claim 1: You Get More Out Than You Put In

For more than 15 years, David Pimentel, Professor of Ecology and Agriculture at Cornell University in New York, and his colleague, Professor Tad Patzek at Berkeley, have published peer-reviewed research showing that biofuels give out less energy when burnt than was used in their manufacture.

By using a 'cradle-to-grave' approach—measuring all the energy inputs to the production of ethanol from the production of nitrogen fertiliser, through to the energy required to clean up the waste from bio-refineries—they have shown that while it takes 6,597 kilocalories of nonrenewable energy to produce a litre of ethanol from corn, that same litre contains only 5,130 kilocalories of energy—a 22 per cent loss.

Their work has been fiercely attacked by the biofuel lobby, who argue that Pimentel and Patzek include too many 'energy input' costs, and fail to give credit to the other, useful 'co-products' created in the process of refining biofuel.

Neither objection stands up under closer scrutiny. In fact, corn uses more herbicides, insecticides and fertiliser than any other crop(3); and 99 per cent of all cornfields used for producing bioethanol are heavily fertilised with nitrogen. Pimentel and Patzek have shown that although the energy costs involved with fertiliser production have fallen, most of the factories producing nitrate fertiliser in the USA today were built in the 1960s and are highly inefficient. As such, they estimate that the energy costs of nitrogen fertiliser manufacture account for over 30 per cent of the total energy needed to grow corn. When the energy costs of labour, machinery, petrol and diesel, other fertilisers, herbicides, insecticides and corn seed production are figured into the equation, merely growing corn using intensive agriculture accounts for 38 per cent of the energy needed to produce a litre of ethanol.

To make their energy costs appear more favourable, proponents of biofuels frequently 'off-set' the energy value of other substances produced during the refining process against the total energy used to produce the fuel. For bioethanol, these co-products include animal feed and carbon dioxide gas. For biodiesel, they include animal feed and glycerine, a component of soap. They argue that, by calculating the energy that would have been required to produce these substances by themselves, the amount of energy accounted for in the biofuel production process can be reduced. In some studies, the energy value of co-products has been calculated at 150 per cent more than the energy required to produce the fuel.

But the energy and monetary value of these co-products is highly subjective. In the UK, the production of glycerine, which biodiesel producers had hoped to sell to cosmetics companies to offset the costs of production, has reached such levels that supply is exceeding demand. Some refiners have been forced to simply burn it. In the US, the value of the grains left over after ethanol distillation has been much touted as an animal feed. But research has shown that this grain contains less energy than normal animal feed (usually made from much less fertiliser-intensive soya), and that production of soya has not fallen as ethanol production has risen, indicating that livestock farmers have been reluctant to change their animals' diet and use the new feed. David Morris, a biofuel lobbyist, has even admitted that it may benefit refiners more to burn the animal feed as fuel than to sell it.

Some ethanol distilleries have bottled the carbon dioxide that is given off during the fermentation process and sold it to carbonated drinks manufacturers, counting the value of the by-product against their overall energy costs. Most, however, have not.

Energy offset benefits can only be counted if the co-products are genuinely used in substitute for another product. Refining ethanol produces roughly equal parts ethanol, carbon dioxide and animal feed. Given that US corn-based ethanol production in 2005 peaked at 16.2 billion litres, this means that an almost equivalent amount of co-products (by volume) must have been produced. If these products are, as market figures suggest, unwanted, then instead of providing a useful 'offset', they are set to become a serious waste problem.

Claim 2: It Makes Economic Sense

In 2006, the American government handed out between $5.1 and $6.8 billion in ethanol subsidies. These include payments made to farmers, tax breaks given to refiners and payments made under carbon reduction programmes. But instead of these subsidies finding their way into farmers' pockets, they are instead swelling the accounts of several large biofuel manufacturers.

One company, Archer Daniel Midlands (ADM, one of the world's largest agribusiness companies), accounted for nearly 28 per cent of the US ethanol industry in 2006. According to attorney Arnold Reitze, Professor of Environmental Law and Director of the Environmental Life Programme at George Washington University Law School, every dollar of ADM's profit has cost US taxpayers $30. To ensure the continuation of ethanol subsidies, the Renewable Fuels Association (of which ADM is a member) had reportedly contributed $772,000 to Republican coffers between 1991 and 1992.

Biofuels have already been taken out of the hands of farmers and turned into big business. Where the demand for ethanol has benefited corn farmers, it has done so only at the expense of cattle farmers, for whom the cost of animal feed has vastly increased. Ethanol production from corn has been estimated to add $1 billion to the cost of beef production.

In the USA, a litre of petrol costs roughly 33 cents to produce; a litre of ethanol can cost up to $1.88. At present, these differentials are disguised behind subsidies, tax breaks, levies and laws. Germany subsidises biofuels to the value of 47 cents per litre, and France to the value of 33 cents per litre.

In his recent pre-Budget report, Gordon Brown reduced the tax on UK blended biofuels from 53 pence per litre to 8 pence per litre. In Brazil, although subsidies of ethanol officially ended in the mid-1990s, a number of 'incentives' still exist. Personal diesel-engined vehicles have been banned, to encourage the uptake of ethanol burning models, despite the greater fuel economies of many diesel cars. In addition, new 'flex-fuel' cars—models that can run on both ethanol and petrol—have been made available at a reduced rate of VAT.

Behind this raft of measures, it is difficult to see whether biofuels could ever compete with fossil fuels without continued subsidies, covert or otherwise. It is important to remember exactly what is being subsidised as well—excessive motor transport. As Michael O'Hare, Professor of Public Policy at UC Berkeley, pointed out in a recent article:

'Driving your car with a gallon of ethanol doesn't do 50 cents worth of good for society, it just does less damage than driving it with gasoline.'

Claim 3: It Is the Solution to Our Energy Problems

Recent figures show that if high-yield bio-energy crops were grown on all the farmland on earth, the resulting fuel would account for only 20 per cent of our current demand. The Organisation for Economic Cooperation and Development (OECD) published research which shows that more than 70 per cent of Europe's farmland would be required for biofuel crops to account for even 10 per cent of road transport fuel.

But there are more basic reasons why biofuels cannot be the answer to our energy problems. A normal petrol engine cannot run on more than a 15 per cent ethanol blend, and it is considered too expensive to modify a car after manufacture. Given that the average life expectancy of a vehicle is 14 years, it would take approximately this long to replace the current petrol fleet. By 2021, however, it could already be too late to make a difference to serious global warming.

The European Union Biofuels Directive requires that all EU member states have a blend of 5.75 per cent biofuel in their road transport fuels by 2010. However, a litre of biodiesel contains 12 per cent less chemical energy than an equivalent litre of mineral diesel, and is five per cent less fuel efficient when burnt in an engine. A litre of ethanol contains 33 per cent less energy than a litre of petrol, and a blend of 85 per cent ethanol to 15 per cent petrol (known as E85) can see vehicle fuel consumption rise by 31 per cent. The UK uses approximately 26 billion litres of petrol each year. If this were to be blended with 5.75 per cent bioethanol, the net energy contained in a litre of pump fuel would drop by approximately 2 per cent. In addition, ethanol blended fuels cannot be transported by pipeline, as the ethanol attracts water, which would render it ineffective as a fuel. It must, therefore, be transported by road. This means that an extra 521.5 million litres of fuel would need to be transported annually to make up for the energy deficit—equivalent to an extra 16,478 tanker journeys in the UK each year, which could increase the carbon emissions involved in distribution from refinery to tanker terminals by 38 per cent.

Claim 4: It's Clean and Safe

The biofuels ethanol and biodiesel are often referred to as 'clean-burning' fuels, and much has been made of their lower emissions of carbon monoxide. However, analyses of exhaust emissions from cars burning ethanol show an increase in nitrogen oxides, acetaldehyde and peroxy-acetyl-nitrate.

Likewise, cars burning biodiesel have been shown to emit higher levels of nitrogen oxides than those burning mineral diesel. Nitrous oxides are powerful greenhouse gases and can lead to the depletion of atmospheric ozone. At low levels they can react with VOCs and create low-level ozone, which can give rise to urban smog and respiratory problems.

When ethanol is blended with gasoline it makes the entire fuel more volatile. This means that it is more likely to evaporate, especially in the summer, through rubber and plastic parts of the fuel system. A study by the California Air Quality Board in 2004 found that blending ethanol with petrol increased fuel evaporation by 14 to 18 per cent. This means a higher quantity of hydrocarbon and nitrogen oxide emissions, as the fuel dissipates from vehicle tanks.

Ethanol is a solvent, and corrodes soft metals including aluminium, zinc, brass and lead. This means that existing underground storage tanks designed for fossil fuels and made from metal or even fibreglass could leak if filled with ethanolblended fuel, leaching pollutants into groundwater. If this happens, there is evidence that pollution would be even more widespread with a petrol-ethanol blend than with petrol alone. The presence of ethanol in the mix increases the persistence of the toxic substances benzene, toluene, ethylbenzene and xylene, and can cause them to travel 2.5 times farther in groundwater than would have been the case with a non-ethanol blended fuel.

Biodiesel is also a natural solvent, whereas mineral diesel is not. This means that parts of the fuel system, particularly in older cars, may start to corrode when biodiesel blends are used. This can lead to a build-up of deposits in the fuel system and engine, which in turn could reduce vehicle performance and increase fuel consumption.

Biodiesel also solidifies at around 4-5°C. This means that it must be preheated on cold winter mornings before it will flow from the tank. One biodiesel information website recommends the use of highly toxic 'anti-gelling' compounds mixed in with the fuel—or a 'heated garage'. It is this kind of solution that typifies the utter dependence of biofuels upon the continuing extravagant use of fossil energy.

Claim 5: It's Good for the Environment

A bio-refinery is an extraordinarily wasteful facility. For every litre of bioethanol produced in a modern refinery, 13 litres of waste water are generated. This waste water contains dead yeast and small amounts of ethanol, and has what is known as a Biological Oxgen Demand (BOD)—which means that the effluent competes with various other organisms in the water for available oxygen.

If effluent with a BOD is discharged into a watercourse, microorganisms in the water use oxygen in the water to break down, or oxidise, the pollutants, thus making the oxygen less available for other species. In extreme cases, fish and other aquatic organisms can suffocate from lack of oxygen.

The BOD of raw sewage is around 600mg per litre; that of bio-refinery waste water can be between 18,000 and 37,000mg per litre. This must be treated before it can leave the refinery, which requires an energy input of around 69,000 kilocalories, roughly equivalent to 306.7 cu ft of natural gas per 1,000 litres of ethanol produced.

In sugarcane ethanol plants, which are particularly common in Brazil, 12 cu ft of a thick, dark red, acid substance called 'vinasse' is left behind for every cubic foot of ethanol that has been produced. It is piped from the refinery to settlement ponds, where it is allowed to cool. If vinasse is left in the pools, anaerobic breakdown will lead to the production of methane, a greenhouse gas.

Some refinery operators have chosen to dilute vinasse at a ratio of up to 1:400 with water for use as a fertiliser on the sugarcane plantations. But it is so potent that the soil has to be carefully monitored to make sure that plants are not scorched or waterways polluted. Some farmers have used vinasse as a 'binding agent' on gravel drives, only to find that it corrodes the underside of vehicles that frequently drive over it.

Ethanol refineries also produce significant amounts of nitrous oxides (a greenhouse gas more than 300 times more potent that CO_2), carbon monoxide and VOCs (also linked to the destruction of the ozone layer and damage to human health). Their emissions are so high that in March 2006, the Environmental Protection Agency in the USA was forced under political pressure from the biofuels lobby to propose raising the threshold for facilities considered to be 'minor source of emissions' from 100 tons per year to 250 tons per year.

POSTSCRIPT

Do Biofuels Enhance Energy Security?

In March 2007, President Bush visited Brazil, which meets much of its need for vehicle fuel with ethanol from sugarcane, and agreed to work with Brazil in developing and promoting biofuels. According to the U.S. State Department, the agreement "reassures small countries in Central America and the Caribbean that they can reduce their dependence on foreign oil." In both Europe and the United States, governments are rushing to encourage the production and use of biofuels. L. Pelmans, et al., "European Biofuels Strategy," *International Journal of Environmental Studies* (June 2007), attempts to classify nations according to their strategies so that "the formulation of a strategy to support the advancement of biofuels and alternative motor fuels in general should become more manageable." Corporations and investors see huge potential for profit, and many environmentalists see benefits for the environment.

However, there *are* problems with biofuels, and those problems are getting a great deal of attention. Robin Maynard, "Against the Grain," *The Ecologist* (March 2007), stresses that when food and fuel compete for farmland, food prices will rise, perhaps drastically. The poor will suffer, as will rain forests. Renton Righelato, "Forests or Fuel," *The Ecologist* (March 2007), reminds us that when forests are cleared, they no longer serve as "carbon sinks"; deforestation thus adds to the global warming problem, and it may take a century for the benefit of biofuels to show itself. Harriet Williams, "How Green Is My Tank?" *The Ecologist* (March 2007), calls biofuels a dangerous distraction. Heather Augustyn, "A Burning Issue," *World Watch* (July/August 2007), describes the impact of forest fires to clear land for oil palm plantations in Indonesia. Palm oil holds great promise as a biofuel, but the plantations displace natural ecosystems and destroy habitat for numerous species, as well as for indigenous peoples.

Laura Venderkam, "Biofuels or Bio-Fools?" *American: A Magazine of Ideas* (May/June 2007), describes the huge amounts of money being invested in companies planning to bring biofuels to market. A great deal of research is also going on, including efforts to use genetic engineering to produce enzymes that can cheaply and efficiently break cellulose into its component sugars (see Matthew L. Wald, "Is Ethanol for the Long Haul?" *Scientific American*, January 2007, and Michael E. Himmel, et al., "Biomass Recalcitrance: Engineering Plants and Enzymes for Biofuels Production," *Science*, February 9, 2007), make bacteria or yeast that can turn a greater proportion of sugar into alcohol (see Francois Torney, et al., "Genetic Engineering Approaches to Improve Bioethanol Production from Maize," *Current Opinion in Biotechnology*, June 2007), and even bacteria that can convert sugar or cellulose into hydrocarbons that can easily be turned into gasoline or diesel fuel (see Neil Savage, "Building Better Biofuels," *Technology Review*, July/August 2007). If these efforts succeed, the price of biofuels may drop drastically, leading investors to abandon the field. Such a price drop would, of course, benefit the consumer and lead to wider use of biofuels. It would also, say C. Ford Runge and Benjamin Senauer, "How Biofuels Could Starve the Poor," *Foreign Affairs* (May/June 2007), ease the impact on food supply.

ISSUE 12

Is It Time to Revive Nuclear Power?

YES: Michael J. Wallace, from "Nuclear Power 2010 Program," Testimony before the United States Senate Committee on Energy & Natural Resources, Hearing on the Department of Energy's Nuclear Power 2010 Program (April 26, 2005)

NO: Karen Charman, from "Brave Nuclear World? Part II," *World Watch* (July/August 2006)

ISSUE SUMMARY

YES: Michael J. Wallace argues that because the benefits of nuclear power include energy supply and price stability, air pollution control, and greenhouse gas reduction, new nuclear power plant construction—with federal support—is essential.

NO: Karen Charman argues that nuclear power's drawbacks and the promise of clean, lower-cost, less dangerous alternatives greatly weaken the case for nuclear power.

T he technology of releasing for human use the energy that holds the atom together got off to an inauspicious start. Its first significant application was military, and the deaths associated with the Hiroshima and Nagasaki explosions have ever since tainted the technology. It did not help that for the ensuing half century, millions of people grew up under the threat of nuclear Armageddon. But almost from the beginning, nuclear physicists and engineers wanted to put nuclear energy to more peaceful uses, largely in the form of power plants. Touted in the 1950s as an astoundingly cheap source of electricity, nuclear power soon proved to be more expensive than conventional sources, largely because safety concerns caused delays in the approval process and prompted elaborate built-in precautions. Many say that safety measures have worked well when needed—Three Mile Island, often cited as a horrific example of what can go wrong with nuclear power, released very little radioactive material to the environment. The Chernobyl disaster occurred when safety measures were ignored. In both cases, human error was more to blame than the technology itself. The related issue of nuclear waste has also raised fears and added expense to the technology.

It is clear that two factors—fear and expense—impede the wide adoption of nuclear power. If both could somehow be alleviated, it might become possible to gain the benefits of the technology. Among those benefits are that nuclear power does not burn oil, coal, nor any other fuel; does not emit air pollution and thus contribute to smog and haze; does not depend on foreign sources of fuel and thus weaken national independence; and does not emit carbon dioxide. The last may be the most important benefit at a time when society is concerned about global warming, and it is the one that prompted James Lovelock, creator of the Gaia Hypothesis and an inspiration to many environmentalists, to say, "If we had nuclear power we wouldn't be in this mess now, and whose fault was it? It was [the antinuclear environmentalists']." See his autobiography, *Homage to Gaia: The Life of an Independent Scientist* (Oxford University Press, 2001). Stewart Brand, "Environmental Heresies," *Technology Review* (May 2005), says that he expects environmentalists to change their minds. The Organisation for Economic Co-operation and Development (OECD's) Nuclear Energy Agency, in "Nuclear Power and Climate Change," (Paris, France, 1998), available at http://www.nea.fr/html/ndd/climate/climate.pdf, found that a greatly expanded deployment of nuclear power to combat global warming was both technically and economically feasible. In 2000 Robert C. Morris published *The Environmental Case for Nuclear Power: Economic, Medical, and Political Considerations* (Paragon House). In August 2000 *USA Today Magazine* published "A Nuclear Solution to Global Warming?" "The time seems right to reconsider the future of nuclear power," say James A. Lake, Ralph G. Bennett, and John F. Kotek, in "Next-Generation Nuclear Power," *Scientific American* (January 2002). See also I. Fells, "Clean and Secure Energy for the Twenty-First Century," *Proceedings of the Institution of Mechanical Engineers, Part A—Power & Energy* (August 1, 2002). David Talbot, "Nuclear Powers Up," *Technology Review* (September 2005), notes that "While the waste problem remains unsolved, current trends favor a nuclear renaissance. Energy needs are growing. Conventional energy sources will eventually dry up. The atmosphere is getting dirtier." Peter Schwartz and Spencer Reiss, "Nuclear Now!" *Wired* (February 2005), argue that nuclear power is the one practical answer to global warming and coming shortages of fossil fuels. Paul Lorenzini, "A Second Look at Nuclear Power," *Issues in Science and Technology* (Spring 2005), says that nuclear power is essential to a sustainable future.

In the following selections, Michael J. Wallace, executive vice president of a major energy company, argues that because the benefits of nuclear power include energy supply and price stability, air pollution control, and greenhouse gas reduction, new nuclear power plant construction is essential, and there is a clear place for federal support. Karen Charman argues that nuclear power's drawbacks—risk, expense, and waste—and the promise of clean, lower-cost, less dangerous alternatives greatly weaken the case for nuclear power.

YES

Michael J. Wallace

Nuclear Power 2010 Program

. . . Constellation Energy, a Fortune 200 company based in Baltimore, is the nation's leading competitive supplier of electricity to large and industrial customers and the nation's largest wholesale power seller. Constellation Energy also manages fuels and energy services on behalf of energy intensive industries and utilities. The company delivers electricity and natural gas through the Baltimore Gas and Electric Company (BGE), its regulated utility in Maryland. We are the owners of 107 generating units at 35 different locations in 11 states, totaling approximately 12,500 megawatts of generation capacity. In 2004, the combined revenues of the integrated energy company totaled more that $12.5 billion and we are the fastest growing Fortune 500 Company over the past two years.

Our portfolio based on electricity produced is approximately 50 percent nuclear, 35 percent coal-fired, 7 percent gas-fired and 5 percent renewables. We own and operate the Calvert Cliffs nuclear plant in Maryland, and the Nine Mile Point and Ginna nuclear stations in New York State.

Constellation is part of the NuStart consortium that is preparing an application to the NRC for a license that would allow us to build and operate a new nuclear plant. Additionally, in December 2004, we submitted a proposal to the Department of Energy (DOE) for studies that could lead to an application to the Nuclear Regulatory Commission for an Early Site Permit as part of the Nuclear Power 2010 program. So, as you can tell, we have a vested interest in the continued success of Nuclear Power 2010, and we're bullish on the future of nuclear power.

Although I am here testifying today on behalf of Constellation, this testimony is supported by our trade association, the Nuclear Energy Institute (NEI).

My statement this morning will address four major issues:

1. The strategic value of our 103 operating nuclear power plants, and the compelling need to build new nuclear plants to preserve our nation's energy security, meet our environmental goals, and sustain our economic growth.
2. The critical importance of the Department of Energy's Nuclear Power 2010 program as a platform from which to launch the next generation of nuclear power plants in the United States.

United States Senate Committee on Energy & Natural Resources Hearing on the Department of Energy's Nuclear Power 2010 Program, April 26, 2005.

3. The need to recognize that the Nuclear Power 2010 program does not address all of the challenges facing companies interested in building new nuclear power plants, and that additional joint investment initiatives by the federal government and the private sector will be necessary.

4. The urgent need for comprehensive energy legislation that squarely addresses the critical need for additional investment in our electricity and energy infrastructure, including advanced nuclear and coal-fired generating capacity, electric and natural gas transmission, and other areas. Construction of the next nuclear power plants in the United States will require some form of investment stimulus, but I know I speak for the entire electric sector when I say that the need for investment stimulus extends well beyond nuclear power. This sector is starved for investment capital, and new federal government policy initiatives are necessary to reverse that trend and place our economy and our future on a sound foundation.

The Strategic Value of Nuclear Power and the Need for New Nuclear Power Plants

The United States has 103 reactors operating today. Nuclear power represented 20 percent of U.S. electricity supply 10 years ago, and it represents 20 percent of our electricity supply today, even though we have six fewer reactors than a decade ago and even though total U.S. electricity supply has increased by 25 percent in the period.

Nuclear power has maintained its market share thanks to dramatic improvements in reliability, safety, productivity and management of our nuclear plants, which today operate, on average, at 90 percent capacity factors, year in and year out. Improved productivity at our nuclear plants satisfied 20 percent of the growth in electricity demand over the last decade.

Due, in part, to excellent plant performance, we've seen steady growth in public support for nuclear energy. The industry has monitored public opinion closely since the early 1980s and two key trends are clear: First, public favorability to nuclear energy has never been higher; and second, the spread between those who support the use of nuclear energy and those opposed is widening steadily: 80 percent of Americans think nuclear power is important for our energy future and 67 percent favor the use of nuclear energy; 71 percent favor keeping the option to build more nuclear power plants. Six in 10 Americans agree that "we should definitely build more nuclear power plants in the future." Sixty-two percent said it would be acceptable to build new plants next to a nuclear power plant already operating.

The operating nuclear plants are such valuable electric generating assets that virtually all companies are planning to renew the operating licenses for these plants, as allowed by law and Nuclear Regulatory Commission regulations, and operate for an additional 20 years beyond their initial 40-year license terms. Sixty-eight U.S. reactors have now renewed their licenses, filed their formal applications, or indicated to the Nuclear Regulatory Commission that they intend to do so. The remaining 35 reactors have not yet declared because most of them are not yet old enough to do so. We believe that virtually all U.S. nuclear plants will renew their licenses and operate for an additional 20 years. At Constellation, we are proud that our Calvert Cliffs station was the first U.S. nuclear plant to renew its license. At the time, the license renewal process was a novel concept. Today, thanks to efficient management of the process by the Nuclear Regulatory Commission, it is a stable and predictable licensing action. Ten years from now, we hope and believe that the issuance of combined construction/operating licenses for new nuclear plants—a novel concept today—will be similarly efficient and predictable.

Although it has not yet started to build new nuclear plants, the industry continues to achieve small but steady increases in generating capability—either through power uprates or the restart of shutdown nuclear capacity. The Tennessee Valley Authority is restarting Unit 1 at its Browns Ferry site in northern Alabama. This is a very complex project—fully as challenging as building a new nuclear plant—and it is on schedule and within budget at the midpoint of the project.

However, despite the impressive gains in reliability and output, there are obviously limits to how much capacity we can derive from our existing nuclear power plants. The time has come to create the business conditions under which we can build new nuclear power plants in the United States. We believe there are compelling public policy reasons for new nuclear generating capacity.

First, new nuclear power plants will continue to contribute to the fuel and technology diversity that is the core strength of the U.S. electric supply system. This diversity is at risk because today's business environment and market conditions in the electric sector make investment in large, new capital-intensive technologies difficult, particularly the advanced nuclear power plants and advanced coal-fired power plants best suited to supply baseload electricity. More than 90 percent of all new electric generating capacity added over the past five years is fueled with natural gas. Natural gas has many desirable characteristics and should be part of our fuel mix, but over-reliance on any one fuel source leaves consumers vulnerable to price spikes and supply disruptions.

Second, new nuclear power plants provide future price stability that is not available from electric generating plants fueled with natural gas. Intense volatility in natural gas prices over the last several years is likely to continue, thanks partly to unsustainable demand for natural gas from the electric sector, and subjects the U.S. economy to potential damage. Although nuclear plants are capital-intensive to build, the operating costs of nuclear power plants are stable and can dampen volatility of consumer costs in the electricity market.

Third, new nuclear plants will reduce the price and supply volatility of natural gas, thereby relieving cost pressures on other users of natural gas that have no alternative fuel source.

And finally, new nuclear power plants will play a strategic role in meeting U.S. clean air goals and the nation's goal of reducing greenhouse gas emissions. New nuclear power plants produce electricity that otherwise would be supplied by oil-, gas- or coal-fired generating capacity, and thus avoid the emissions associated with that fossil-fueled capacity.

In summary, nuclear energy represents a unique value proposition: new nuclear power plants would provide large volumes of electricity—cleanly, reliably, safely and affordably. They would provide future price stability and serve as a hedge against price and supply volatility. New nuclear plants also have valuable environmental attributes. These characteristics demonstrate why new nuclear plant construction is such an imperative in the United States.

The Critical Value of the Nuclear Power 2010 Program

As I said earlier, the Department of Energy's Nuclear Power 2010 program is an essential foundation in the joint government/industry partnership to build new nuclear power plants. This committee and, in particular, you, Mr. Chairman, deserve great credit for your leadership in ensuring adequate funding for this program in the 2005 Fiscal Year.

Nuclear Power 2010 is designed to demonstrate the various components of the new licensing system for nuclear power plants, including the process of obtaining early site permits (ESPs) and combined construction/operating licenses (COLs), sharing the cost of the detailed design and engineering work necessary to prepare COLs, and resolving generic licensing issues. This work is an essential risk-management exercise because it allows industry and the NRC staff to identify and resolve scores of technical and regulatory issues that must be settled before companies can undertake high-risk, capital-intensive construction projects like new nuclear plant construction.

The Nuclear Power 2010 program is the springboard that launched a tangible and visible industry commitment to new plant construction. The industry's commitment to Nuclear Power 2010 includes a planned investment of $650 million over the next several years on design, engineering, and licensing work, which will create a business foundation for decisions to build. Three companies have applications for early site permits under review at NRC. In addition to these three, Constellation and possibly one other company are also considering ESP applications. The industry is developing at least three applications for construction/operating licenses; the first will be filed in 2007, the second and third in 2008.

As you know, the administration has proposed $56 million for the Nuclear Power 2010 program in the 2006 fiscal year. The $56 million funding proposed for 2006 is sufficient for the ESP and COL demonstration projects already underway. It is not adequate, however, to cover more recent expressions of interest from Constellation and others, and additional resources will be needed to ensure this program is viable into the future.

It is also important to recognize that Nuclear Power 2010 is a multi-year undertaking. Certainty of future funding and program stability are a big concern for industry. However, our biggest frustration with the Nuclear Power 2010 program involves the time it has taken the DOE to award the grants. In the case of NuStart, we submitted our application in April 2004 and we were not notified that we received the grant until November 2004. As for Constellation's ESP application, we submitted it almost four months ago and have yet to hear from DOE.

To support the ESP and COL demonstration projects currently underway and future projects, we anticipate that the Department of Energy will need to significantly increase funding for Nuclear Power 2010 over FY 2006 levels.

The process of developing the first COL applications, certifying new designs and completing NRC review of the first ESP and COL applications will take some time. We are looking for ways to accelerate that process, and the Congress may be able to help there—by ensuring sufficient funding for Nuclear Power 2010 and even accelerating that funding; and by providing NRC sufficient resources to ensure that the commission has adequate manpower to conduct licensing reviews and meet aggressive but realistic schedules.

The Nuclear Power 2010 Program Does Not Address All the Challenges Facing New Nuclear Plant Construction

The Department of Energy's Nuclear Power 2010 program is a necessary, but not sufficient, step toward new nuclear plant construction. We must address other challenges as well.

Our industry is not yet at the point where we can announce specific decisions to build. We are not yet at the point where we can take a $1.5 billion to $2 billion investment decision to our boards of directors. We do yet not have fully certified designs that are competitive, for example. We do not know the licensing process will work as intended: That is why we are working systematically through the ESP and COL processes. We must identify and contain the risks to make sure that nothing untoward occurs after we start building. We cannot make a $1.5–$2 billion investment decision and end up spending twice that because the licensing process failed us.

The industry believes federal investment is necessary and appropriate to offset some of the risks I've mentioned. We recommend that the federal government's investment include the incentives identified by the Secretary of Energy Advisory Board's Nuclear Energy Task Force in its recent report. That investment stimulus includes:

1. secured loans and loan guarantees;
2. transferable investment tax credits that can be taken as money is expended during construction;
3. transferable production tax credits;
4. accelerated depreciation.

This portfolio of incentives is necessary because it's clear that no single financial incentive is appropriate for all companies, because of differences in company-specific business attributes or differences in the marketplace—namely, whether the markets they serve are open to competition or are in a regulated rate structure.

The next nuclear plants might be built as unregulated merchant plants, or as regulated rate-base projects. The next nuclear plants could be built by single entities, or by consortia of companies. Business environment and project structure have a major impact on which financial incentives work best. Some companies prefer tax-related incentives. Others expect that construction loans or loan guarantees will enable them to finance the next nuclear plants.

It is important to preserve both approaches. We must maintain as much flexibility as possible.

It's important to understand why federal investment stimulus and investment protection is necessary and appropriate.

Federal investment stimulus is necessary to offset the higher first-time costs associated with the first few nuclear plants built.

Federal investment protection is necessary to manage and contain the one type of risk that we cannot manage, and that's the risk of some kind of regulatory failure (including court challenges) that delays construction or commercial operation.

The new licensing process codified in the 1992 Energy Policy Act is conceptually sound. It allows for public participation in the process at the time when that participation is most effective—before designs and sites are approved and construction begins. The new process is designed to remove the uncertainties inherent in the Part 50 process that was used to license the nuclear plants operating today. In principle, the new licensing process is intended to reduce the risk of delay in construction and commercial operation and thus the risk of unanticipated cost increases. The goal is to provide certainty before companies begin construction and place significant investment at risk.

In practice, until the process is demonstrated, the industry and the financial community cannot be assured that licensing will proceed in a disciplined manner, without unfounded intervention and delay. Only the successful licensing and commissioning of several new nuclear plants (such as proposed by the NuStart and Dominion-led consortia) can demonstrate that the licensing issues discussed above have been adequately resolved. Industry and investor concern over these potential regulatory impediments may require techniques like the standby default coverage and standby interest coverage contained in S. 887, introduced by Senators Hagel, Craig and others.

Let me also be clear on two other important issues:

1. The industry is not seeking a totally risk-free business environment. It is seeking government assistance in containing those risks that are beyond the private sector's control. The goal is to ensure that the level of risk associated with the next nuclear plants built in the U.S. generally approaches what the electric industry would consider normal commercial risks. The industry is fully prepared to accept construction management risks and operational risks that are properly within the private sector's control.

2. The industry's financing challenges apply largely to the first few plants in any series of new nuclear reactors. As capital costs decline to the "nth-of-a-kind" range, as investors gain confidence that the licensing process operates as intended and does not represent a source of unpredictable risk, follow-on plants can be financed more conventionally, without the support necessary for the first few projects. What is needed [is] limited federal investment in a limited

number of new plants for a limited period of time to overcome the financial and economic hurdles facing the first few plants built.

In summary, we believe the industry and the federal government should work together to finance the first-of-a-kind design and engineering work and to develop an integrated package of financial incentives to stimulate construction of new nuclear power plants. Any such package must address a number of factors, including the licensing/regulatory risks; the investment risks; and the other business issues that make it difficult for companies to undertake capital-intensive projects. Such a cooperative industry/government financing program is a necessary and appropriate investment in U.S. energy security.

I hope this Committee can find a place for this type of investment stimulus in the comprehensive energy legislation now being developed.

In addition, I would be remiss if I did not thank the Chairman for his support for three additional programs/provisions that will assist in the construction of new nuclear power plants in the United States:

1. Sustained progress with the Yucca Mountain project is essential. This includes the funding necessary to maintain the schedule, ensure timely filing of the license application, and access to the full receipts of the Nuclear Waste Fund.
2. Renewal of the Price-Anderson Act, which provides the framework for the industry's self-funded liability insurance. I am pleased to note that this is included in the recently House-passed energy bill.
3. Updated tax treatment of decommissioning funds that would provide comparable treatment for unregulated merchant generating companies and regulated companies. This provision, included in the energy tax legislation passed recently by the House, would allow all companies to establish qualified decommissioning funds and ensure that annual contributions to those funds are treated appropriately as a deductible business expense.

The U.S. electricity business and our nation are paying the price today for our inability to strike an appropriate balance between what was expedient and easy in the short-term, and what was prudent and more difficult in the long-term. We are paying the price today for 10 to 15 years of neglect of longer-term imperatives and the oversupply of base-load generation in the 1990s.

The United States faces a critical need for investment in energy infrastructure, including the capital-intensive, long-lead-time advanced nuclear and coal-fired power plants that represent the backbone of the U.S. electricity supply system.

While some may not realize it, the United States faces an imminent energy crisis today.

Electric power sales represent three to four percent of our gross domestic product. But the other 96 to 97 percent of our $11-trillion-a-year economy depends on that three to four percent. We cannot afford to gamble with something as fundamental as energy supply, and the biggest problem we face with nuclear energy is not having enough of it.

Karen Charman

 **NO**

Brave Nuclear World?

This year marks the 20th anniversary of the world's most notorious nuclear disaster. At 1:23 a.m. on April 26, 1986, the Number Four reactor at the Chornobyl* nuclear plant in northern Ukraine exploded and burned uncontrolled for 10 days, releasing over 100 times more radiation into the atmosphere than the Hiroshima and Nagasaki bombs combined. At least 19 million hectares were heavily contaminated in Belarus, Ukraine, and Russia. Prevailing winds and rain sent radioactive fallout over much of Europe, and it was measured as far away as Alaska. Approximately 7 million people lived in the contaminated zones in the former Soviet Union at the time of the accident (over 5 million still do). More than 350,000 were evacuated, and 2,000 villages were demolished. Radioactive foodstuffs from Belarus and Ukraine continue to show up in the markets of Moscow, and farmers on 375 properties in Wales, Scotland, and England still must grapple with restrictions due to radioactive contamination from Chornobyl.

The operating crew and the 600 men in the plant's fire service who first responded to the disaster received the highest doses of radiation, between 0.7 and 13 Sieverts (Sv). According to chernobyl.info, a United Nations Internet-based information clearinghouse, this is 700 to 13,000 times more radiation in just a few hours than the maximum dose of 1 millisievert that the European Union says people living near a nuclear power plant should be exposed to in one year. Thirty-one of those first on the scene died within three months. A total of 800,000 "liquidators"—mainly military conscripts from all over the former Soviet Union—were involved in the cleanup until 1989, and government agencies in Belarus, Ukraine, and Russia have reported that 25,000 have since died.

By any measure, Chornobyl was a horrific catastrophe and has become the icon of nuclear power's satanic side. Yet controversy has dogged the environmental and health impacts of Chornobyl from the beginning. The Soviet leadership first hoped nobody would notice the accident and then did their best to conceal and minimize the damage. As a result, a full and accurate assessment of the consequences has proved impossible. Historian and Chornobyl expert David

---

*In this article we use the Ukrainian spelling of "Chornobyl." The word may appear as "Chernobyl" in the formal names of organizations.

From *World Watch Magazine*, July/August 2006, pp. 12–18. Copyright © 2006 by Worldwatch Institute. Reprinted by permission. www.worldwatch.org

Marples wrote that authorities in the former Soviet Union classified all medical information related to the accident while denying that illnesses among cleanup workers resulted from their radiation exposure. Independent researchers have had difficulty locating significant numbers of evacuees and those who worked on the cleanup, and they have had to piece together their conclusions from interviews with medical providers, citizens, officials in the contaminated areas, others involved, and those cleanup workers they could find.

In September 2005, a report on the health impacts of Chornobyl by the UN Chernobyl Forum (seven UN agencies plus the World Bank and officials from Belarus, Ukraine, and Russia) said only 50 deaths could be attributed to Chornobyl and ultimately 4,000 will die as a result of the accident. The Chernobyl Forum report acknowledges that nine children died from thyroid cancer and that 4,000 children contracted the disease, but puts the survival rate at 99 percent. It denies any link with fertility problems and says that the most significant health problems are due to poverty, lifestyle (e.g., smoking, poor diet), and emotional problems, especially among evacuees. Marples notes that the overall assessment of the Chernobyl Forum is "a reassuring message."

The reality on the ground offers a different picture. In Gomel, a city of 700,000 in Belarus less than 80 kilometers from the destroyed reactor and one of the most severely contaminated areas, the documentary film *Chernobyl Heart* reports the incidence of thyroid cancer is 10,000 times higher than before the accident and by 1990 had increased 30-fold throughout Belarus, which received most of the radioactive fallout. Chernobyl.info states that congenital birth defects in Gomel have jumped 250 percent since the accident, and infant mortality is 300 percent higher than in the rest of Europe. A doctor interviewed in *Chernobyl Heart* says just 15 to 20 percent of the babies born at the Gomel Maternity Hospital are healthy. Chernobyl Children's Project International executive director Adi Roche says it's impossible to prove that Chornobyl caused the problems: "All we can say is the defects are increasing, the illnesses are increasing, the genetic damage is increasing." Referring to a facility for abandoned children, she adds, "places like this didn't exist before Chornobyl, so it speaks for itself." Marples, who has made numerous trips to the Chornobyl region over the past 20 years, reports the health crisis in Belarus today is so serious that there are open discussions of a "demographic doomsday."

The long-lived nature of the radionuclides and the fact that they are migrating through the contaminated regions' ecosystems into the groundwater and food chain further complicate the task of predicting the full impact of the disaster. But as the global campaign to build new reactors gains momentum, it bears asking whether a Chornobyl could happen elsewhere.

It Can't Happen Here

Nobody wants any more Chornobyls. The question is, can that outcome be ensured without phasing out nuclear power altogether? The Nuclear Energy Institute (NEI), the trade association and lobbying arm of the American nuclear power industry, says a Chornobyl-type accident is highly unlikely in the United States because of "key differences in U.S. reactor design, regulation, and emergency preparedness." Safety is assured, NEI says, by the strategy of "defense in depth," which relies on a combination of multiple, redundant, independently operating safety systems; physical barriers such as the steel reactor vessel and the typically three- to four-foot steel-reinforced concrete containment dome that would stop radiation from escaping; ongoing preventive and corrective maintenance; ongoing training of technical staff; and extensive government oversight. A key argument for nuclear power these days is the claim that nuclear reactors are safe and reliable.

The U.S. nuclear fleet has substantially increased its "capacity factor" (for a given period, the output of a generating unit as a percentage of total possible output if run at full power) since 1980. However, David Lochbaum, director of the Nuclear Safety Project at the Union of Concerned Scientists (UCS), points out that since the Three Mile Island accident in central Pennsylvania in 1979, 45 reactors (out of 104 operating U.S. units) have been shut down longer than one year to restore safety margins. A nuclear engineer by training, Lochbaum left the industry after 17 years when he and a co-worker were unable to get their employer or the Nuclear Regulatory Commission (NRC) to address safety issues at the Susquehanna plant in northeastern Pennsylvania. (The problem at that plant and others across the country was corrected after they testified before Congress.) For the last 10 years Lochbaum has been at UCS monitoring the safety of the nation's nuclear power plants and raising concerns with the NRC. He does not share the industry's confidence in the safety of the current fleet.

Nuclear power plants are incredibly complex systems that perform a relatively simple task: heating water to create steam that spins a turbine and generates electricity. Lochbaum explains that nuclear plant safety problems tend to follow a bathtub curve; the greatest number come at the beginning of a reactor's life, then after a few years when the plant is "broken in" and staff are familiar with its specific needs, problems drop and level off until the plant begins to age.

Most of the current U.S. fleet is either in or entering its twilight years, and since the late 1990s the NRC has allowed reactors to increase the amount of electricity they generate by up to 20 percent, which exceeds what the plants were designed to handle. Such "power uprates" push greater volumes of cooling water through the plant, causing more wear and tear on pipes and other equipment. The agency has also granted 20-year license extensions to 39 reactors, and most of the rest are expected to apply before their initial 40-year licenses expire. At the same time, Lochbaum says, the NRC is cutting back on the amount and frequency of safety tests and inspections. Tests that were carried out quarterly are now performed annually, and once-annual tests are now done when reactors are shut down for refueling, about every two years.

The NRC maintains that it is providing adequate oversight to keep the public safe and prevent serious reactor accidents. Gary Holahan, an official in the NRC's Office of Nuclear Reactor Regulation, explains that extended power uprates, which raise the power output of a reactor between 7 and 20 percent, require modifications to the plant that involve upgrading or replacing equipment like high pressure turbines, pumps, motors, main generators, and transformers. Before a power uprate is granted, he says, the NRC must make a finding that it complies with federal regulations and that there's "a reasonable assurance" that the health and safety of the public will not be endangered.

Lochbaum says the NRC's handling of the large power uprates illustrates the problems with its oversight. In an issue brief entitled "Snap, Crackle, & Pop: The BWR Power Uprate Experiment," he says the Quad Cities Unit 2 reactor in Illinois "literally began shaking itself apart at the higher power level" after operating for nearly 30 years at its originally licensed power level. After the uprate was approved, the steam dryer developed a 2.7 meter crack, and the component was replaced in May 2005. In early April of this year, he says Quad Cities staff found a 1.5 meter crack in the new steam dryer, and they still don't know exactly what is causing the problem. After the problem was first reported, manufacturer General Electric (GE) surveyed 15 of its other boiling water reactors around the world that had been granted 20-percent power uprates and reported problems—all vibration related—in 13.

Despite objections from the Vermont Public Service Board and one of its own commissioners, the NRC recently granted a 20-percent power uprate to the 33-year-old Vermont Yankee reactor. Stuart Richards, deputy director of the NRC's Division of Inspection, says the commission approved the power uprate after a first-time pilot engineering inspection that included an 11,000-manhour technical review failed to find any significant safety issues. "It's not the age of the plant but the physical condition of the components and how well the facility maintains the plant" that is important, he says. In addition, the power is being increased in NRC-monitored stages. But none of this reassures Lochbaum, who points out that this single-unit plant was badly maintained for much of its operating life, making it an especially poor candidate for a practice known to stress reactors. Applications for extended power uprates at six reactors are pending, and the NRC expects nine more through 2011.

The NRC says it is doing a smarter job of regulating the industry today by pinpointing areas likely to need more attention. "The agency and the industry as a whole over the last 10 to 15 years have developed better and better tools to determine what is risk-significant and what is less risk-significant," Richards explains. "So in some cases where in the past we have required more maintenance or surveillance, now those requirements are less stringent, because the components have been demonstrated to be less significant." In other cases, he says, performing too much maintenance can be detrimental, because the components are needed to do their job, and they can be tested "to the point where it causes them to have degradation."

Lochbaum says the flaw in that logic is well illustrated by a near miss at the Davis-Besse plant in Ohio. In 2002 it was discovered that boric acid escaping from the reactor for several years had eaten a 15-centimeter hole in the reactor vessel's steel lid, leaving a thin layer of stainless steel bulging outward from the pressure. Boric acid had been observed on the vessel head in 1996, 1998, and again in 2000, and NRC staff drafted an order in November 2001 to shut Davis-Besse down for a safety inspection. NRC nevertheless allowed the reactor to continue operating until February 2002, when plant workers almost accidentally found the hole. If the reactor head had burst, the reactor would likely have melted down.

Lochbaum and former NRC commissioner Peter Bradford say the Davis-Besse incident and numerous others indicate that the agency seems to be more interested in the short-term economic interest of the nuclear industry than in carrying out its mission to protect public health and safety. Bradford points to an internal NRC survey in 2002 revealing that nearly half of all NRC employees thought they would be retaliated against if they raised safety concerns, and that of those who did report problems, one-third said they suffered harassment as a result. Several critics say the safety culture of the commission changed after Senator Pete Domenici—perhaps the nuclear industry's biggest champion in Congress—told the NRC chairman in 1998 that he would cut the agency's budget by a third if it didn't reverse its "adversarial attitude" toward the industry.

Given the regulatory environment and an aging fleet of reactors, Lochbaum fears that another serious accident is inevitable. He uses the analogy of a slot machine, but instead of oranges, bananas, and cherries, the winning combination is an initiating event, like a broken pipe or a fire; equipment failure; and human error. "As the plants get older, we're starting to see the wheels come up more often, which suggests it's only a matter of time before all three come up at once," he says.

Nuclear proponents claim the new advanced designs are much safer. Unlike current plants with their multiple back-up systems, the new "passive safety" designs, such as Westinghouse's AP1000 pressurized water reactor (PWR) and GE's ABWR (Advanced Boiling Water Reactor) and ESBWR (Economic Simplified Boiling Water Reactor), rely on gravity rather than an army of pumps to push the water up into the reactor vessel and through the cooling system. Because the systems are smaller, there are fewer components to break.

Physicist Ed Lyman, a colleague of Lochbaum's at UCS who has been studying the new designs, is skeptical of the safety claims of the passive designs. He explains that slashing costs, particularly of piping and the enormously expensive steel-reinforced rebar concrete, motivated the new LWR designs, not safety. It was thought that if the power output of the reactors was lower, a gravity-driven system could dump water into the reactor core without the need for forced circulation and its miles of pipes and accompanying equipment.

Numerous tests of the gravity-driven water system for the AP600, the smaller predecessor to the AP 1000, showed the system worked, and NRC certified the design. However, the current trend in reactors is for larger units with higher output. The cost of the AP600 wasn't low enough to offset the loss in generation capacity, so none sold. The AP600 then morphed into the AP1000. GE's new "passive safety" designs followed a similar trajectory beginning with a 600-megawatt design, the SBWR (Simplified Boiling Water Reactor). The company's next design, the ABWR, was 1,350 megawatts, and its ESBWR is 1,560.

The NRC recently certified the AP1000. Lyman is concerned the agency is relying on computer modeling rather than experimental data to demonstrate that gravity-driven cooling will work in these much larger designs. He's also troubled that the containment structures of the new PWRs are less robust than those in the current fleet. NRC's Gary Holahan acknowledges that the agency relied on the tests from the AP600 and computer modeling for the AP 1000, but says that after extensive review by the commission's technical staff and the Advisory Committee on Reactor Safeguards, it determined that additional testing was not necessary. Nor does the NRC have any concerns about the thickness of the AP1000's containment dome compared to those of existing PWRs.

Increasing numbers of nuclear proponents and news reports are describing new reactor designs, such as the pebble bed modular reactor, as "accident-proof" or "fail-safe"—so safe, in fact, that the pebble bed doesn't need (or have) a containment structure. Lyman disagrees. The pebble bed is moderated by helium instead of water and uses uranium fuel pellets encased in silicon carbide, ceramic material, and graphite. He says experiments conducted at the AVR demonstration reactor in Germany, the first one ever built, have shown that the models underestimated how hot the pellets could get. The pellets degrade quickly upon reaching the critical temperature, which could lead to a large release of radiation. "So, they just don't have the predictive capacity or the understanding of how these reactors or the fuel technology work to say it's meltdown-proof," he says.

Going to Waste

In the light-water reactors that make up the majority of the world's reactor fleet, uranium fuel is loaded into the reactor, then bombarded by neutrons to trigger the nuclear fission chain reaction. After awhile all of the fissionable material in the uranium fuel is used up, or "spent." But the neutron bombardment makes the fuel two-and-a-half million times more radioactive, according to Marvin Resnikoff, a nuclear physicist with Radioactive Waste Management Associates in New York. By 2035, American nuclear power plants will have created an estimated 105,000 metric tons of spent fuel that is so deadly it must be completely isolated from the environment for tens or even hundreds of thousands of years. A Nevada state agency report put the toxicity in perspective: even after 10 years out of the reactor, an unshielded spent fuel assembly would emit enough radiation to kill somebody standing a meter away from it in less than three minutes.

No country has yet successfully dealt with its high-level nuclear waste from the first generation of reactors, let alone made plans for the added waste from a vast expansion of nuclear power. Most agree that deep geologic burial is the safest and cheapest disposal method, and countries are in various stages of picking and developing their sites. Steve Frishman of the Nevada Agency for Nuclear Projects thinks the Finns are furthest along, having chosen a permanent repository at a crystalline bedrock site at Olkiluoto that already hosts two operating reactors and one under construction. The site has been tested extensively to ensure it will effectively isolate the waste 420–520 meters down. The repository is expected to open in 2020.

The Swedes also plan to construct their repository in a deep underground granite site, though they have not yet picked the final location. They will encapsulate the spent nuclear fuel in copper canisters surrounded by bentonite clay, which swells up and makes its own watertight seal when exposed to water. Frishman says that's an extra precaution, because while they will probably find some water 500 meters underground where they plan to put the canisters, the water there is not oxygenated and would probably not corrode the canisters even if it did come in contact with them. The Swedish approach is enormously expensive, but they say results, not costs, are guiding their decisions.

These approaches seem reasonably cautious and thus offer some hope that the waste problem—which must be solved no matter what happens to nuclear power—might not be intractable. The U.S. approach, however, is less reassuring. Politics, rather than science-determined suitability, led the U.S. Department of Energy (DOE) to Yucca Mountain, a ridge of volcanic tuff on the edge of the U.S. Nuclear Test Site in the Nevada desert about 145 kilometers northwest of Las Vegas. Nevada was designated by default in an amendment (later tagged as the "Screw Nevada Bill") to the 1982 Nuclear Waste Policy Act that prohibited DOE from considering any sites in granite.

Aside from being located in the third most seismically active region in the country, Yucca Mountain is so porous that after just 50 years isotopes from atmospheric atom bomb tests have already seeped down into the underlying aquifer. But since the mountain was designated as the nation's only repository site, Frishman says DOE has been trying to engineer its way around the problems, and when it can't do that, change the rules. The latest attempt is legislation proposed by the Bush administration that among other things would raise the repository's current legal limit of 70,000 metric tons of high-level waste, remove the nuclear waste fund (money collected over the years from ratepayers by nuclear utilities to build a repository) from federal budgetary oversight, and exempt metals in the underground metal containers from regulation, leaving chromium, molybdenum, and zinc free to contaminate the area's groundwater.

On the basis of the geological instability of the site, Nevada is aggressively fighting the repository. In 2004 a federal court ruled that an Environmental Protection Agency (EPA) health standard that applied for the first 10,000 years was inadequate because the National Academy of Sciences determined that peak doses would likely occur at least 200,000 years after the waste was placed in the site. NRC therefore could not license the site. EPA has since proposed another health standard, which appears to ignore the court ruling by allowing radiation exposure to residents of the nearby Amargossa Valley to jump from a mean of 15 millirems per year for the first 10,000 years to a median value of 350 millirems per year subsequently.

Ultimately, Frishman does not believe Yucca Mountain can meet any real health-based standard. Furthermore, he points out, whatever standard is finally adopted is irrelevant once a licensing decision is made and the waste is placed in the repository: "The site is the standard."

Reprocessing

The nuclear power industry did not expect Nevada's legal challenges to be so successful, and U.S. nuclear proponents have begun to think beyond Yucca Mountain. They maintain that the development of fast breeder reactors, which create nuclear fuel by producing more fissile material than they consume, along with reprocessing the spent fuel (separating out the still-usable plutonium and uranium) will reduce the volume of waste and negate the need for geologic disposal.

Since it was originally assumed that reprocessing would be part of the nuclear fuel cycle, commercial reactors were not designed to house all of the waste they would create during their operational lives. Three commercial reprocessing facilities were built in the United States, though only one, at West Valley in western New York state, ever operated. After six years of troubled operation marked by accidents, mishandling of high-level wastes, and contamination of nearby waterways, it was shut down in 1972. In 1977 the Carter administration banned reprocessing due to concerns about nuclear weapons proliferation after India stunned the world by testing its first atomic bomb, which was made with plutonium from its reprocessing facility. According to UCS, approximately 240 metric tons of separated plutonium—enough for 40,000 nuclear weapons— was in storage worldwide as of the end of 2003. Reprocessing the U.S. spent fuel inventory would add more than 500 metric tons.

France, Britain, Russia, India, and Japan currently reprocess spent fuel, and the Bush administration is pushing to revive reprocessing in the United States. It has allocated $130 million to begin developing an "integrated spent fuel cycle," and recently announced another $250 million, primarily to develop UREX+, a technology said to address proliferation concerns by leaving the separated plutonium too radioactive for potential thieves to handle. In addition, the U.S. Congress has directed the administration to prepare a plan by 2007 to pick a technology to reprocess all of the spent fuel from commercial nuclear reactors and start building an engineering-scale demonstration plant.

UCS's Ed Lyman says it is "a myth" that reprocessing spent nuclear fuel reduces the volume of nuclear waste: "All reprocessing does is take spent fuel that's compact, and it spreads—smears—it out into dozens of different places." Current reprocessing technology uses nitric acid to dissolve the fuel assemblies and separate out plutonium and uranium. But it also leaves behind numerous extremely radioactive fission products as well as high-level liquid waste that is typically solidified in glass. In the process, a lot of radioactive gas is discharged into the environment, and there is additional liquid waste that's too expensive to isolate, he says: "So, that's just dumped into the ocean—that's the practice in France and the U.K."

Matthew Bunn, acting director of Harvard University's Project on Managing the Atom, has laid out a number of additional arguments against reprocessing. First, reprocessing spent fuel doesn't negate the need for or reduce the space required in a permanent repository, because a repository's size is determined by the heat output of the waste, not its volume. Second, reprocessing would substantially increase the cost of managing nuclear waste and wouldn't make sense economically unless uranium topped US$360 per kilogram, a price he says is not likely for several decades, if ever. Third, in this new era of heightened violence and terrorism, the proliferation risks—which would not be addressed by the new reprocessing technologies—take on even greater urgency. Fourth, reprocessing is also a dangerous technology with a track record of terrible accidents, including the world's worst pre-Chornobyl nuclear accident (a 1957 explosion at a reprocessing plant near Khystym in Russia) and other incidents in Russia and Japan as recently as the 1990s. Fifth, the new "advanced" reprocessing technologies, UREX+ and pyroprocessing, are complex, expensive, in their infancy, and unlikely to yield substantial improvements over existing reprocessing methods. Finally, Bunn argues, the Bush administration's rush to embrace reprocessing spent nuclear fuel is premature and unnecessary, since the spent fuel can remain in dry casks at nuclear power plants for decades while better solutions are sought.

Solution in Search of Problem

In the end, the case for nuclear power hinges on an evaluation of its costs and benefits compared with those of the alternatives. Many observers expect a growing ecological, social, and economic crisis unless we figure out how to retard and ultimately reverse climate change by weaning ourselves off increasingly scarce, expensive, and conflict-ridden fossil fuels. Nuclear power, until recently a pariah due to its enormous cost and demonstrated potential for serious accidents, is now touted as an indispensable solution. Nuclear power's dark side—its environmental legacy, high cost, and danger of accidents and the spread of atomic weapons—is currently downplayed. No energy system is without costs, but alternatives that avoid these particularly grave drawbacks do exist.

Space limitations preclude a comprehensive review of the alternatives, but their prospects have never been brighter. For instance, a 2005 report by the New Economics Foundation (NEF) says a broad mix of renewable energy sources that includes micro, small-, medium- and large-scale technologies applied flexibly could "more than meet all our needs." Besides solar and wind power, the mix includes tidal, wave, small-scale hydro, geothermal, biomass, and landfill gas. Rather than relying exclusively on large baseload suppliers of electricity like nuclear plants, or single sources of renewable energy that are not always available, the foundation says the key is setting up an extensive, diverse, and decentralized network of power sources, which would also be much less susceptible to widespread power outages. The total capital cost of setting up such a system has not been calculated and would vary greatly depending on whether it was implemented all at once or incrementally, building on transition technologies. According to the NEF report, a nuclear-generated kilowatthour of electricity—factoring in construction and operating costs but not waste management, insurance against accidents, or preventing nuclear weapons proliferation—costs up to 15.6 U.S. cents, significantly higher than other sources.

Governments and markets are beginning to recognize the potential of renewable energy and its use is growing rapidly. According to Worldwatch Institute's *Renewables 2005,* global investment in renewable energy in 2004 was about US$30 billion. The report points out that renewable sources generated 20 percent of the amount of electricity produced by the world's 443 operating nuclear reactors in 2004. Renewables now account for 20–25 percent of global power sector investment, and the Organisation for Economic Co-operation and

Development predicts that over the next 30 years one-third of the investment in new power sources in OECD countries will be for renewable energy.

Alternative energy guru Amory Lovins says the investment in alternatives is currently "an order of magnitude" greater than that now being spent on building new nuclear plants. Lovins has been preaching lower-cost alternatives, including energy conservation, for more than three decades, and the realization of his vision of sustainable, renewable energy is perhaps closer than ever. He argues that the current moves to re-embrace nuclear power are a huge step backwards, and that contrary to claims that we need to consider all options to deal with global warming, nuclear power would actually hinder the effort because of the high cost and the long time it would take to get enough carbon-displacing nuclear plants up and running. "In practice, keeping nuclear power alive means diverting private and public investment from the cheaper market winners—cogeneration, renewables, and efficiency—to the costly market loser. Its higher cost than competitors, per unit of net CO_2 displaced, means that every dollar invested in nuclear expansion will *worsen* climate change," he writes in his 2005 paper "Nuclear Power: Economics and Climate-Protection Potential."

[D]oubling the world's current nuclear energy output would reduce global carbon emissions by just one-seventh of the amount required to avoid the worst impacts of global warming. Researchers at the Massachusetts Institute of Technology point out that achieving even this inadequate result would require siting a permanent repository the size of Yucca Mountain every three to four years to deal with the additional waste—an enormous and expensive challenge. Given nuclear power's drawbacks, and the growth and promise of clean, lower cost, less dangerous alternatives, the case for nuclear power wobbles badly. Stripped of the pretext that nuclear power is the answer to climate change, the case essentially collapses.

POSTSCRIPT

Is It Time to Revive Nuclear Power?

Christine Laurent, in "Beating Global Warming with Nuclear Power?" *UNESCO Courier* (February 2001), notes that "For several years, the nuclear energy industry has attempted to cloak itself in different ecological robes. Its credo: nuclear energy is a formidable asset in battle against global warming because it emits very small amounts of greenhouse gases. This stance, first presented in the late 1980s when the extent of the phenomenon was still the subject of controversy, is now at the heart of policy debates over how to avoid droughts, downpours and floods." Laurent adds that it makes more sense to focus on reducing carbon emissions by reducing energy consumption. Robert Evans, "Nuclear Power: Back in the Game," *Power Engineering* (October 2005), reports that a number of power companies are now considering new nuclear power plants. See also Eliot Marshall, "Is the Friendly Atom Poised for a Comeback?", Daniel Clery, "Nuclear Industry Dares to Dream of a New Dawn," *Science* (August 19, 2005), and Josh Goodman, "The Nuclear Option," *Governing* (November 2006). Nuclear momentum is growing, says Charles Petit, "Nuclear Power: Risking a Comeback," *National Geographic* (April 2006), thanks in part to new technologies. John Geddes, "Harper Embraces the Nuclear Future," *Maclean's* (May 7, 2007), notes that the Canadian Prime Minister has endorsed the expanded use of nuclear power as a way to reduce fossil fuel use and greenhouse gas emissions. Karen Charman, "Brave Nuclear World? Part I" *World Watch* (May/June 2006), objects that producing nuclear fuel uses huge amounts of electricity derived from fossil fuels, so going nuclear can hardly prevent all releases of carbon dioxide (although using electricity derived from nuclear power would reduce the problem). She also notes that "Although no comprehensive and integrated study comparing the collateral and external costs of energy sources globally has been done, all currently available energy sources have them. . . . Burning coal—the single largest source of air pollution in the United States—causes global warming, acid rain, soot, smog, and other toxic air emissions and generates waste ash, sludge, and toxic chemicals. Landscapes and ecosystems are completely destroyed by mountaintop removal mining, while underground mining imposes high fatality, injury, and sickness rates. Even wind energy kills birds, can be noisy, and, some people complain, blights landscapes."

Michael J. Wallace tells us that there are 103 nuclear reactors operating in the United States today. Stephen Ansolabehere, et al., "The Future of Nuclear Power," *An Interdisciplinary MIT* Study (MIT, 2003), note that in 2000 there were 352 in the developed world as a whole, and a mere 15 in developing nations, and that even a very large increase in the number of nuclear power plants—to 1,000 to 1,500—will not stop all releases of carbon dioxide.

In fact, if carbon emissions double by 2050 as expected, from 6,500 to 13,000 million metric tons per year, the 1,800 million metric tons not emitted because of nuclear power will seem relatively insignificant. Nevertheless, say John M. Deutch and Ernest J. Moniz, "The Nuclear Option," *Scientific American* (September 2006), such a cut in carbon emissions would be "significant." However, says Jose Goldemberg, "The Limited Appeal of Nuclear Energy," *Scientific American* (July 2007), nuclear power is likely to be of limited value to developing nations, which will need to find other options.

The debate over the future of nuclear power is likely to remain vigorous for some time to come. But as Richard A. Meserve says in a *Science* editorial ("Global Warming and Nuclear Power," *Science* [January 23, 2004]), "For those who are serious about confronting global warming, nuclear power should be seen as part of the solution. Although it is unlikely that many environmental groups will become enthusiastic proponents of nuclear power, the harsh reality is that any serious program to address global warming cannot afford to jettison any technology prematurely. . . . The stakes are large, and the scientific and educational community should seek to ensure that the public understands the critical link between nuclear power and climate change." Paul Lorenzini, "A Second Look at Nuclear Power," *Issues in Science and Technology* (Spring 2005), argues that the goal must be energy "sufficiency for the foreseeable future with minimal environmental impact." Nuclear power can be part of the answer, but making it happen requires that we shed ideological biases. "It means ceasing to deceive ourselves about what might be possible."

Alvin M. Weinberg, former director of the Oak Ridge National Laboratory, notes in "New Life for Nuclear Power," *Issues in Science and Technology* (Summer 2003), that to make a serious dent in carbon emissions would require perhaps four times as many reactors as suggested in the MIT study. The accompanying safety and security problems would be challenging. If the challenges can be met, says John J. Taylor, retired vice president for nuclear power at the Electric Power Research Institute, in "The Nuclear Power Bargain," *Issues in Science and Technology* (Spring 2004), there are a great many potential benefits. Are new reactor technologies needed? Richard K. Lester, "New Nukes," *Issues in Science and Technology* (Summer 2006), says that better centralized waste storage is what is needed, at least in the short term.

Environmental groups such as Friends of the Earth are adamantly opposed, saying "Those who back nuclear over renewables and increased energy efficiency completely fail to acknowledge the deadly radioactive legacy nuclear power has created and continues to create" ("Nuclear Power Revival Plan Slammed," Press Release, April 18, 2004, http://www.foe-scotland.org.uk/press/pr20040408.html). However, there are signs that some environmentalists do not agree; see William M. Welch, "Some Rethinking Nuke Opposition," *USA Today* (March 23, 2007).

Internet References . . .

The Population Council

Established in 1952, the Population Council "is an international, nonprofit institution that conducts research on three fronts: biomedical, social science, and public health. This research—and the information it produces—helps change the way people think about problems related to reproductive health and population growth." Many of the Council's publications are available on-line.

http://www.popcouncil.org/

United Nations Population Division

The United Nations Population Division is responsible for monitoring and appraising a broad range of areas in the field of population. This site offers a wealth of recent data and links.

http://www.un.org/esa/population/unpop.htm

The Agriculture Network Information Center

The Agriculture Network Information Center is a guide to quality agricultural (including biotechnology) information on the Internet as selected by the National Agricultural Library, Land-Grant Universities, and other institutions.

http://www.agnic.org/

Agriculture: Genetic Resources and GMOs

The Agriculture portion of the European Union's portal Web site (EUROPA) provides information on different subjects associated with the genetic base for agricultural activities.

http://www.europa.eu/agriculture/res/index_en.htm

EarthTrends

The World Resources Institute offers data on biodiversity, fisheries, agriculture, population, and a great deal more.

http://earthtrends.wri.org

USDA National Organic Program

The USDA's National Organic Standards Board, comprised of farmers, environ-mentalists, scientists, and consumer advocates, helps the USDA develop standards for substances used in "organic" farming.

http://www.ams.usda.gov/nop/indexNet.htm

Organic Farming Research Foundation

The Organic Farming Research Foundation sponsors research related to organic farming practices, disseminates research results, and educates the public and decision-makers about organic farming issues.

http://www.ofrf.org/

Food and Population

*T*o many, *"sustainability" means arranging things so that the natural world—plants and animals, forests and coral reefs, fresh water and landscapes—can continue to exist more or less (mostly less) as it did before human beings multiplied, developed technology, and began to cause extinctions, air and water pollution, soil erosion, desertification, climate change, and so on. To others, "sustainability" means arranging things so that humankind can continue to survive and thrive, even keeping up its history of growth, technological development, and energy use—as if the environment and its resources were infinite.*

Because the two visions of "sustainability" are logically incompatible, we must struggle to find some sort of middle ground. Must we reduce the numbers of people on the planet? Their use of technology? Their standard of living? If we do, will human well-being be lessened? If we do not, how can we continue to feed everyone? Is genetic engineering the answer? Is organic farming? Both issues provoke considerable debate.

- Do Falling Birthrates Pose a Threat to Human Welfare?

- Is Genetic Engineering the Answer to Hunger?

- Is a Large-Scale Shift to Organic Farming the Best Way to Increase World Food Supply?

ISSUE 13

Do Falling Birthrates Pose a Threat to Human Welfare?

YES: Michael Meyer, from "Birth Dearth," *Newsweek* (September 27, 2004)

NO: David Nicholson-Lord, from "The Fewer the Better," *New Statesman* (November 8, 2004)

ISSUE SUMMARY

YES: Michael Meyer argues that when world population begins to decline after about 2050, economies will no longer continue to grow, government benefits will decline, young people will have to support ever more elders, and despite some environmental benefits, quality of life will suffer.

NO: David Nicholson-Lord argues that the economic problems of population decline all have straightforward solutions. A less crowded world will not suffer from the environmental ills attendant on overcrowding and will, overall, be a roomier, gentler, less materialistic place to live, with cleaner air and water.

In 1798 the British economist Thomas Malthus published his *Essay on the Principle of Population.* In it, he pointed with alarm at the way the human population grew geometrically (a hockey-stick curve of increase) and at how agricultural productivity grew only arithmetically (a straight-line increase). It was obvious, he said, that the population must inevitably outstrip its food supply and experience famine. Contrary to the conventional wisdom of the time, population growth was not necessarily a good thing. Indeed, it led inexorably to catastrophe. For many years, Malthus was something of a laughingstock. The doom he forecast kept receding into the future as new lands were opened to agriculture, new agricultural technologies appeared, new ways of preserving food limited the waste of spoilage, and the developed nations underwent a "demographic transition" from high birthrates and high death rates to low birthrates and low death rates.

Demographers initially attributed the demographic transition to increasing prosperity and predicted that as prosperity increased in countries whose populations were rapidly growing, birthrates would surely fall. Later, some scholars analyzed the historical data and concluded that the transition had actually preceded prosperity. The two views have contrasting implications for public policy designed to slow population growth—economic aid or family planning aid—but neither has worked very well. In 1994 the UN Conference on Population and Development, held in Cairo, Egypt, concluded that better results would follow from improving women's access to education and health care.

Should we be trying to slow or reverse population growth? In the 1968 book *The Population Bomb* (Ballantine Books), Paul R. Ehrlich warned that unrestricted population growth would lead to both human and environmental disaster. But some religious leaders oppose population control because, they say, family planning is against God's will. Furthermore, minority groups and developing nations contend that they are unfairly targeted by family planning programs.

The world's human population has grown tremendously. In Malthus's time, there were about 1 billion human beings on earth. By 1950 there were a little over 2.5 billion. In 2007 the tally passed 6.7 billion. By 2050, the United Nations expects world population to be about 9.2 billion (see *World Population Prospects: The 2006 Revision;* http://www.un.org/esa/population/publications/wpp2006/wpp2006.htm; United Nations, 2007).

While global agricultural production has also increased, it has not kept up with rising demand, and—because of the loss of topsoil to erosion, the exhaustion of aquifers for irrigation water, and the high price of energy for making fertilizer (among other things)—the prospect of improvement seems exceedingly slim to many observers. Paul R. Ehrlich and Anne H. Ehrlich argue in "The Population Explosion: Why We Should Care and What We Should Do About It," *Environmental Law* (Winter 1997) that "population growth may be the paramount force moving humanity inexorably towards disaster." They therefore maintain that it is essential to reduce the impact of population in terms of both numbers and resource consumption. See also William G. Moseley, "A Population Remedy Is Right Here at Home: U.S. Overconsumption Is a Bigger Issue than Fertility," *The Philadelphia Inquirer* (July 11, 2007).

Was Malthus wrong? Both environmental scientists and many economists now say that if population continues to grow, problems are inevitable. But earlier predictions of a world population of 10 or 12 billion by 2050 are no longer looking very likely. The UN's population statistics show a slowing of growth, to be followed by an actual decline in population. Stewart Brand, "Environmental Heresies," *Technology Review* (May 2005), credits much of the slowing to the increasing urbanization of the world's population (city families are smaller).

Some people worry that such a decline will not be good for human welfare. Michael Meyer argues that shrinking population will mean that the economic growth which has meant constantly increasing standards of living, must come to an end, government programs (from war to benefits for the poor and elderly) will no longer be affordable, young people will have to support ever more elders, and despite some environmental benefits, quality of life will suffer. David Nicholson-Lord argues that the economic problems of population decline all have straightforward solutions. A less crowded world will not suffer from the environmental ills attendant on overcrowding and will, overall, be a roomier, gentler, less materialistic place to live, with cleaner air and water.

YES

Michael Meyer

Birth Dearth

Everyone knows there are too many people in the world. Whether we live in Lahore or Los Angeles, Shanghai or Sao Paulo, our lives are daily proof. We endure traffic gridlock, urban sprawl and environmental depredation. The evening news brings variations on Ramallah or Darfur—images of Third World famine, poverty, pestilence, war, global competition for jobs and increasingly scarce natural resources.

Just last week the United Nations warned that many of the world's cities are becoming hopelessly overcrowded. Lagos alone will grow from 6.5 million people in 1995 to 16 million by 2015, a miasma of slums and decay where a fifth of all children will die before they are 5. At a conference in London, the U.N. Population Fund weighed in with a similarly bleak report: unless something dramatically changes, the world's 50 poorest countries will triple in size by 2050, to 1.7 billion people.

Yet this is not the full story. To the contrary, in fact. Across the globe, people are having fewer and fewer children. Fertility rates have dropped by half since 1972, from six children per woman to 2.9. And demographers say they're still falling, faster than ever. The world's population will continue to grow—from today's 6.4 billion to around 9 billion in 2050. But after that, it will go sharply into decline. Indeed, a phenomenon that we're destined to learn much more about—depopulation—has already begun in a number of countries. Welcome to the New Demography. It will change everything about our world, from the absolute size and power of nations to global economic growth to the quality of our lives.

This revolutionary transformation will be led not so much by developed nations as by the developing ones. Most of us are familiar with demographic trends in Europe, where birthrates have been declining for years. To reproduce itself, a society's women must each bear 2.1 children. Europe's fertility rates fall far short of that, according to the 2002 U.N. population report. France and Ireland, at 1.8, top Europe's childbearing charts. Italy and Spain, at 1.2, bring up the rear. In between are countries such as Germany, whose fertility rate of 1.4 is exactly Europe's average. What does that mean? If the U.N. figures are right, Germany could shed nearly a fifth of its 82.5 million people over the next 40 years—roughly the equivalent of all of east Germany, a loss of population not seen in Europe since the Thirty Years' War.

From *Newsweek*, vol. 144, issue 13, September 27, 2004, pp. 54–61. Copyright © 2004 by Newsweek. Reprinted by permission via PARS International. www.newsweek.com

And so it is across the Continent. Bulgaria will shrink by 38 percent, Romania by 27 percent, Estonia by 25 percent. "Parts of Eastern Europe, already sparsely populated, will just empty out," predicts Reiner Klingholz, director of the Berlin Institute for Population and Development. Russia is already losing close to 750,000 people yearly. (President Vladimir Putin calls it a "national crisis.") So is Western Europe, and that figure could grow to as much as 3 million a year by midcentury, if not more.

The surprise is how closely the less-developed world is following the same trajectory. In Asia it's well known that Japan will soon tip into population loss, if it hasn't already. With a fertility rate of 1.3 children per woman, the country stands to shed a quarter of its 127 million people over the next four decades, according to U.N. projections. But while the graying of Japan (average age: 42.3 years) has long been a staple of news headlines, what to make of China, whose fertility rate has declined from 5.8 in 1970 to 1.8 today, according to the U.N.? Chinese census data put the figure even lower, at 1.3. Coupled with increasing life spans, that means China's population will age as quickly in one generation as Europe's has over the past 100 years, reports the Center for Strategic and International Studies in Washington. With an expected median age of 44 in 2015, China will be older on average than the United States. By 2019 or soon after, its population will peak at 1.5 billion, then enter a steep decline. By midcentury, China could well lose 20 to 30 percent of its population every generation.

The picture is similar elsewhere in Asia, where birthrates are declining even in the absence of such stringent birth-control programs as China's. Indeed, it's happening despite often generous official incentives to procreate. The industrialized nations of Singapore, Hong Kong, Taiwan and South Korea all report subreplacement fertility, says Nicholas Eberstadt, a demographer at the American Enterprise Institute in Washington. To this list can be added Thailand, Burma, Australia and Sri Lanka, along with Cuba and many Caribbean nations, as well as Uruguay and Brazil. Mexico is aging so rapidly that within several decades it will not only stop growing but will have an older population than that of the United States. So much for the cliche of those Mexican youths swarming across the Rio Grande? "If these figures are accurate," says Eberstadt, "just about half of the world's population lives in subreplacement countries."

There are notable exceptions. In Europe, Albania and the outlier province of Kosovo are reproducing energetically. So are pockets of Asia: Mongolia, Pakistan and the Philippines. The United Nations projects that the Middle East will double in population over the next 20 years, growing from 326 million today to 649 million by 2050. Saudi Arabia has one of the highest fertility rates in the world, 5.7, after Palestinian territories at 5.9 and Yemen at 7.2. Yet there are surprises here, too. Tunisia has tipped below replacement. Lebanon and Iran are at the threshold. And though overall the region's population continues to grow, the increase is due mainly to lower infant mortality; fertility rates themselves are falling faster than in developed countries, indicating that over the coming decades the Middle East will age far more rapidly than other regions of the world. Birthrates in Africa remain high, and despite the AIDS epidemic its population is projected to keep growing. So is that of the United States.

We'll return to American exceptionalism, and what that might portend. But first, let's explore the causes of the birth dearth, as outlined in a pair of new books on the subject. "Never in the last 650 years, since the time of the Black Plague, have birth and fertility rates fallen so far, so fast, so low, for so long, in so many places," writes the sociologist Ben Wattenberg in "Fewer: How the New Demography of Depopulation Will Shape Our Future." Why? Wattenberg suggests that a variety of once independent trends have conjoined to produce a demographic tsunami. As the United Nations reported last week, people everywhere are leaving the countryside and moving to cities, which will be home to more than half the world's people by 2007. Once there, having a child becomes a cost rather than an asset. From 1970 to 2000, Nigeria's urban population climbed from 14 to 44 percent. South Korea went from 28 to 84 percent. So-called megacities, from Lagos to Mexico City, have exploded seemingly overnight. Birthrates have fallen in inverse correlation.

Other factors are at work. Increasing female literacy and enrollment in schools have tended to decrease fertility, as have divorce, abortion and the worldwide trend toward later marriage. Contraceptive use has risen dramatically over the past decade; according to U.N. data, 62 percent of married or "in union" women of reproductive age are now using some form of nonnatural birth control. In countries such as India, now the capital of global HIV, disease has become a factor. In Russia, the culprits include alcoholism, poor public health and industrial pollution that has whacked male sperm counts. Wealth discourages childbearing, as seen long ago in Europe and now in Asia. As Wattenberg puts it, "Capitalism is the best contraception."

The potential consequences of the population implosion are enormous. Consider the global economy, as Phillip Longman describes it in another recent book, "The Empty Cradle: How Falling Birthrates Threaten World Prosperity and What to Do About It." A population expert at the New America Foundation in Washington, he sees danger for global prosperity. Whether it's real estate or consumer spending, economic growth and population have always been closely linked. "There are people who cling to the hope that you can have a vibrant economy without a growing population, but mainstream economists are pessimistic," says Longman. You have only to look at Japan or Europe for a whiff of what the future might bring, he adds. In Italy, demographers forecast a 40 percent decline in the working-age population over the next four decades—accompanied by a commensurate drop in growth across the Continent, according to the European Commission. What happens when Europe's cohort of baby boomers begins to retire around 2020? Recent strikes and demonstrations in Germany, Italy, France and Austria over the most modest pension reforms are only the beginning of what promises to become a major sociological battle between Europe's older and younger generations.

That will be only a skirmish compared with the conflict brewing in China. There market reforms have removed the cradle-to-grave benefits of the planned economy, while the Communist Party hasn't constructed an adequate social safety net to take their place. Less than one quarter of the population is covered by retirement pensions, according to CSIS. That puts the burden of elder care almost entirely on what is now a generation of only children. The one-child policy has led to the so-called 4-2-1 problem, in which each child will be potentially responsible for caring for two parents and four grandparents.

Incomes in China aren't rising fast enough to offset this burden. In some rural villages, so many young people have fled to the cities that there may be nobody left to look after the elders. And the aging population could soon start to dull China's competitive edge, which depends on a seemingly endless supply of cheap labor. After 2015, this labor pool will begin to dry up, says economist Hu Angang. China will have little choice but to adopt a very Western-sounding solution, he says: it will have to raise the education level of its work force and make it more productive. Whether it can is an open question. Either way, this much is certain: among Asia's emerging economic powers, China will be the first to grow old before it gets rich.

Equally deep dislocations are becoming apparent in Japan. Akihiko Matsutani, an economist and author of a recent best seller, "The Economy of a Shrinking Population," predicts that by 2009 Japan's economy will enter an era of "negative growth." By 2030, national income will have shrunk by 15 percent. Speculating about the future is always dicey, but economists pose troubling questions. Take the legendarily high savings that have long buoyed the Japanese economy and financed borrowing worldwide, especially by the United States. As an aging Japan draws down those assets in retirement, will U.S. and global interest rates rise? At home, will Japanese businesses find themselves competing for increasingly scarce investment capital? And just what will they be investing in, as the country's consumers grow older, and demand for the latest in hot new products cools off? What of the effect on national infrastructure? With less tax revenue in state coffers, Matsutani predicts, governments will increasingly be forced to skimp on or delay repairs to the nation's roads, bridges, rail lines and the like. "Life will become less convenient," he says. Spanking-clean Tokyo might come to look more like New York City in the 1970s, when many urban dwellers decamped for the suburbs (taking their taxes with them) and city fathers could no longer afford the municipal upkeep. Can Japanese cope? "They will have to," says Matsutani. "There's no alternative."

Demographic change magnifies all of a country's problems, social as well as economic. An overburdened welfare state? Aging makes it collapse. Tensions over immigration? Differing birthrates intensify anxieties, just as the need for imported labor rises—perhaps the critical issue for the Europe of tomorrow. A poor education system, with too many kids left behind? Better fix it, because a shrinking work force requires higher productivity and greater flexibility, reflected in a new need for continuing job training, career switches and the health care needed to keep workers working into old age.

In an ideal world, perhaps, the growing gulf between the world's wealthy but shrinking countries and its poor, growing ones would create an opportunity. Labor would flow from the overpopulated, resource-poor south to the depopulating north, where jobs would continue to be plentiful. Capital and remittance income from the rich nations would flow along the reverse path, benefiting all. Will it happen? Perhaps, but that presupposes considerable labor mobility. Considering the resistance Europeans display toward large-scale immigration from North Africa, or Japan's almost zero-immigration policy, it's hard to be optimistic. Yes, attitudes are changing. Only a decade ago, for instance, Europeans also spoke of zero immigration. Today they recognize the need and, in bits and pieces, are beginning to plan for it. But will it happen on the scale required?

A more probable scenario may be an intensification of existing tensions between peoples determined to preserve their beleaguered national identities on the one hand, and immigrant groups on the other seeking to escape overcrowding and lack of opportunity at home. For countries such as the Philippines—still growing, and whose educated work force looks likely to break out of low-status jobs as nannies and gardeners and move up the global professional ladder—this may be less of a problem. It will be vastly more serious for the tens of millions of Arab youths who make up a majority of the population in the Middle East and North Africa, at least half of whom are unemployed.

America is the wild card in this global equation. While Europe and much of Asia shrinks, the United States' indigenous population looks likely to stay relatively constant, with fertility rates hovering almost precisely at replacement levels. Add in heavy immigration, and you quickly see that America is the only modern nation that will continue to grow. Over the next 45 years the United States will gain 100 million people, Wattenberg estimates, while Europe loses roughly as many.

This does not mean that Americans will escape the coming demographic whammy. They, too, face the problems of an aging work force and its burdens. (The cost of Medicare and Social Security will rise from 4.3 percent of GDP in 2000 to 11.5 percent in 2030 and 21 percent in 2050, according to the Congressional Budget Office.) They, too, face the prospect of increasing ethnic tensions, as a flat white population and a dwindling black one become gradually smaller minorities in a growing multicultural sea. And in our interdependent era, the troubles of America's major trading partners—Europe and Japan—will quickly become its own. To cite one example, what becomes of the vaunted "China market," invested in so heavily by U.S. companies, if by 2050 China loses an estimated 35 percent of its workers and the aged consume an ever-greater share of income?

America's demographic "unipolarity" has profound security implications as well. Washington worries about terrorism and failing states. Yet the chaos of today's fragmented world is likely to prove small in comparison to what could come. For U.S. leaders, Longman in "The Empty Cradle" sketches an unsettling prospect. Though the United States may have few military competitors, the technologies by which it projects geopolitical power—from laser-guided missiles and stealth bombers to a huge military infrastructure—may gradually become too expensive for a country facing massively rising social entitlements in an era of slowing global economic growth. If the war on terrorism turns out to be the "generational struggle" that national-security adviser Condoleezza Rice says it is, Longman concludes, then the United States might have difficulty paying for it.

None of this is writ, of course. Enlightened governments could help hold the line. France and the Netherlands have instituted family-friendly policies that help women combine work and motherhood, ranging from tax credits for kids to subsidized day care. Scandinavian countries have kept birthrates up with generous provisions for parental leave, health care and part-time employment. Still, similar programs offered by the shrinking city-state of Singapore—including a state-run dating service—have done little to reverse the birth dearth. Remember, too, that such prognoses have been wrong in the past. At the cusp of the postwar baby boom, demographers predicted a sharp fall in fertility and a global birth dearth. Yet even if this generation of seers turns out to be right, as seems likely, not all is bad. Environmentally, a smaller world is almost certainly a better world, whether in terms of cleaner air or, say, the return of wolves and rare flora to abandoned stretches of the East German countryside. And while people are living longer, they are also living healthier—at least in the developed world. That means they can (and probably should) work more years before retirement.

Yes, a younger generation will have to shoulder the burden of paying for their elders. But there will be compensations. As populations shrink, says economist Matsutani, national incomes may drop—but not necessarily per capita incomes. And in this realm of uncertainty, one mundane thing is probably sure: real-estate prices will fall. That will hurt seniors whose nest eggs are tied up in their homes, but it will be a boon to youngsters of the future. Who knows? Maybe the added space and cheap living will inspire them to, well, do whatever it takes to make more babies. Thus the cycle of life will restore its balance. . . .

David Nicholson-Lord  **NO**

The Fewer the Better

This is a story of two Britains and two futures. In the first Britain, the work culture dominates; the talk is of economic growth and dynamism and competing with the rest of the world. Labour is young, cheap and biddable, and driven by the urge to "succeed"—to make it in material and career terms, with the consumer goods and lifestyles to match. In the cities of this 24/7 society, population densities rise: so do crime, violence and antisocial behaviour. Outside the cities, urbanisation spreads, along with noise, congestion, the creep of human clutter and development. Unspoilt places are increasingly hard to find. Pollution gets steadily worse.

The second Britain is a quieter place. The age profile is older, the values less strident and materialistic. People work longer—they are not pensioned off in their fifties—but they save more and spend less, at least on ephemera and gadgets. They drink much less, too, and don't get involved in fights. Work is important but so are hobbies, family and community life. Cities are more spacious, roads emptier, the countryside more rural. The air and water are cleaner and there is hope of getting the weather back to normal because the planet is no longer warming so rapidly.

Which future would we prefer? The first—let us call it "UK plc"—with its economic engine revving at full speed? Or the second, where quality of life matters more: not so much a plc as a community enterprise, with the emphasis on community rather than enterprise? Most people would plump for the second. Yet we seem to be heading for the first.

What we do not admit is that the difference between the two futures is largely one of human numbers. Population is a subject we don't like to mention. In September Michael Howard, the Conservative leader, pointed out that, over the next 30 years, Britain's population would grow by 5.6 million—an increase of nearly 10 percent on the current 59 million. Immigration, running at an average net inflow of 158,000 a year in the past five years, accounts for 85 percent of this increase. Because population is forecast to rise, the government plans an extra 3.8 million houses in England over the next 20 years. But that plan is based on net immigration of 65,000 a year. If it continues at 158,000 a year, we will need 4.85 million new homes.

Howard went on to quote, with approval, the conclusions of the government's Community Cohesion Panel, which said in July that people "need

From *New Statesman*, November 8, 2004, pp. 24–26. Copyright © 2004 by New Statesman, Ltd. Reprinted by permission. All rights reserved.

sufficient time to come to terms with and accommodate incoming groups, regardless of their ethnic origin. The 'pace of change' . . . is simply too great . . . at present."

Alarmist? Electioneering? Playing the race card? In so far as these parts of Howard's speech were reported at all, that was how the left-liberal media interpreted them. Yet his figures understate the contribution of immigration to housing forecasts, because they ignore the changes in fertility and household formation resulting from a younger population. According to the Optimum Population Trust, a continuation of the 2001–2003 growth rate of 0.4 percent would result in a UK population of 71 million in 2050 and 100 million by the end of the century.

The implications of this bear examination. Given a population of "only" 65–66 million by mid-century, for example, we would need an extra nine or ten million houses by 2050—more than twice the numbers Howard was talking about, and an increase of nearly 50 percent on the current English housing stock. Should this worry us? Clearly many people think so; the government's housing plans have been a source of controversy ever since they were published. Examine this controversy in greater depth, and you will find a developing awareness of what ecologists call "carrying capacity": the balance (or lack of it) between a physical environment and the numbers it can support.

About all this, the environmental lobby is now silent. The last time such issues were deemed fit for public debate was in 1973, when a government population panel said Britain must accept "that its population cannot go on increasing indefinitely." The progressive-minded believe, on the one hand, in liberal multiculturalism and, on the other, in sustainability. They cannot resolve the conflict. The field has thus been abandoned to the political right.

The demographic facts are undeniable, however. Before the start of the current immigration surge in the 1990s, Britain's population, like that of many other developed countries, was heading for decline—as early as 2013, according to some forecasts. British women are having 1.7 children each, on average, above Germany (1.4) and Japan (1.3) but below the replacement level of 2.1. If this had been allowed to continue, with no immigration, we would be down to 30 million by 2120.

What would it be like to live in a country where population halved in the space of three or four generations? Environmentally, the case for population decline is unanswerable—less pollution, less strain on natural systems, greater national self-sufficiency, a reduction in fossil-fuel emissions, the freeing up of land for other species and higher-order human uses, such as wilderness. Psychologically, what the economist Fred Hirsch called "positional goods"—a view, an unspoilt beach, a piece of heritage—would be freer of the crowds and queues that now, for most people, mar them. Applied to social and economic life, this might reduce the awful sense of competitiveness that is a relatively recent feature of cultural life, for jobs, places at school or university, or entry to prized social institutions or niches.

Given the close association between crowding, densities, congestion and stress, and the greater distances available between people, we would also probably see less casual public aggression: less of the "rage" that emerged in the late 1980s. And, because young people are more likely to commit crimes, the ageing of society that would result from population decline would reinforce these trends. A Britain of 30 million people would almost certainly be a kindlier, more easygoing, more socially concerned place—exactly the sort of Britain that many readers of the *New Statesman* would like to see.

⋯⊙⋯

Most of the argument so far, however, has focused on the perils of decline: economic and social stagnation, the decrease in the support ratio (of workers to pensioners), emerging labour shortages and so on. Given that all of these "problems" are either illusory, fantastical or soluble it is instructive to ask why they obsess us. Why were the Tories, for example, thinking until recently about encouraging people to have babies and why does the government still envisage no upper limit on immigration? There are two answers. First, population growth is such a feature of the past two centuries—although not of preceding ages—that it has become synonymous in our minds with progress. Second, economic growth is how politicians and economists measure national success. And having more people is the quickest and easiest way to boost gross domestic product.

Much is made, therefore, of the impact of immigration on economic growth. Yet the growth comes almost entirely from additions to the national headcount. The increased wealth per person may be as little as 0.1 percent a year, according to US research.

More important is what happens when "immigrants" become "natives." This is the central fallacy of the demographic "timebomb" argument. Immigrants eventually become pensioners, and pensioners keep living longer. The only way to preserve a support ratio regarded as optimal is thus to have permanently high levels of immigration—and a population permanently, indeed infinitely, growing. David Coleman, professor of demography at Oxford University, has calculated that to keep the support ratio between pensioners and those of working age at roughly current levels would require a UK population in 2100 of approximately 300 million and rising. He calls it "the incredible in pursuit of the implausible."

And what about the world as a whole? How are developing countries, presumably expected to provide the young immigrants to the UK and other western countries, supposed to support their own old people?

New figures from the US Population Reference Bureau suggest a world population of roughly 9.3 billion in 2050, against 6.4 billion now. Studies such as the WWF's *Living Planet Report* say that by that time, humanity's footprint will be up to 220 percent of the earth's biological capacity. We would need, in other words, another couple of planets to survive. But if we manage to control global population (and it looks increasingly likely that we can) numbers will start to decline, possibly around 2070. What is the world supposed to do then? Import extra-planetary aliens to maintain the support ratio?

Even within Britain, it is hard to make a case for labour shortages when unemployment is three times the number of vacancies and economic inactivity, notably among the over-fifties, is at an all-time high. It is also hard, morally at least, to argue that we should deliberately cream off the skilled and educated workers of poorer countries—little different from people-trafficking, according to a National Health Service overseas recruiter addressing this year's Royal College of Nursing conference—or that we should bring people in because there is nobody else to sweep our streets and clean our toilets.

⋅⨀⋅

The solutions to the "problems" of population decline, in fact, lie safely within the range of realistic policy options. They include: people saving more and consuming less; governments investing more in preventative health measures, to lengthen illness-free old age; better labour productivity; a higher retirement age; drawing the economically inactive back into economic activity (with penalties for ageism); and restructuring hard-to-fill jobs to make them more attractive. Population decline creates a (relative) shortage of workers and therefore shifts power from capital to labour and raises pay rates generally, as happened after the Black Death. Isn't the left supposed to be in favour of such an outcome and against the use of immigrants to create a US-style low-wage economy?

Yet the argument is not primarily about economics. Those who advocate increases in immigration and population do so largely on the grounds that they are good for GDP. They forget, as most economists do, that they are often bad for the environment and society. Economic growth, after all, is ethically undiscriminating: the wages earned from clearing up the effects of a car crash or a pollution mishap count towards GDP in the same way as those earned from making a loaf of bread. All over the world, Britain included, population growth is generating an extraordinary range of negative effects, from climate change and resource exhaustion to the destruction of species and habitats and the poisoning of the biosphere. Deliberately boosting Britain's population, either through large-scale net immigration or by telling people to have more babies, will ultimately make it a much worse place to live in.

How big should Britain's population be? That depends which sums you do—but some calculations from the Optimum Population Trust suggest 20 million or fewer.

Before you throw up your hands in disbelief at this idea, consider the view of a liberal from another generation, John Stuart Mill, who in his *Principles of Political Economy* (1848) acknowledged the economic potential for a "great increase in population" but confessed he could see little reason for desiring it. "The density of population necessary to enable mankind to obtain . . . all the advantages both of co-operation and of social intercourse," he wrote, "has, in all the most populous countries, been attained." In 1848, the world contained just over a billion people and the population of Britain was 21 million.

POSTSCRIPT

Do Falling Birthrates Pose a Threat to Human Welfare?

Resources and population come together in the concept of "carrying capacity," defined very simply as the size of the population that the environment can support, or "carry," indefinitely, through both good years and bad. It is not the size of the population that can prosper in good times alone, for such a large population must suffer catastrophically when droughts, floods, hurricanes, or blights arrive or the climate warms or cools. It is a long-term concept, where "long-term" means not decades or generations, nor even centuries, but millennia or more. See Mark Nathan Cohen, "Carrying Capacity," *Free Inquiry* (August/September 2004). It is worth noting that many see the threat of global warming and its accompanying changes in rainfall patterns, sea level, and disease distribution as the beginning of a long run of bad years. See Sarah DeWeerdt, "Climate Change, Coming Home," *World Watch* (May/June 2007), and Jeffrey D. Sachs, "Climate Change Refugees," *Scientific American* (June 2007).

What is Earth's carrying capacity for human beings? It is surely impossible to set a precise figure on the number of human beings the world can support for the long run. As Joel E. Cohen discusses in *How Many People Can the Earth Support?* (W. W. Norton, 1995), estimates of Earth's carrying capacity range from under a billion to over a trillion. The precise number depends on our choices of diet, standard of living, level of technology, willingness to share with others at home and abroad, and desire for an intact physical, chemical, and biological environment, as well as on whether or not our morality permits restraint in reproduction and our political or religious ideology permits educating and empowering women. The key, Cohen stresses, is human choice, and the choices are ones we must make within the next 50 years. See also Joel E. Cohen, "Human Population Grows Up," *Scientific American* (September 2005). Phoebe Hall, "Carrying Capacity," *E Magazine* (March/April 2003), notes that even countries with large land areas and small populations, such as Australia and Canada, can be overpopulated in terms of resource availability. The critical resource appears to be food supply; see Russell Hopfenberg, "Human Carrying Capacity Is Determined by Food Availability," *Population & Environment* (November 2003). Jared Diamond's *Collapse: How Societies Choose to Fail or Succeed* (Viking, 2005) is an excellent presentation of how past societies have run afoul of the need to match resources and population.

Andrew R. B. Ferguson, in "Perceiving the Population Bomb," *World Watch* (July/August 2001), sets the maximum sustainable human population at about 2 billion. Sandra Postel, in the Worldwatch Institute's *State of the*

World 1994 (W. W. Norton, 1994), says, "As a result of our population size, consumption patterns, and technology choices, we have surpassed the planet's carrying capacity. This is plainly evident by the extent to which we are damaging and depleting natural capital" (including soil and water).

If population growth is now declining and world population will actually begin to decline during this century, there is clearly hope. But the question of carrying capacity remains. Most estimates of carrying capacity put it at well below the current world population size, and it will take a long time for global population to fall far enough to reach such levels. We seem to be moving in the right direction, but it remains an open question whether our numbers will decline far enough soon enough, i.e., before environmental problems become critical. On the other hand, Jeroen Van den Bergh and Piet Rietveld, "Reconsidering the Limits to World Population: Meta-analysis and Meta-prediction," *Bioscience* (March 2004), set their best estimate of human global carrying capacity at 7.7 billion, which is distinctly reassuring, at least for the environment.

How high a level will population actually reach? Fertility levels are definitely declining in many developed nations; see Alan Booth and Ann C. Crouter (eds.), *The New Population Problem: Why Families in Developed Countries Are Shrinking and What It Means* (Lawrence Erlbaum Associates, 2005). Does this mean an actual "birth dearth"? Not according to Doug Moss, "What Birth Dearth?" *E Magazine* (November/December 2006), who reminds us that there is still a large surplus of births—and therefore a growing population—in the less developed world. If we think globally, there is no shortage of people. Developed nations worried about falling fertility should not try to raise fertility levels but rather should increase immigration. Still, the end of population growth will have severe social and economic impacts; see Wolfgang Lutz, Warren C. Sanderson, and Sergei Scherbov, eds., *The End of World Population Growth in the 21st Century: New Challenges for Human Capital Formation and Sustainable Development* (Earthscan, 2004).

ISSUE 14

Is Genetic Engineering the Answer to Hunger?

YES: Gerald D. Coleman, from "Is Genetic Engineering the Answer to Hunger?" *America* (February 21, 2005)

NO: Sean McDonagh, from "Genetic Engineering Is Not the Answer," *America* (May 2, 2005)

ISSUE SUMMARY

YES: Gerald D. Coleman argues that genetically engineered crops are useful, healthful, and nonharmful, and although caution may be justified, such crops can help satisfy the moral obligation to feed the hungry.

NO: Sean McDonagh argues that those who wish to feed the hungry would do better to address land reform, social inequality, lack of credit, and other social issues.

In the early 1970s scientists first discovered that it was technically possible to move genes—the biological material that determines a living organism's physical traits—from one organism to another and thus (in principle) to give bacteria, plants, and animals new features. Most researchers in molecular genetics were excited by the potentialities that suddenly seemed within their reach. However, a few researchers—as well as many people outside the field—were disturbed by the idea; they thought that genetic mix-and-match games might spawn new diseases, weeds, and pests. Researchers in support of genetic experimentation responded by declaring a moratorium on their own work until suitable safeguards (in the form of government regulations) could be devised.

A 1987 National Academy of Sciences report said that genetic engineering posed no unique hazards. And, despite continuing controversy, by 1989 the technology had developed tremendously: researchers could obtain patents for mice with artificially added genes ("transgenic" mice); firefly genes had been added to tobacco plants to make them glow (faintly) in the dark; and growth hormone produced by genetically engineered bacteria was being used to grow low-fat pork and increase milk production in cows. The growing biotechnology industry promised more productive crops that made their own fertilizer and pesticide. Proponents argued that genetic engineering was in no significant way different from traditional selective breeding. Critics argued that genetic engineering was unnatural and violated the rights of both plants and animals to their "species integrity"; that expensive, high-tech, tinkered animals gave the competitive advantage to big agricultural corporations and drove small farmers out of business; and that putting human genes into animals, plants, or bacteria was downright offensive. Trey Popp, "God and the New Foodstuffs," *Science & Spirit* (March/April 2006), discusses objections to genetically modified rice containing human genes. For a summary of events related to the development of agricultural biotechnology, see "Biotechnology Timeline: Chronology of Key Events," *International Debates* (March 2006).

In 1992 the U.S. Office of Science and Technology issued guidelines to bar regulations that are based on the assumption that genetically engineered crops pose greater risks than similar crops produced by traditional breeding methods. The result was the rapid commercial introduction of crops that were genetically engineered to make the bacterial insecticide Bt and to resist herbicides and disease, among other things. Rice has now been engineered for herbicide and stress resistance, attempts have been made to improve nutrient content and yield, and "Transgenic rice can serve as a biofactory for the production of molecules of pharmaceutical and industrial utility. The drive to apply transgenic rice for public good as well as commercial gains has fueled research to an all time high"; see Hitesh Kathuria, et al., "Advances in Transgenic Rice Biotechnology," *Critical Reviews in Plant Sciences* (2007). In 2006, some 70 engineered crop varieties were grown on over 102 million hectares (252 million acres) in 22 countries (see GMO-Compass.org, http://www.gmo-compass.org/eng/agri_biotechnology/gmo_planting/). Commercial GM crops are chiefly soybeans, corn, cotton, and canola. Sales of genetically engineered crop products are expected to reach $210 billion by 2014.

Skepticism about the benefits remains, but in 2000, the national academies of science of the United States, the United Kingdom, China, Brazil, India, Mexico, and the third world recognized that though the use of genetically modified crops has some worrisome potentials that deserve further research and continuing caution, those crops hold the potential to feed the world during the twenty-first century while also protecting the environment (Royal Society of London, et al., "Transgenic Plants and World Agriculture," July 2000; http://books.nap.edu/catalog.php?record_id=9889). Genetic engineering is also poised to improve turfgrass and help produce biofuels (see Derek Burke, "Biofuels—Is There a Role for GM?" *Biologist*, February 2007).

The following selections illustrate the different current perspectives from within the Roman Catholic Church. Gerald D. Coleman argues that genetically engineered crops are useful, healthful, and nonharmful, and though caution may be justified, such crops can help satisfy the moral obligation to feed the hungry. Sean McDonagh, a priest who writes frequently on environmental matters, argues that ethical and environmental problems connected to the implementation of genetically engineered crops stand in the way of their promise. Those who wish to feed the hungry would do better to address land reform, social inequality, lack of credit, and other social issues.

YES

Gerald D. Coleman

Is Genetic Engineering the Answer to Hunger?

Both the developed and developing worlds are facing a critical moral choice in the controversial issue of genetically modified food, also known as genetically modified organisms and genetically engineered crops. Critics of these modifications speak dismissively of biotech foods and genetic pollution. On the other hand, proponents like Nina Federoff and Nancy Marie Brown, authors of *Mendel in the Kitchen: A Scientist's View of Genetically Modified Foods* (2004), promote genetically modified organisms (GMs or GMOs) as "the miracle of seed science and fertilizers."

To mark the 20th anniversary of U.S. diplomatic relations with the Holy See, the U.S. Embassy to the Holy See, in cooperation with the Pontifical Academy of Sciences, hosted a conference last fall at Rome's Gregorian University on "Feeding a Hungry World: The Moral Imperative of Biotechnology." Archbishop Renato Martino, who heads the Pontifical Council for Justice and Peace and has been a strong and outspoken proponent of GMOs, told Vatican Radio: "The problem of hunger involves the conscience of every man. For this reason the Catholic Church follows with special interest and solicitude every development in science to help the solution of a plight that affects . . . humanity."

Americans have grown accustomed, perhaps unwittingly, to GMO products. In the United States, for example, 68 percent of the soybeans, 70 percent of the cotton crop, 26 percent of corn and 55 percent of canola are genetically engineered. GMOs represent an estimated 60 percent of all American processed foods. A recent study by the National Center for Food and Agriculture found that farmers in the United States investing in biotech products harvested 5.3 billion additional pounds of crops and realized $22 billion in increased income. Most of the world's beer and cheese is made with GMOs, as are hundreds of medications. In an article published last October, James Nicholson, then U.S. Ambassador to the Vatican and an aggressive promoter of U.S. policy in Vatican circles, wrote that "millions of Americans, Canadians, Australians, Argentines and other people have been eating genetically modified food for nearly a decade—without one proven case of an illness, allergic reaction or even the hiccups. . . . Mankind has been genetically altering food throughout human history." And according to its supporters, biotechnology helps the environment by reducing the use of pesticides and tilling.

The World Health Association recently reported that more than 3.7 billion people around the world are now malnourished, the largest number in history. To this, opponents of GMOs reply that the "real problems" causing hunger, especially in the developing world, are poverty, lack of education and training, unequal land distribution and lack of access to markets. The moral point they advance is that distribution, not production, is the key to solving hunger.

Another significant moral issue relates to "intellectual property policies" and the interest of companies in licensing potentially valuable discoveries. The Rev. Giulio Albanese, head of the Missionary News Agency, insists that unless the problem of intellectual property is resolved in favor of the poor, it represents a "provocation" to developing countries: "The concern of many in the missionary world over the property rights to GM seeds . . . cannot but accentuate the dependence of the poor nations on the rich ones." In response to this concern, a proposal was made recently (reported in *Science* magazine on March 19) that research universities cooperate to seek open licensing provisions that would allow them to share their intellectual property through a "developing-country license." Universities would still retain rights for research and education and maintain negotiating power with the biotechnology and pharmaceutical industries. Catholic social ethics would support this type of proposal, since it places the good of people over amassing profit.

Three moral paths suggest themselves:

1. *Favor the use of GMOs.* Nobel Prize winner Norman Borlaug, who developed the Green Revolution wheat and rice strains, recently wrote: "Biotechnology absolutely should be part of Africa's agricultural reform. African leaders would be making a grievous error if they turn their back on it." Proponents at the Rome conference agreed, arguing that the use of GMOs decreases pesticide use, creates more nutrient-filled crops that require less water and have greater drought resistance, produces more food at a lower cost and uses less land. One small-scale South African farmer concluded, "We need this technology. We don't want always to be fed food aid. We want access to this technology so that one day we can also become commercial farmers."

   This position concludes that the use of GMOs amounts to a moral obligation.

2. *Condemn the use of GMOs.* Many Catholic bishops take an opposing stance. Perhaps the clearest statement comes from the National Conference of Bishops of Brazil and their Pastoral Land Commission. Their argument is threefold: the use of GMOs involves potential risks to human health; a small group of large corporations will be the greatest beneficiaries, with grave damage to the family farmers; and the environment will be gravely damaged.

   The bishops of Botswana, South Africa and Swaziland agree: "We do not believe that agro-companies or gene technologies will help our farmers to produce the food that is needed in the 21st century." Roland Lesseps and Peter Henriot, two Jesuits working in

Zambia who are experts on agriculture in the developing world, state their opposition on principle: "Nature is not just useful to us as humans, but is valued and loved in itself, for itself, by God in Christ. . . . The right to use other creatures does not give us the right to abuse them."

In a similar but distinct criticism, the executive director of the U.S. National Catholic Rural Conference, David Andrews, C.S.C., feels that "the Pontifical Academy of Sciences has allowed itself to be subordinated to the U.S. government's insistent advocacy of biotechnology and the companies which market it." Sean McDonagh states: "With patents [on genetically engineered food], farmers will never own their own food. . . ." He believes that "corporate greed" is at the heart of the GMO controversy. Biowatch's Elfrieda Pschorn-Strauss agrees: "With GM crops, small-scale farmers will become completely reliant on and controlled by big foreign companies for their food supply."

This position concludes that the use of GMOs is morally irresponsible.

3. *Approach the use of GMOs with caution.* Two years ago Pope John Paul II declared that GMO agriculture could not be judged solely on the basis of "short-term economic interests," but needed to be subject to "a rigorous scientific and ethical process of verification." This cautionary stance has been adopted by the Catholic Bishops Conference of the Philippines in urging its government to postpone authorization of GMO corn until comprehensive studies have been made: "We have to be careful because, once it is there, how can we remedy its consequences?"

In 2003 the Rural Life Committee of the North Dakota Conference of Churches also called for "rigorous examination" to understand fully the outcomes of the use of GMOs. This document endorses the "Precautionary Principle" formulated in 1992 by the United Nations Conference on Environment and Development in order to avoid "potential harm and unforeseen and unintended consequences."

This view mandates restraint and places the fundamental burden on demonstrating safety. The arguments are based on three areas of concern: the impact on the natural environment, the size of the benefit to the small farmer if the owners and distributors are giant companies like Bristol-Myers and Monsanto and the long-term effects of GMOs on human and animal health and nutrition.

This position concludes that the use of GMOs should be approached with caution.

While the "Precautionary Principle" seems prudent, there is simultaneously a strong moral argument that a war on hunger is a grave, universal need. Last year, 10 million people died of starvation. Every 3.6 seconds someone dies from hunger—24,000 people each day. Half of sub-Saharan Africans are malnourished, and this number is expected to increase to 70 percent by 2010. It was a moral disgrace that in 2002 African governments gave in to GMO opponents and returned to the World Food Program tons of GMO corn simply because it was produced in the U.S. by biotechnology.

The Roman conference gives solid reasons that GMOs are useful, healthful and nonharmful. After all, organisms have been exchanging genetic information for centuries. The tomato, corn and potato would not exist today if human engineering had not transferred genes between species.

The *Catechism of the Catholic Church* teaches that we have a duty to "make accessible to each what is needed to lead a truly human life." The very first example given is food. In *Populorum Progressio* (1967), *Sollicitude Rei Socialis* (1987) and *Centesimus Annus* (1991), Paul VI and later John Paul II forcefully insisted that rich countries have an obligation to help the poor, just as global economic interdependence places us on a moral obligation to be in solidarity with poor nations. Likewise, The Challenge of Faithful Citizenship, published by the U.S. bishops in 2004, argues that the church's preferential option for the poor entails "a moral responsibility to commit ourselves to the common good at all levels."

At the same time, it is critical that farmers in developing countries not become dependent on GMO seeds patented by a small number of companies. Intellectual knowledge must be considered the common patrimony of the entire human family. As the U.S. bishops have stated, "Both public and private entities have an obligation to use their property, including intellectual and scientific property, to promote the good of all people" (*For I Was Hungry and You Gave Me Food, 2003*).

The Catholic Church sees deep sacramental significance in wheat and bread, and insists on the absolute imperative to feed and care for the poor of the world. A vital way to promote and ensure the dignity of every human being is to enable them to have their daily bread.

Sean McDonagh

NO

Genetic Engineering Is Not the Answer

In 1992 the then-chief executive of Monsanto, Robert Shapiro, told the *Harvard Business Review* that genetically modified crops will be necessary to feed a growing world population. He predicted that if population levels were to rise to 10 billion, humanity would face two options: either open up new land for cultivation or increase crop yields. Since the first choice was not feasible, because we were already cultivating marginal land and in the process creating unprecedented levels of soil erosion, we would have to choose genetic engineering. This option, Shapiro argued, was merely a further improvement on the agricultural technologies that gave rise to the Green Revolution that saved Asia from food shortages in the 1960s and 1970s.

Genetically engineered crops might seem an ideal solution. Yet both current data and past examples show problems and provoke doubts as to their necessity.

The Green Revolution

The Green Revolution involved the production of hybrid seeds by crossing two genetically distant parents, which produced an offspring plant that gave increased yield. Critics of genetic engineering question the accepted wisdom that its impact has been entirely positive. Hybrid seeds are expensive and heavily reliant on fertilizers and pesticides. And because they lose their vigor after the first planting, the farmer must purchase new seeds for each successive planting.

In his book *Geopolitics and the Green Revolution*, John H. Perkins describes the environmentally destructive and socially unjust aspects of the Green Revolution. One of its most important negative effects, he says, is that it has contributed to the loss of three-quarters of the genetic diversity of major food crops and that the rate of erosion continues at close to 2 percent per year. The fundamental importance of genetic diversity is illustrated by the fact that when a virulent fungus began to destroy wheat fields in the United States and Canada in 1950, plant breeders staved off disaster by cross-breeding five Mexican wheat varieties with 12 imported ones. In the process they created a new strain that was able to resist so-called "stem rust." The loss of these varieties would have been a catastrophe for wheat production globally.

The Terminator Gene

The development by a Monsanto-owned company of what is benignly called a Technology Protection System—a more apt name is terminator technology—is another reason for asserting that the feed-the-world argument is completely spurious. Genetically engineered seeds that contain the terminator gene self-destruct after the first crop. Once again, this forces farmers to return to the seed companies at the beginning of each planting season. If this technology becomes widely used, it will harm the two billion subsistence farmers who live mainly in the poor countries of the world. Sharing seeds among farmers has been at the very heart of subsistence farming since the domestication of staple food crops 11,000 years ago. The terminator technology will lock farmers into a regime of buying genetically engineered seeds that are herbicide tolerant and insect resistant, tethering them to the chemical treadmill.

On an ethical level, a technology that, according to Professor Richard Lewontin of Harvard University, "introduces a 'killer' transgene that prevents the germ of the harvested grain from developing" must be considered grossly immoral. It is a sin against the poor, against previous generations who freely shared their knowledge of plant life with us, against nature itself and finally against the God of all creativity. To set out deliberately to create seeds that self-destruct is an abomination no civilized society should tolerate. Furthermore, there is danger that the terminator genes could spread to neighboring crops and to the wild and weedy relatives of the plant that has been engineered to commit suicide. This would jeopardize the food security of many poor people.

The current situation promoting genetically modified organisms also means supporting the patenting of living organisms—both crops and animals. I find it difficult to understand the support that Cardinal Renato Martino, prefect of the Pontifical Council for Justice and Peace, seems to be giving to genetically modified organisms, given the Catholic Church's strong pro-life position. In my book *Patenting Life? Stop!* I argue that "patenting life is a fundamental attack on the understanding of life as interconnected, mutually dependent and a gift of God to be shared with everyone. Patenting opts for an atomized, isolated understanding of life." The Indian scientist and activist Dr. Vandana Shiva believes that patented crops will lead to food dictatorship by a handful of northern transnational corporations. This would certainly be a recipe for hunger and starvation—in conflict with Catholic social teaching on food and agriculture.

No Higher Yield, No Reduction in Chemicals

Early in 2003 a researcher at the Institute of Development Studies at Sussex University in England published an analysis of the GMO crops that biotech companies are developing for Africa. Among the plants studied were cotton, maize and the sweet potato. The GMO research on the sweet potato is now approaching its 12th year and has involved the work of 19 scientists; to date it has cost $6 million. Results indicated that yield has increased by 18 percent. On the other hand, conventional sweet potato breeding, working with a small budget, has produced a virus-resistant variety with a 100 percent yield increase.

Claims that GMOs lead to fewer chemicals in agriculture are also being challenged. A comprehensive study using U.S. government data on the use of chemicals on genetically engineered crops was carried out by Charles Benbrook, head of the Northwest Science and Environmental Policy Center in Sandpoint, Idaho. He found that when GMOs were first introduced, they needed 25 percent fewer chemicals for the first three years. But in 2001, 5 percent more chemicals were sprayed compared with conventional crop varieties. Dr. Benbrook stated: "The proponents of biotechnology claim GMO varieties substantially reduce pesticide use. While true in the first few years of widespread planting, it is not the case now. There's now clear evidence that the average pound of herbicide applied per acre planted to herbicide-tolerant varieties have increased compared to the first few years."

Toward a Solution

Hunger and famine around the world have more to do with the absence of land reform, social inequality, bias against women farmers and the scarcity of cheap credit and basic agricultural tools than with lack of agribusiness super-seeds. This fact was recognized by those who attended the World Food Summit in Rome in November 1996. People are hungry because they do not have access to food production processes or the money to buy food. Brazil, for example, is the third largest exporter of food in the world, yet one-fifth of its population, over 30 million people, do not have enough food to eat. Clearly hunger there is not due to lack of food but to the unequal distribution of wealth and the fact that a huge number of people are landless.

Do the proponents of genetically engineered food think that agribusiness companies will distribute such food free to the hungry poor who have no money? There was food in Ireland during the famine in the 1840s, for example, but those who were starving had no access to it or money to buy it.

As a Columban missionary in the Philippines, I saw something similar during the drought caused by El Niño in 1983. There was a severe food shortage among the tribal people in the highlands of Mindanao. The drought destroyed their cereal crops, and they could no longer harvest food in the tropical forest because it had been cleared during the previous decades. Even during the height of the drought, an agribusiness corporation was exporting tropical fruit from the lowlands. There was also sufficient rice and corn in the lowlands, but the tribal people did not have the money to buy it. Had it not been for food aid from nongovernmental organizations, many of the tribal people would have starved.

In 1990 the World Food Program at Brown University calculated that if the global food harvests over the previous few years were distributed equitably among all the people of the world, it could provide a vegetarian diet for over

6 billion people. In contrast, a meat-rich diet, favored by affluent countries and currently available to the global elite, could feed only 2.6 billion people. Human society is going to be faced with the option of getting protein from plants or from animals. If we opt for animal protein, the consequence will be a much less equitable world, with increasing levels of human misery.

Those who wish to banish hunger should address the social and economic inequalities that create poverty and not claim that a magic-bullet technology will solve all the problems.

POSTSCRIPT

Is Genetic Engineering
the Answer to Hunger?

Kathleen Hart, in *Eating in the Dark* (Pantheon, 2002), expresses horror at the fact that "Frankenfood" is not labeled and that U.S. consumers are not as alarmed by genetically modified foods as European consumers are. The worries—and the scientific evidence to support them—are summarized by Kathryn Brown, in "Seeds of Concern," and Karen Hopkin, in "The Risks on the Table," *Scientific American* (April 2001). In the same issue, Sasha Nemecek poses the question "Does the World Need GM Foods?" to two prominent figures in the debate: Robert B. Horsch, vice president of the Monsanto Corporation and recipient of the 1998 National Medal of Technology for his work on modifying plant genes, says yes; Margaret Mellon, of the Union of Concerned Scientists, says no, adding that much more work needs to be done with regard to safety. The May 2002 U.S. General Accounting Office Report to Congressional Requesters, "Genetically Modified Foods: Experts View Regimen of Safety Tests as Adequate, but FDA's Evaluation Process Could Be Enhanced," urges more attention to verifying safety testing performed by biotechnology companies. Carl F. Jordan, in "Genetic Engineering, the Farm Crisis, and World Hunger," *Bioscience* (June 2002), says that a major problem is already apparent in the way agricultural biotechnology is widening the gap between the rich and the poor.

Charles Mann, in "Biotech Goes Wild," *Technology Review* (July/August 1999), discusses the continuing "lack of a rigorous regulatory framework to sort out the risks inherent in agricultural biotech." Margaret Kriz, in "Global Food Fight," *National Journal* (March 4, 2000), describes the January 2000 Montreal meeting, in which representatives of 130 countries reached "an agreement that requires biotechnology companies to ask permission before importing genetically altered seeds [and] forces food companies to clearly identify all commodity shipments that may contain genetically altered grain." Also see "Environmental Effects of Transgenic Plants: The Scope and Adequacy of Regulation," a report of the National Research Council's Committee on Environmental Impacts Associated With Commercialization of Transgenic Crops (National Academy Press, 2002).

Gregory Conko and C. S. Prakash, in "The Attack on Plant Biotechnology," in Ronald Bailey, ed., *Global Warming and Other Eco-Myths: How the Environmental Movement Uses False Science to Scare Us to Death* (Prima, 2002), say that genetically engineered crops have successfully increased yields and decreased pesticide usage, have not had notable bad environmental side effects, and will be essential for feeding the world's growing population. Jerry D. Glover, Cindy M. Cox, and John P. Reganold, "Future Farming: A Return to Roots?" *Scientific*

American (August 2007), argue that current work aimed at using biotechnology to turn present annual crops such as corn, wheat, and rice into perennial crops will boost the food supply while greatly reducing soil erosion and water pollution. Lee Silver, "Why GM Is Good for Us," *Newsweek (Atlantic Edition)* (March 20, 2006), reports that genetically modified foods may actually have fewer allergy-related problems than organic foods. On the other hand, Jeffrey M. Smith, "Frankenstein Peas," *The Ecologist* (April 2006), reports on a study that genetically modified peas fed to mice caused serious immune system problems.

The UN Food and Agriculture Organization's 2004 annual report, *The State of Food and Agriculture 2003–2004*, FAO Agriculture Series No. 35 (Rome, 2004), maintains that the biggest problem with genetically engineered crops is that the technology has focused so far on crops of interest to large commercial firms. GM crops have not spread fast enough to small farmers, although where they have been introduced into developing countries, they have yielded economic gains and reduced the use of toxic chemicals. The report concludes that there have been no adverse health or environmental consequences so far. Continued safety will require more research and governmental regulation and monitoring. Jerry Cayford notes in "Breeding Sanity into the GM Food Debate," *Issues in Science and Technology* (Winter 2004) that the issue is one of social justice as much as it is one of science. Who will control the world's food supply? Which philosophy—democratic competition or technocratic monopoly—will prevail? Theodoros H. Varzakas, et al., "The Politics and Science Behind GMO Acceptance," *Critical Reviews in Food Science & Nutrition* (June 2007), notes that in Europe a number of food scares (including Mad Cow disease) have alarmed consumers, and GMOs "despite the intense reactions from Non Governmental Organizations and consumer organizations have entered our lives with inadequate legislative measures to protect consumers from their consumption."

ISSUE 15

Is a Large-Scale Shift to Organic Farming the Best Way to Increase World Food Supply?

YES: **Brian Halweil**, from "Can Organic Farming Feed Us All?" *World Watch* (May/June 2006)

NO: **John J. Miller**, from "The Organic Myth: A Food Movement Makes a Pest of Itself," *National Review* (February 9, 2004)

ISSUE SUMMARY

YES: Brian Halweil, senior researcher at the Worldwatch Institute, argues that organic agriculture is potentially so productive that it could sustainably increase world food supply, although the future may be more likely to see a mix of organic and nonorganic techniques.

NO: John J. Miller argues that organic farming is not productive enough to feed today's population, much less larger future populations, it is prone to dangerous biological contamination, and it is not sustainable.

$\mathbf{T}$here was a time when all farming was organic. Fertilizer was compost and manure. Fields were periodically left fallow (unfarmed) to recover soil moisture and nutrients. Crops were rotated to prevent nutrient exhaustion. Pesticides were nonexistent. And farmers were at the mercy of periodic droughts (despite irrigation) and insect infestations.

As population grew, so did the demand for food. In Europe and America, the concomitant demand for fertilizer led in the nineteenth century to a booming trade in guano mined from Caribbean and Pacific islands where deposits of seabird dung could be a hundred and fifty feet thick. When the guano deposits were exhausted, there was an agricultural crisis that was relieved only by the invention of synthetic nitrogen-containing fertilizers early in the twentieth century. See Jimmy Skaggs, *The Great Guano Rush: Entrepreneurs and American Overseas Expansion* (St. Martin's, 1994), and G. J. Leigh, *The World's Greatest Fix: A History of Nitrogen and Agriculture* (Oxford, 2004). Unfortunately, synthetic fertilizers do not maintain the soil's content of organic matter (humus). This deficit can be amended by tilling in sewage sludge, but the public is not usually very receptive to the idea, partly because of the "yuck factor," but also because sewage sludge may contain human pathogens and chemical contaminants.

Synthetic pesticides, beginning with DDT, came into use in the 1940s. Their history is nicely outlined in Keith S. Delaplane, *Pesticide Usage in the United States: History, Benefits, Risks, and Trends* (University of Georgia Extension Bulletin 1121) (http://pubs.caes.uga.edu/caespubs/pubcd/B1121.htm). When they turned out to have problems—target species quickly became resistant, and when the chemicals reached human beings and wildlife on food and in water, they proved to be toxic—some people sought alternatives. These alternatives to synthetic fertilizers and pesticides (among other things) are what is usually meant by "organic farming." Proponents of organic farming have called it holistic, biodynamic, ecological, and natural and claimed a number of advantages for its practice. They say it both preserves the health of the soil and provides healthier food for people. They also argue that it should be used more, even to the point of replacing chemical-based "industrial" agriculture. Because proponents of chemicals hold that fertilizers and pesticides are essential to produce food in the quantities that a world population of over six billion people requires, and to hold food prices down to affordable levels, one strand of debate has been over whether organic agriculture can do the job.

In the following selections, Worldwatch researcher Brian Halweil argues that organic agriculture is potentially so productive that it could sustainably increase world food supply beyond current levels while strengthening the economic position of small farmers. However, the realities of the marketplace are such that many farmers are likely to use a mixture of organic and nonorganic techniques known as "low-input" farming. John J. Miller argues that organic farming is not productive enough to feed today's population, much less larger future populations, it is prone to dangerous biological contamination, and it is not sustainable. "Wishful thinking is at the heart of the organic-food movement."

YES

Brian Halweil

Can Organic Farming Feed Us All?

The only people who think organic farming can feed the world are delusional hippies, hysterical moms, and self-righteous organic farmers. Right?

Actually, no. A fair number of agribusiness executives, agricultural and ecological scientists, and international agriculture experts believe that a large-scale shift to organic farming would not only *increase* the world's food supply, but might be the only way to eradicate hunger.

This probably comes as a surprise. After all, organic farmers scorn the pesticides, synthetic fertilizers, and other tools that have become synonymous with high-yield agriculture. Instead, organic farmers depend on raising animals for manure, growing beans, clover, or other nitrogen-fixing legumes, or making compost and other sources of fertilizer that cannot be manufactured in a chemical plant but are instead grown—which consumes land, water, and other resources. (In contrast, producing synthetic fertilizers consumes massive amounts of petroleum.) Since organic farmers can't use synthetic pesticides, one can imagine that their fields suffer from a scourge of crop-munching bugs, fruitrotting blights, and plant-choking weeds. And because organic farmers depend on rotating crops to help control pest problems, the same field won't grow corn or wheat or some other staple as often.

As a result, the argument goes, a world dependent on organic farming would have to farm more land than it does today—even if it meant less pollution, fewer abused farm animals, and fewer carcinogenic residues on our vegetables. "We aren't going to feed 6 billion people with organic fertilizer," said Nobel Prize-winning plant breeder Norman Borlaug at a 2002 conference. "If we tried to do it, we would level most of our forest and many of those lands would be productive only for a short period of time." Cambridge chemist John Emsley put it more bluntly: "The greatest catastrophe that the human race could face this century is not global warming but a global conversion to 'organic firming'—an estimated 2 billion people would perish."

In recent years, organic farming has attracted new scrutiny, not just from critics who fear that a large-scale shift in its direction would cause billions to starve, but also from farmers and development agencies who actually suspect that such a shift could *better* satisfy hungry populations. Unfortunately, no one had ever systematically analyzed whether in fact a widespread shift to organic farming would run up against a shortage of nutrients and a lack of yields—until recently. The results are striking.

From *World Watch Magazine*, May/June 2006, pp. 18–24. Copyright © 2006 by Worldwatch Institute. Reprinted by permission. www.worldwatch.org

High-Tech, Low-Impact

There are actually myriad studies from around the world showing that organic farms can produce about as much, and in some settings much more, than conventional farms. Where there is a yield gap, it tends to be widest in wealthy nations, where farmers use copious amounts of synthetic fertilizers and pesticides in a perennial attempt to maximize yields. It is true that farmers converting to organic production often encounter lower yields in the first few years, as the soil and surrounding biodiversity recover from years of assault with chemicals. And it may take several seasons for farmers to refine the new approach.

But the long-standing argument that organic farming would yield just one-third or one-half of conventional farming was based on biased assumptions and lack of data. For example, the often-cited statistic that switching to organic farming in the United States would only yield one-quarter of the food currently produced there is based on a U.S. Department of Agriculture study showing that all the manure in the United States could only meet one-quarter of the nation's fertilizer needs—even though organic farmers depend on much more than just manure.

More up-to-date research refutes these arguments. For example, a recent study by scientists at the Research Institute for Organic Agriculture in Switzerland showed that organic farms were only 20 percent less productive than conventional plots over a 21-year period. Looking at more than 200 studies in North America and Europe, Per Pinstrup Andersen (a Cornell professor and winner of the World Food Prize) and colleagues recently concluded that organic yields were about 80 percent of conventional yields. And many studies show an even narrower gap. Reviewing 154 growing seasons' worth of data on various crops grown on rain-fed and irrigated land in the United States, University of California–Davis agricultural scientist Bill Liebhardt found that organic corn yields were 94 percent of conventional yields, organic wheat yields were 97 percent, and organic soybean yields were 94 percent. Organic tomatoes showed no yield difference.

More importantly, in the world's poorer nations where most of the world's hungry live, the yield gaps completely disappear. University of Essex researchers Jules Pretty and Rachel Hine looked at over 200 agricultural projects in the developing world that converted to organic and ecological approaches, and found that for all the projects—involving 9 million farms on nearly 30 million hectares—yields increased an average of 93 percent. A seven-year study from Maikaal District in central India involving 1,000 farmers cultivating 3,200 hectares found that average yields for cotton, wheat, chili, and soy were as much as 20 percent higher on the organic farms than on nearby conventionally managed ones. Farmers and agricultural scientists attributed the higher yields in this dry region to the emphasis on cover crops, compost, manure, and other practices that increased organic matter (which helps retain water) in the soils. A study from Kenya found that while organic farmers in "high-potential areas" (those with above-average rainfall and high soil quality) had lower maize yields than nonorganic farmers, organic farmers in areas with poorer resource endowments consistently outyielded conventional growers.

(In both regions, organic farmers had higher net profits, return on capital, and return on labor.)

Contrary to critics who jibe that it's going back to farming like our grandfathers did or that most of Africa already farms organically and it can't do the job, organic farming is a sophisticated combination of old wisdom and modern ecological innovations that help harness the yield-boosting effects of nutrient cycles, beneficial insects, and crop synergies. It's heavily dependent on technology—just not the technology that comes out of a chemical plant.

High-Calorie Farms

So could we make do without the chemical plants? Inspired by a field trip to a nearby organic farm where the farmer reported that he raised an amazing 27 tons of vegetables on six-tenths of a hectare in a relatively short growing season, a team of scientists from the University of Michigan tried to estimate how much food could be raised following a global shift to organic farming. The team combed through the literature for any and all studies comparing crop yields on organic farms with those on nonorganic farms. Based on 293 examples, they came up with a global dataset of yield ratios for the world's major crops for the developed and the developing world. As expected, organic farming yielded less than conventional farming in the developed world for most food categories, while studies from the developing world showed organic farming boosting yields. The team then ran two models. The first was conservative in the sense that it applied the yield ratio for the developed world to the entire planet, i.e., they assumed that every farm regardless of location would get only the lower developed-country yields. The second applied the yield ratio for the developed world to wealthy nations and the yield ratio for the developing world to those countries.

"We were all surprised by what we found," said Catherine Badgley, a Michigan paleoecologist who was one of the lead researchers. The first model yielded 2,641 kilocalories ("calories") per person per day, just under the world's current production of 2,786 calories but significantly higher than the average caloric requirement for a healthy person of between 2,200 and 2,500. The second model yielded 4,381 calories per person per day, 75 percent greater than current availability—and a quantity that could theoretically sustain a much larger human population than is currently supported on the world's farmland. (It also laid to rest another concern about organic agriculture; see sidebar . . . .)

The team's interest in this subject was partly inspired by the concern that a large-scale shift to organic farming would require clearing additional wild areas to compensate for lower yields—an obvious worry for scientists like Badgley, who studies present and past biodiversity. The only problem with the argument, she said, is that much of the world's biodiversity exists in close proximity to farmland, and that's not likely to change anytime soon. "If we simply try to maintain biodiversity in islands around the world, we will lose most of it," she said. "It's very important to make areas between those islands friendly to biodiversity. The idea of those areas being pesticide-drenched fields is just going to be a disaster for biodiversity, especially in the

ENOUGH NITROGEN TO GO AROUND?

In addition to looking at raw yields, the University of Michigan scientists also examined the common concern that there aren't enough available sources of non-synthetic nitrogen—compost, manure, and plant residues—in the world to support large-scale organic farming. For instance, in his book *Enriching the Earth: Fritz Haber, Carl Bosch, and the Transformation of World Food Production*, Vaclav Smil argues that roughly two-thirds of the world's food harvest depends on the Harber-Bosch process, the technique developed in the early 20th century to synthesize ammonia fertilizer from fossil fuels. (Smil admits that he largely ignored the contribution of nitrogen-fixing crops and assumed that some of them, like soybeans, are net users of nitrogen, although he himself points out that on average half of all the fertilizer applied globally is wasted and not taken up by plants.) Most critics of organic farming as a means to feed the world focus on how much manure—and how much related pastureland and how many head of livestock—would be needed to fertilize the world's organic farms. "The issue of nitrogen is different in different regions," says Don Lotter, an agricultural consultant who has published widely on organic farming and nutrient requirements. "But lots more nitrogen comes in as green manure than animal manure."

Looking at 77 studies from the temperate areas and tropics, the Michigan team found that greater use of nitrogen-fixing crops in the world's major agricultural regions could result in 58 million metric tons more nitrogen than the amount of synthetic nitrogen currently used every year. Research at the Rodale Institute in Pennsylvania showed that red clover used as a winter cover in an oat/wheat–corn–soy rotation, with no additional fertilizer inputs, achieved yields comparable to those in conventional control fields. Even in arid and semi-arid tropical regions like East Africa, where water availability is limited between periods of crop production, drought-resistant green manures such as pigeon peas or groundnuts could be used to fix nitrogen. In Washington state, organic wheat growers have matched their non-organic neighbor's wheat yields using the same field pea rotation for nitrogen. In Kenya, farmers using leguminous tree crops have doubled or tripled corn yields as well as suppressing certain stubborn weeds and generating additional animal fodder.

The Michigan results imply that no additional land area is required to obtain enough biologically available nitrogen, even without including the potential for intercropping (several crops grown in the same field at the same time), rotation of livestock with annual crops, and inoculation of soil with *Azobacter, Azospirillum,* and other free-living nitrogen-fixing bacteria.

tropics. The world would be able to sustain high levels of biodiversity much better if we could change agriculture on a large scale."

Badgley's team went out of the way to make its assumerions as conservative as possible: most of the studies they used looked at the yields of a single crop, even though many organic farms grow more than one crop in a field at the same time, yielding more total food even if the yield of any given crop may be lower. Skeptics may doubt the team's conclusions—as ecologists, they are likely to be sympathetic to organic farming—but a second recent study of the potential of a global shift to organic farming, led by Niels Halberg of the Danish Institute of Agricultural Sciences, came to very similar conclusions, even though the authors were economists, agronomists, and international development experts.

Like the Michigan team, Halberg's group made an assumption about the differences in yields with organic farming for a range of crops and then plugged those numbers into a model developed by the World Bank's International Food Policy Research Institute (IFPRI). This model is considered the definitive algorithm for predicting food output, farm income, and the number of hungry people throughout the world. Given the growing interest in organic farming among consumers, government officials, and agricultural scientists, the researchers wanted to assess whether a large-scale conversion to organic farming in Europe and North America (the world's primary food exporting regions) would reduce yields, increase world food prices, or worsen hunger in poorer nations that depend on imports, particularly those people living in the Third World's swelling megacities. Although the group found that total food production declined in Europe and North America, the model didn't show a substantial impact on world food prices. And because the model assumed, like the Michigan study, that organic farming would boost yields in Africa, Asia, and Latin America, the most optimistic scenario even had hunger-plagued sub-Saharan Africa exporting food surpluses.

"Modern non-certified organic farming is a potentially sustainable approach to agricultural development in areas with low yields due to poor access to inputs or low yield potential because it involves lower economic risk than comparative interventions based on purchased inputs and may increase farm level resilience against climatic fluctuations," Halberg's team concluded. In other words, studies from the field show that the yield increases from shifting to organic farming are highest and most consistent in exactly those poor, dry, remote areas where hunger is most severe. "Organic agriculture could be an important part of increased food security in sub-Saharan Africa," says Halberg.

That is, if other problems can be overcome. "A lot of research is to try to kill prejudices," Halberg says—like the notion that organic farming is only a luxury, and one that poorer nations cannot afford. "I'd like to kill this once and for all. The two sides are simply too far from each other and they ignore the realities of the global food system." Even if a shift toward organic farming boosted yields in hungry African and Asian nations, the model found that nearly a billion people remained hungry, because any surpluses were simply exported to areas that could best afford it.

Wrong Question?

These conclusions about yields won't come as a surprise to many organic farmers. They have seen with their own eyes and felt with their own hands how productive they can be. But some supporters of organic farming shy away from even asking whether it can feed the world, simply because they don't think it's the most useful question. There is good reason to believe that a global conversion to organic farming would not proceed as seamlessly as plugging some yield ratios into a spreadsheet.

To begin with, organic farming isn't as easy as farming with chemicals. Instead of choosing a pesticide to prevent a pest outbreak, for example, a particular organic farmer might consider altering his crop rotation, planting a crop that will repel the pest or one that will attract its predators—decisions that require some experimentation and long-term planning. Moreover, the IFPRI study suggested that a large-scale conversion to organic farming might require that most dairy and beef production eventually "be better integrated in cereal and other cash crop rotations" to optimize use of the manure. Bringing cows back to one or two farms to build up soil fertility may seem like a no-brainer, but doing it wholesale would be a challenge—and dumping ammonia on depleted soils still makes for a quicker fix.

Again, these are just theoretical assumptions, since a global shift to organic farming could take decades. But farmers are ingenious and industrious people and they tend to cope with whatever problems are at hand. Eliminate nitrogen fertilizer and many farmers will probably graze cows on their fields to compensate. Eliminate fungicides and farmers will look for fungus-resistant crop varieties. As more and more farmers begin to farm organically, everyone will get better at it. Agricultural research centers, universities, and agriculture ministries will throw their resources into this type of farming—in sharp contrast to their current neglect of organic agriculture, which partly stems from the assumption that organic farmers will never play a major role in the global food supply.

So the problems of adopting organic techniques do not seen insurmountable. But those problems may not deserve most of our attention; even if a mass conversion over, say, the next two decades, dramatically increased food production, there's little guarantee it would eradicate hunger. The global food system can be a complex and unpredictable beast. It's hard to anticipate how China's rise as a major importer of soybeans for its feedlots, for instance, might affect food supplies elsewhere. (It's likely to drive up food prices.) Or how elimination of agricultural subsidies in wealthy nations might affect poorer countries. (It's likely to boost farm incomes and reduce hunger.) And would less meat eating around the world free up food for the hungry? (It would, but could the hungry afford it?) In other words, "Can organic farming feed the world?" is probably not even the right question, since feeding the world depends more on politics and economics than any technological innovations.

"'Can organic farming feed the world' is indeed a bogus question," says Gene Kahn, a long-time organic farmer who founded Cascadian Farms organic foods and is now vice president of sustainable development for General Mills. "The real question is, can we feed the world? Period. Can we fix the disparities in human nutrition?" Kahn notes that the marginal difference in today's organic yields and the yields of conventional agriculture wouldn't matter if food surpluses were redistributed.

But organic farming will yield other benefits that are too numerous to name. Studies have shown, for example, that the "external" costs of organic farming—erosion, chemical pollution to drinking water, death of birds and other wildlife—are just one-third those of conventional farming. Surveys from every continent show that organic farms support many more species of birds, wild plants, insects, and other wildlife than conventional farms. And tests by several governments have shown that organic foods carry just a tiny fraction of the pesticide residues of the nonorganic alternatives, while completely banning growth hormones, antibiotics, and many additives allowed in many conventional foods. There is even some evidence that crops grown organically have considerably higher levels of health-promoting antioxidants.

There are social benefits as well. Because organic farming doesn't depend on expensive inputs, it might help shift the balance towards smaller farmers in hungry nations. A 2002 report from the UN Food and Agriculture Organization noted that "organic systems can double or triple the productivity of traditional systems" in developing nations but suggested that yield comparisons offer a "limited, narrow, and often misleading picture" since farmers in these countries often adopt organic farming techniques to save water, save money, and reduce the variability of yields in extreme conditions. A more recent study by the International Fund for Agricultural Development found that the higher labor requirements often mean that "organic agriculture can prove particularly effective in bringing redistribution of resources in areas where the labour force is underemployed. This can help contribute to rural stability."

Middle Earth

These benefits will come even without a complete conversion to a sort of organic Utopia. In fact, some experts think that a more hopeful, and reasonable, way forward is a sort of middle ground, where more and more farmers adopt the principles of organic farming even if they don't follow the approach religiously. In this scenario, both poor farmers and the environment come out way ahead. "Organic agriculture is not going to do the trick," says Roland Bunch, an agricultural extensionist who has worked for decades in Africa and the Americas and is now with COSECHA (Association of Consultants for a Sustainable, Ecological, and People-Centered Agriculture) in Honduras. Bunch knows first-hand that organic agriculture can produce more than conventional farming among poorer farmers. But he also knows that these farmers cannot get the premium prices paid for organic produce elsewhere, and that they are often unable, and unwilling, to shoulder some of the costs and risks associated with going completely organic.

Instead, Bunch points to "a middle path," of eco-agriculture, or low-input agriculture that uses many of the principles of organic farming and depends on just a small fraction of the chemicals. "These systems can immediately produce two or three times what smallholder farmers are presently producing" Bunch says. "And furthermore, it is attractive to smallholder farmers because it is less costly per unit produced." In addition to the immediate gains in food production, Bunch suggests that the benefits for the environment of this middle path will be far greater than going "totally organic," because "something like five to ten times as many smallholder farmers will adopt it

FOOD VERSUS FUEL

Sometimes, when humans try to solve one problem, they end up creating another. The global food supply is already under serious strain: more than 800 million people go hungry every day, the world's population continues to expand, and a growing number of people in the developing world are changing to a more Western, meat-intensive diet that requires more grain and water per calorie than traditional diets do. Now comes another potential stressor: concern about climate change means that more nations are interested in converting crops into biofuels as an alternative to fossil fuels. But could this transition remove land from food production and further intensify problems of world hunger?

For several reasons, some analysts say no, at least not in the near future. First, they emphasize that nearly 40 percent of global cereal crops are fed to livestock, not humans, and that global prices of grains and oil seeds do not always affect the cost of food for the hungry, who generally cannot participate in formal markets anyway.

Second, at least to date, hunger has been due primarily to inadequate income and distribution rather than absolute food scarcity. In this regard, a biofuels economy may actually help to reduce hunger and poverty. A recent UN Food and Agriculture Organization report argued that increased use of biofuels could diversify agricultural and forestry activities, attract investment in new small and medium-sized enterprises, and increase investment in agricultural production, thereby increasing the incomes of the world's poorest people.

Third, biofuel refineries in the future will depend less on food crops and increasingly on organic wastes and residues. Producing biofuels from corn stalks, rice hulls, sawdust, or waste paper is unlikely to affect food production directly. And there are drought-resistant grasses, fast-growing trees, and other energy crops that will grow on marginal lands unsuitable for raising food.

Nonetheless, with growing human appetites for both food and fuel, biofuels' long-run potential may be limited by the priority given to food production if bioenergy systems are not harmonized with food systems. The most optimistic assessments of the long-term potential of biofuels have assumed that agricultural yields will continue to improve and that world population growth and food consumption will stabilize. But the assumption about population may prove to be wrong. And yields, organic or otherwise, may not improve enough if agriculture in the future is threatened by declining water tables or poor soil maintenance.

per unit of extension and training expense, because it behooves them economically. They aren't taking food out of their kids' mouths. If five farmers eliminate half their use of chemicals, the effect on the environment will be two and one-half times as great as if one farmer goes totally organic."

And farmers who focus on building their soils, increasing biodiversity, or bringing livestock into their rotation aren't precluded from occasionally turning to biotech crops or synthetic nitrogen or any other yield-enhancing innovations in the future, particularly in places where the soils are heavily depleted. "In the end, if we do things right, we'll build a lot of organic into conventional systems," says Don Lotter, the agricultural consultant. Like Bunch, Lotter notes that such an "integrated" approach often out-performs both a strictly organic and chemical-intensive approach in terms of yield, economics, and environmental benefits. Still, Lotter's not sure we'll get there tomorrow, since the world's farming is hardly pointed in the organic direction—which could be the real problem for the world's poor and hungry. "There is such a huge area in sub-Saharan Africa and South America where the Green Revolution has never made an impact and it's unlikely that it will for the next generation of poor farmers," argues Niels Halberg, the Danish scientist who lead the IFPR] study. "It seems that agro-ecological measures for some of these areas have a beneficial impact on yields and food insecurity. So why not seriously try it out?"

John J. Miller

The Organic Myth: A Food Movement Makes a Pest of Itself

Somewhere in the cornfields of Britain, a hungry insect settled on a tall green stalk and decided to have a feast. It chewed into a single kernel of corn, filled its little belly, and buzzed off—leaving behind a tiny hole that was big enough to invite a slow decay. The agent of the decomposition was a fungus known to biologists as *Fusarium*. Farmers have a much simpler name for it: corn ear rot.

As the mold spread inside the corn, it left behind a cancer causing residue called fumonisin. This sequence repeated itself thousands and thousands of times until the infested corn was harvested and sold last year as Fresh and Wild Organic Maize Meal, Infinity Foods Organic Maize Meal, and several other products.

Consuming trace amounts of fumonisin is harmless, but large doses can be deadly. Last fall, the United Kingdom's Food Standards Agency detected alarming concentrations of the toxin in all six brands of organic corn meal subjected to testing—for a failure rate of 100 percent. The average level of contamination was almost 20 times higher than the safety threshold Europeans have set for fumonisin. The tainted products were immediately recalled from the food chain. In contrast, inspectors determined that 20 of the 24 non-organic corn meal products they examined were unquestionably safe to eat.

Despite this, millions of people continue to assume that organic foods are healthier than non-organic ones, presumably because they grow in pristine settings free from icky chemicals and creepy biotechnology. This has given birth to an energetic political movement. In 2002, activists in Oregon sponsored a ballot initiative that essentially would have required the state to slap biohazard labels on anything that wasn't produced in ways deemed fit by anti-biotech agitators. Voters rejected it, but the cause continues to percolate. Hawaiian legislators are giving serious thought to banning biotech crop tests in their state. In March, California's Mendocino County may outlaw biotech plantings altogether.

Beneath it all lurks the belief that organic food is somehow better for us. In one poll, two-thirds of Americans said that organic food is healthier. But they're wrong: It's no more nutritious than food fueled by industrial fertilizers, sprayed with synthetic pesticides, and genetically altered in science labs. And the problem isn't limited to the fungal infections that recently cursed organic

From *National Review*, February 9, 2004, pp. 35–37. Copyright © 2004 by National Review, Inc., 215 Lexington Avenue, New York, NY 10016. Reprinted by permission.

corn meal in Britain; bacteria are a major source of disease in organic food as well. To complicate matters further, organic farming is incredibly inefficient. If its appeal ever grew beyond the boutique, it would pose serious threats to the environment. Consumers who go shopping for products emblazoned with the USDA's "organic" seal of approval aren't really helping themselves or the planet—and they're arguably hurting both.

No Fear

Here's the good news: At no point in human history has food been safer than it is today, despite occasional furors like the recent one over an isolated case of mad-cow disease here in the U.S. People still get sick from food—each year, about 76 million Americans pick up at least a mild illness from what they put in their mouths—but modern agricultural methods have sanitized our fare to the point where we may eat without fear. This is true for all food, organic or otherwise.

And that raises a semantic question: What is it about organic food that makes it "organic"? The food we think of as nonorganic isn't really *in*organic, as if it were composed of rocks and minerals. In truth, everything we eat is organic—it's just not "organic" the way the organic-food movement has come to define the word.

About a decade ago, the federal government decided to wade into this semantic swamp. There was no compelling reason for this, but Congress nonetheless called for the invention of a National Organic Rule. It became official in 2002. Organic food, said the bureaucrats, is produced without synthetic fertilizers, conventional pesticides, growth hormones, genetic engineering, or germ-killing radiation. There are also varying levels of organic-ness: Special labels are available for products that are "made with organic ingredients" (which means the food is 70 percent organic), "organic" (which means 95 percent organic), and "100 percent organic." It's not at all clear what consumers are supposed to do with this information. As the Department of Agriculture explains on its website, the "USDA makes no claims that organically produced food is safer or more nutritious than conventionally produced food."

It doesn't because it can't: There's no scientific evidence whatsoever showing that organic food is healthier. So why bother with a National Organic Rule? When the thing was in development, the Clinton administration's secretary of agriculture, Dan Glickman, offered an answer: "The organic label is a marketing tool. It is not a statement about food safety." In other words, those USDA labels are intended to give people warm fuzzies for buying pricey food.

And herein lies one of the dirty secrets of organic farming: It's big business. Although the organic movement has humble origins, today most of its food isn't produced on family farms in quaint villages or even on hippie communes in Vermont. Instead, the industry has come to be dominated by large corporations that are normally the dreaded bogeymen in the minds of many organic consumers. A single company currently controls about 70 percent of the market in organic milk. California grows about $400 million per year in organic produce—and about half of it comes from just five farms. The membership list of the

Organic Trade Association includes the biggest names in agribusiness, such as Archer Daniels Midland, Gerber, and Heinz. Even Nike is a member. When its capitalist slavedrivers aren't exploiting child labor in Third World sweatshops (as they do in the fevered imaginations of campus protesters), they're promoting Nike Organics, a clothing line made from organic cotton.

The Yum Factor

There are, in fact, good reasons to eat organic food. Often it's yummier—though this has nothing to do with the fact that it's "organic." If an organic tomato tastes better than a non-organic one, the reason is usually that it has been grown locally, where it has ripened on the vine and taken only a day or two to get from the picking to the selling. Large-scale farming operations that ship fruits and vegetables across the country or the world can't compete with this kind of homegrown quality, even though they do make it possible for people in Minnesota to avoid scurvy by eating oranges in February. Conventional produce is also a good bargain because organic foods can be expensive—the profit margins are quite high, relative to the rest of the food industry.

Unfortunately, money isn't always the sole cost. Although the overwhelming majority of organic foods are safe to eat, they aren't totally risk-free. Think of it this way: Organic foods may be fresh, but they're also fresh from the manure fields.

Organic farmers aren't allowed to enrich their soils the way most non-organic farmers do, which is with nitrogen fertilizers produced through an industrial process. In their place, many farmers rely on composted manure. When they spread the stuff in their fields, they create luscious breeding grounds for all kinds of nasty microbes. Take the dreaded *E. coli,* which is capable of killing people who ingest it. A study by the Center for Global Food Issues found that although organic foods make up about 1 percent of America's diet, they also account for about 8 percent of confirmed *E. coli* cases. Organic food products also suffer from more than eight times as many recalls as conventional ones.

Some of this problem would go away if organic farmers used synthetic sprays—but this, too, is off limits. Conventional wisdom says that we should avoid food that's been drenched in herbicides, pesticides, and fungicides. Half a century ago, there was some truth in this: Sprays were primitive and left behind chemical deposits that often survived all the way to the dinner table. Today's sprays, however, are largely biodegradable. They do their job in the field and quickly break down into harmless molecules. What's more, advances in biotechnology have reduced the need to spray. About one-third of America's corn crop is now genetically modified. This corn includes a special gene that produces a natural toxin that's safe for every living creature to eat except caterpillars with alkaline guts, such as the European corn borer, a moth larva that can ravage whole harvests. This kind of biotech innovation has helped farmers reduce their reliance on pesticides by about 50 million pounds per year.

Organic farmers, of course, don't benefit from any of this. But they do have some recourse against the bugs, weeds, and fungi that can devastate a crop: They spray their plants with "natural" pesticides. These are less effective than synthetic ones and they're certainly no safer. In rat tests, rotenone—an insecticide extracted from the roots of tropical plants—has been shown to cause the symptoms of Parkinson's disease. The Environmental Protection Agency has described pyrethrum, another natural bug killer, as a human carcinogen. Everything is lethal in massive quantities, of course, and it takes huge doses of pyrethrum to pose a health hazard. Still, the typical organic farmer has to douse his crops with it as many as seven times to have the same effect as one or two applications of a synthetic compound based on the same ingredients. Then there's one of the natural fungicides preferred by organic coffee growers in Guatemala: fermented urine. Think about that the next time you're tempted to order the "special brew" at your local organic java hut.

St. Anthony's Fire, etc.

Fungicides are worth taking seriously—and not just because they might have prevented Britain's corn meal problem a few months ago. Before the advent of modern farming, when all agriculture was necessarily "organic," food-borne fungi were a major problem. One of the worst kinds was ergot, which affects rye, wheat, and other grains. During the Middle Ages, ergot poisoning caused St. Anthony's Fire, a painful contraction of blood vessels that often led to gangrene in limb extremities and sometimes death. Hallucinations also were common, as ergot contains lysergic acid, which is the crucial component of LSD. Historians of medieval Europe have documented several episodes of towns eating large batches of ergot-polluted food and falling into mass hysteria. There is some circumstantial evidence suggesting that ergot was behind the madness of the Salem witches: The warm and damp weather just prior to the infamous events of 1692 would have been ideal for an outbreak. Today, however, chemical sprays have virtually eradicated this affliction.

The very worst thing about organic farming requires the use of a word that doomsaying environmentalists have practically trademarked: It's not *sustainable*. Few activities are as wasteful as organic farming. Its yields are about half of what conventional farmers expect at harvest time. Norman Borlaug, who won the Nobel Peace Prize in 1970 for his agricultural innovations, has said, "You couldn't feed more than 4 billion people" on an all-organic diet.

If organic-food consumers think they're making a political statement when they eat, they're correct: They're declaring themselves to be not only friends of population control, but also enemies of environmental conservation. About half the world's land area that isn't covered with ice or sand is devoted to food production. Modern farming techniques have enabled this limited supply to produce increasing quantities of food. Yields have fattened so much in the last few decades that people refer to this phenomenon as the "Green Revolution," a term that has nothing to do with enviro-greenies and everything to do with improvements in breeding, fertilization, and irrigation. Yet even greater challenges lie ahead, because demographers predict that world population will rise to 9 billion by 2050. "The key is to produce more food," says Alex Avery of CGFI. "Growing more per acre leaves more land for nature." The alternative is to chop down rainforests so that we may dine on organic soybeans.

There's one more important reason that organics can't feed the world: There just isn't enough cow poop to go around. For fun, pretend that U.N. secretary-general Kofi Annan chowed on some ergot rye, decreed that all of humanity must eat nothing but organic food, and that all of humanity responded by saying, "What the heck, we'll give it a try." Forget about the population boom ahead. The immediate problem would be generating enough manure to fertilize all the brand-new, low-yield organic crop fields. There are a little more than a billion cattle in the world today, and each bovine needs between 3 and 30 acres to support it. Conservative estimates say it would take around 7 or 8 billion cattle to produce sufficient heaps of manure to sustain our all-organic diets. The United States alone would need about a billion head (or rear, to be precise). The country would be made up of nothing but cities and manure fields—and the experiment would give a whole new meaning to the term "fruited plains."

This is the sort of future the organic-food movement envisions—and its most fanatical advocates aren't planning to win any arguments on the merits or any consumers on the quality of organic food. In December, when a single U.S. animal was diagnosed with mad-cow disease, nobody was more pleased than Ronnie Cummins of the Organic Consumers Association, who has openly hoped for a public scare that would spark a "crisis of confidence" in American food. No such thing happened, but Cummins should be careful about what he wishes for: Germany's first case of mad-cow disease surfaced at a slaughterhouse that specializes in organic beef.

But then wishful thinking is at the heart of the organic-food movement. Its whole market rationale depends on the misperception that organic foods are somehow healthier for both consumers and Mother Earth. Just remember: Nature's Valley can't be found on any map. It's a state of mind.

POSTSCRIPT

Is a Large-Scale Shift to Organic Farming the Best Way to Increase World Food Supply?

Is organic food better or safer for the consumer than non-organic food? Faidon Magkos, et al., "Organic Food: Buying More Safety or Just Peace of Mind? A Critical Review of the Literature," *Critical Reviews in Food Science & Nutrition* (January 2006), report that the quality and safety of organic food are largely a matter of perception. "Relevant scientific evidence . . . is scarce, while anecdotal reports abound." Pesticide and herbicide residues may be lower, but even on non-organic food they are low. Environmental contaminants are likely to affect both organic and non-organic foods. "'Organic' does not automatically equal 'safe.'"

Is organic farming better for the environment? Soil fertility is in decline in many parts of the world; see Alfred E. Herteminck, "Assessing Soil Fertility Decline in the Tropics Using Soil Chemical Data," *Advances in Agronomy* (2006). In Africa, the situation is extraordinarily serious. According to the International Center for Soil Fertility and Agricultural Development (http://www.ifdc.org/New_Design/Whats_New/africasfailingagriculture033006.pdf), "About 75 percent of the farmland in sub-Saharan Africa is plagued by severe degradation, losing basic soil nutrients needed to grow the crops that feed Africa, according to a new report . . . on the precipitous decline in African soil health from 1980 to 2004. Africa's crisis in food production and battle with hunger are largely rooted in this 'soil health crisis.'" Proponents of organic farming argue that it is essential to relieving the crisis, but one study of changes in soil fertility, as indicated by crop yield, earthworm numbers, and soil properties, after converting from conventional to organic practices, found that different soils responded differently, with some improving and some not; see Anne Kjersti Bakken, et al., "Soil Fertility in Three Cropping Systems after Conversion from Conventional to Organic Farming, "*Acta Agriculturae Scandinavica: Section B, Soil & Plant Science* (June 2006). Richard Wood, et al., "A Comparative Study of Some Environmental Impacts of Conventional and Organic Farming in Australia," *Agricultural Systems* (September 2006), find in a comparison of organic and conventional farms "that direct energy use, energy related emissions, and greenhouse gas emissions are higher for the" former.

According to Catherine M. Cooney, "Sustainable Agriculture Delivers the Crops," *Environmental Science & Technology* (February 15, 2006), "Sustainable agriculture, such as crop rotation, organic farming, and genetically modified seeds, increased crop yields by an average of 79 percent" while also improving the lives of farmers in developing countries. Cong Tu, et al., "Responses of Soil

Microbial Biomass and N Availability to Transition Strategies from Conventional to Organic Farming Systems," *Agriculture, Ecosystems & Environment* (April 2006), note that a serious barrier to changing from conventional to organic farming, despite soil improvements, is an initial reduction in yield and increase in pests. David Pimentel, et al., "Environmental, Energetic, and Economic Comparisons of Organic and Conventional Farming Systems," *Bioscience* (July 2005), find that organic farming uses less energy, improves soil, and has yields comparable to those of conventional farming but crops probably cannot be grown as often (because of fallowing), which then reduces long-term yields. However, "because organic foods frequently bring higher prices in the marketplace, the net economic return per [hectare] is often equal to or higher than that of conventionally produced crops." This is one of the factors that prompts Craig J. Pearson to argue in favor of shifting "from conventional open or leaky systems to more closed, regenerative systems" in "Regenerative, Semiclosed Systems: A Priority for Twenty-First-Century Agriculture," *Bioscience* (May 2007).

Better economic return means that organic farming is currently good for the organic farmer. However, the advantage would disappear if the world converted to organic farming. Initial declines in yield mean the conversion would be difficult, but the longer we wait and the more population grows, the more difficult it will be. Whether the conversion is essential depends on the availability of alternative solutions to the problem, and it is worth noting that high energy prices make chemical fertilizers increasingly expensive. Sadly, Stacey Irwin, "Battle High Fertilizer Costs," *Farm Industry News* (January 2006), does not mention the possibility of using organic methods.

Internet References . . .

The Pesticide Action Network

The Pesticide Action Network North America (PANNA) challenges the global proliferation of pesticides.

http://www.panna.org/

Africa Fighting Malaria

Africa Fighting Malaria is an NGO (non-governmental organization) which seeks to educate people about the scourge of malaria and the political economy of malaria control.

http://www.fightingmalaria.org/

e.hormone

e.hormone is hosted by the Center for Bioenvironmental Research at Tulane and Xavier Universities in New Orleans. It provides accurate, timely information about environmental hormones and their impacts.

http://e.hormone.tulane.edu/

The Silicon Valley Toxics Coalition

The Silicon Valley Toxics Coalition (SVTC) was formed to engage in research, advocacy, and organizing associated with environmental and human problems caused by the rapid growth of the high-tech electronics industry.

http://svtc.etoxics.org/site/PageServer

Superfund

The U.S. Environmental Protection Agency provides a great deal of information on the Superfund program, including material on environmental justice.

http://www.epa.gov/superfund/

Yucca Mountain

The EPA also provides information on the proposed Yucca Mountain permanent nuclear waste repository.

http://www.epa.gov/radiation/yucca/

The La Hague Nuclear Reprocessing Plant

The AREVA NC La Hague site, located on the western tip of the Cotentin Peninsula in Normandy, reprocesses spent power reactor fuel to recycle reusable energy materials—uranium and plutonium—and to condition the waste into suitable final form.

http://www.cogemalahague.com/servlet/ContentServer?pagename=cogema_en/Page/page_html_libre_full_template&c=Page&cid=1039482706828

Toxic Chemicals

A great many of today's environmental issues have to do with industrial development, which expanded greatly during the twentieth century. Just since World War II, many thousands of synthetic chemicals—pesticides, plastics, and antibiotics—have flooded the environment. We have become dependent on the production and use of energy, particularly in the form of fossil fuels. We have discovered that industrial processes generate huge amounts of waste, much of it toxic. Air and water pollution have become global problems. And we have discovered that our actions may change the world for generations to come.

This section deals with two prominent controversies concerning toxic chemicals and two concerning hazardous wastes. There are others in both categories.

- Should DDT Be Banned Worldwide?

- Do Environmental Hormone Mimics Pose a Potentially Serious Health Threat?

- Is the Superfund Program Successfully Protecting Human Health from Hazardous Materials?

- Should the United States Reprocess Spent Nuclear Fuel?

ISSUE 16

Should DDT Be Banned Worldwide?

YES: Anne Platt McGinn, from "Malaria, Mosquitoes, and DDT," *World Watch* (May/June 2002)

NO: Donald R. Roberts, from Statement before the U.S. Senate Committee on Environment & Public Works, Hearing on the Role of Science in Environmental Policy-Making (September 28, 2005)

ISSUE SUMMARY

YES: Anne Platt McGinn, a senior researcher at the Worldwatch Institute, argues that although DDT is still used to fight malaria, there are other, more effective and less environmentally harmful methods. She maintains that DDT should be banned or reserved for emergency use.

NO: Donald R. Roberts argues that the scientific evidence regarding the environmental hazards of DDT has been seriously misrepresented by anti-pesticide activists. The hazards of malaria are much greater and, properly used, DDT can prevent them and save lives.

DDT is a crucial element in the story of environmentalism. The chemical was first synthesized in 1874. Swiss entomologist Paul Mueller was the first to notice that DDT has insecticidal properties, which, it was quickly realized, implied that the chemical could save human lives. It had long been known that more soldiers died during wars because of disease than because of enemy fire. During World War I, for example, some 5 million lives were lost to typhus, a disease carried by body lice. DDT was first deployed during World War II to halt a typhus epidemic in Naples, Italy. It was a dramatic success, and DDT was soon used routinely as a dust for soldiers and civilians. During and after the war, DDT was also deployed successfully against the mosquitoes that carry malaria and other diseases. In the United States cases of malaria fell from 120,000 in 1934 to 72 in 1960, and cases of yellow fever dropped from 100,000 in 1878 to none. In 1948 Mueller received the Nobel Prize for medicine and physiology because DDT had saved so many civilian lives.

DDT was by no means the first pesticide. But its predecessors—arsenic, strychnine, cyanide, copper sulfate, and nicotine—were all markedly toxic to humans. DDT was not only more effective as an insecticide, it was also less hazardous to users. It is therefore not surprising that DDT was seen as a beneficial substance. It was soon applied routinely to agricultural crops and used to control mosquito populations in American suburbs. However, insects quickly became resistant to the insecticide. (In any population of insects, some will be more resistant than others; when the insecticide kills the more vulnerable members of the population, the resistant ones are left to breed and multiply. This is an example of natural selection.) In *Silent Spring* (Houghton Mifflin, 1962), marine scientist Rachel Carson demonstrated that DDT was concentrated in the food chain and affected the reproduction of predators such as hawks and eagles. In 1972 the U.S. Environmental Protection Agency banned almost all uses of DDT (it could still be used to protect public health). Other developed countries soon banned it as well, but developing nations, especially those in the tropics, saw it as an essential tool for fighting diseases such as malaria. Roger Bate, director of Africa Fighting Malaria, argues in "A Case of the DDTs," *National Review* (May 14, 2001) that DDT remains the cheapest and most effective way to combat malaria and that it should remain available for use.

It soon became apparent that DDT is by no means the only pesticide or organic toxin with environmental effects. As a result, on May 24, 2001, the United States joined 90 other nations in signing the Stockholm Convention on Persistent Organic Pollutants (POPs). This treaty aims to eliminate from use the entire class of chemicals to which DDT belongs, beginning with the "dirty dozen," pesticides DDT, aldrin, dieldrin, endrin, chlordane, heptachlor, mirex, and toxaphene, and the industrial chemicals polychlorinated biphenyls (PCBs), hexachlorobenzene (HCB), dioxins, and furans. Since then, 59 countries, not including the United States and the European Union (EU), have formally ratified the treaty. It took effect in May 2004. Fiona Proffitt, in "U.N. Convention Targets Dirty Dozen Chemicals," *Science* (May 21, 2004), notes, "About 25 countries will be allowed to continue using DDT against malaria-spreading mosquitoes until a viable alternative is found."

In the following selection, Anne Platt McGinn, granting that malaria remains a serious problem in the developing nations of the tropics, especially Africa, contends that although DDT is still used to fight malaria in these nations, it is far less effective than it used to be. She argues that the environmental effects are also serious concerns and that DDT should be banned or reserved for emergency use. In the second selection, Professor Donald R. Roberts argues that the scientific evidence regarding the environmental hazards of DDT has been seriously misrepresented by anti-pesticide activists. The hazards of malaria are much greater and, properly used, DDT can prevent them and save lives. Efforts to prevent the use of DDT have produced a "global humanitarian disaster."

YES

Anne Platt McGinn

Malaria, Mosquitoes, and DDT

This year, like every other year within the past couple of decades, uncountable trillions of mosquitoes will inject malaria parasites into human blood streams billions of times. Some 300 to 500 million full-blown cases of malaria will result, and between 1 and 3 million people will die, most of them pregnant women and children. That's the official figure, anyway, but it's likely to be a substantial underestimate, since most malaria deaths are not formally registered, and many are likely to have escaped the estimators. Very roughly, the malaria death toll rivals that of AIDS, which now kills about 3 million people annually.

But unlike AIDS, malaria is a low-priority killer. Despite the deaths, and the fact that roughly 2.5 billion people (40 percent of the world's population) are at risk of contracting the disease, malaria is a relatively low public health priority on the international scene. Malaria rarely makes the news. And international funding for malaria research currently comes to a mere $150 million annually. Just by way of comparison, that's only about 5 percent of the $2.8 billion that the U.S. government alone is considering for AIDS research in fiscal year 2003.

The low priority assigned to malaria would be at least easier to understand, though no less mistaken, if the threat were static. Unfortunately it is not. It is true that the geographic range of the disease has contracted substantially since the mid-20th century, but over the past couple of decades, malaria has been gathering strength. Virtually all areas where the disease is endemic have seen drug-resistant strains of the parasites emerge—a development that is almost certainly boosting death rates. In countries as various as Armenia, Afghanistan, and Sierra Leone, the lack or deterioration of basic infrastructure has created a wealth of new breeding sites for the mosquitoes that spread the disease. The rapidly expanding slums of many tropical cities also lack such infrastructure; poor sanitation and crowding have primed these places as well for outbreaks—even though malaria has up to now been regarded as predominantly a rural disease.

What has current policy to offer in the face of these threats? The medical arsenal is limited; there are only about a dozen antimalarial drugs commonly in use, and there is significant malaria resistance to most of them. In the absence of a reliable way to kill the parasites, policy has tended to focus on

From *World Watch Magazine*, vol. 15, no. 3, May/June 2002. Copyright © 2002 by Worldwatch Institute. Reprinted by permission. www.worldwatch.org

killing the mosquitoes that bear them. And that has led to an abundant use of synthetic pesticides, including one of the oldest and most dangerous: dichlorodiphenyl trichloroethane, or DDT.

DDT is no longer used or manufactured in most of the world, but because it does not break down readily, it is still one of the most commonly detected pesticides in the milk of nursing mothers. DDT is also one of the "dirty dozen" chemicals included in the 2001 Stockholm Convention on Persistent Organic Pollutants [POPs]. The signatories to the "POPs Treaty" essentially agreed to ban all uses of DDT except as a last resort against disease-bearing mosquitoes. Unfortunately, however, DDT is still a routine option in 19 countries, most of them in Africa. (Only 11 of these countries have thus far signed the treaty.) Among the signatory countries, 31—slightly fewer than one-third—have given notice that they are reserving the right to use DDT against malaria. On the face of it, such use may seem unavoidable, but there are good reasons for thinking that progress against the disease is compatible with *reductions* in DDT use.

꿍

Malaria is caused by four protozoan parasite species in the genus *Plasmodium*. These parasites are spread exclusively by certain mosquitoes in the genus *Anopheles*. An infection begins when a parasite-laden female mosquito settles onto someone's skin and pierces a capillary to take her blood meal. The parasite, in a form called the *sporozoite*, moves with the mosquito's saliva into the human bloodstream. About 10 percent of the mosquito's lode of sporozoites is likely to be injected during a meal, leaving plenty for the next bite. Unless the victim has some immunity to malaria—normally as a result of previous exposure—most sporozoites are likely to evade the body's immune system and make their way to the liver, a process that takes less than an hour. There they invade the liver cells and multiply asexually for about two weeks. By this time, the original several dozen sporozoites have become millions of *merozoites*—the form the parasite takes when it emerges from the liver and moves back into the blood to invade the body's red blood cells. Within the red blood cells, the merozoites go through another cycle of asexual reproduction, after which the cells burst and release millions of additional merozoites, which invade yet more red blood cells. The high fever and chills associated with malaria are the result of this stage, which tends to occur in pulses. If enough red blood cells are destroyed in one of these pulses, the result is convulsions, difficulty in breathing, coma, and death.

As the parasite multiplies inside the red blood cells, it produces not just more merozoites, but also *gametocytes*, which are capable of sexual reproduction. This occurs when the parasite moves back into the mosquitoes; even as they inject sporozoites, biting mosquitoes may ingest gametocytes if they are feeding on a person who is already infected. The gametocytes reproduce in the insect's gut and the resulting eggs move into the gut cells. Eventually, more sporozoites emerge from the gut and penetrate the mosquito's salivary glands, where they await a chance to enter another human bloodstream, to begin the cycle again.

Of the roughly 380 mosquito species in the genus *Anopheles,* about 60 are able to transmit malaria to people. These malaria vectors are widespread throughout the tropics and warm temperate zones, and they are very efficient at spreading the disease. Malaria is highly contagious, as is apparent from a measurement that epidemiologists call the "basic reproduction number," or BRN. The BRN indicates, on average, how many new cases a single infected person is likely to cause. For example, among the nonvectored diseases (those in which the pathogen travels directly from person to person without an inter-mediary like a mosquito), measles is one of the most contagious. The BRN for measles is 12 to 14, meaning that someone with measles is likely to infect 12 to 14 other people. (Luckily, there's an inherent limit in this process: as a pathogen spreads through any particular area, it will encounter fewer and fewer susceptible people who aren't already sick, and the outbreak will eventually subside.) HIV/AIDS is on the other end of the scale: it's deadly, but it burns through a population slowly. Its BRN is just above 1, the minimum necessary for the pathogen's survival. With malaria, the BRN varies considerably, depending on such factors as which mosquito species are present in an area and what the temperatures are. (Warmer is worse, since the parasites mature more quickly.) But malaria can have a BRN in excess of 100: over an adult life that may last about a week, a single, malaria-laden mosquito could conceivably infect more than 100 people.

Seven Years, Seven Months

"Malaria" comes from the Italian "mal'aria." For centuries, European physicians had attributed the disease to "bad air." Apart from a tradition of associating bad air with swamps—a useful prejudice, given the amount of mosquito habitat in swamps—early medicine was largely ineffective against the disease. It wasn't until 1897 that the British physician Ronald Ross proved that mosquitoes carry malaria.

The practical implications of Ross's discovery did not go unnoticed. For example, the U.S. administration of Theodore Roosevelt recognized malaria and yellow fever (another mosquito-vectored disease) as perhaps the most serious obstacles to the construction of the Panama Canal. This was hardly a surprising conclusion, since the earlier and unsuccessful French attempt to build the canal—an effort that predated Ross's discovery—is thought to have lost between 10,000 and 20,000 workers to disease. So the American workers draped their water supplies and living quarters with mosquito netting, attempted to fill in or drain swamps, installed sewers, poured oil into standing water, and conducted mosquito-swatting campaigns. And it worked: the incidence of malaria declined. In 1906, 80 percent of the workers had the disease; by 1913, a year before the Canal was completed, only 7 percent did. Malaria could be suppressed, it seemed, with a great deal of mosquito netting, and by eliminating as much mosquito habitat as possible. But the labor involved in that effort could be enormous.

That is why DDT proved so appealing. In 1939, the Swiss chemist Paul Müller discovered that this chemical was a potent pesticide. DDT was first used during World War II, as a delousing agent. Later on, areas in southern Europe, North Africa, and Asia were fogged with DDT, to clear malaria-laden mosquitoes from the paths of invading Allied troops. DDT was cheap and it seemed to be harmless to anything other than insects. It was also long-lasting: most other insecticides lost their potency in a few days, but in the early years of its use, the effects of a single dose of DDT could last for up to six months. In 1948, Müller won a Nobel Prize for his work and DDT was hailed as a chemical miracle.

A decade later, DDT had inspired another kind of war—a general assault on malaria. The "Global Malaria Eradication Program," launched in 1955, became one of the first major undertakings of the newly created World Health Organization [WHO]. Some 65 nations enlisted in the cause. Funding for DDT factories was donated to poor countries and production of the insecticide climbed.

The malaria eradication strategy was not to kill every single mosquito, but to suppress their populations and shorten the lifespans of any survivors, so that the parasite would not have time to develop within them. If the mosquitoes could be kept down long enough, the parasites would eventually disappear from the human population. In any particular area, the process was expected to take three years—time enough for all infected people either to recover or die. After that, a resurgence of mosquitoes would be merely an annoyance, rather than a threat. And initially, the strategy seemed to be working. It proved especially effective on islands—relatively small areas insulated from reinfestation. Taiwan, Jamaica, and Sardinia were soon declared malaria-free and have remained so to this day. By 1961, arguably the year at which the program had peak momentum, malaria had been eliminated or dramatically reduced in 37 countries.

One year later, Rachel Carson published *Silent Spring*, her landmark study of the ecological damage caused by the widespread use of DDT and other pesticides. Like other organochlorine pesticides, DDT bioaccumulates. It's fat soluble, so when an animal ingests it—by browsing contaminated vegetation, for example—the chemical tends to concentrate in its fat, instead of being excreted. When another animal eats that animal, it is likely to absorb the prey's burden of DDT. This process leads to an increasing concentration of DDT in the higher links of the food chain. And since DDT has a high chronic toxicity—that is, long-term exposure is likely to cause various physiological abnormalities—this bioaccumulation has profound implications for both ecological and human health.

With the miseries of malaria in full view, the managers of the eradication campaign didn't worry much about the toxicity of DDT, but they were greatly concerned about another aspect of the pesticide's effects: resistance. Continual exposure to an insecticide tends to "breed" insect populations that are at least partially immune to the poison. Resistance to DDT had been reported as early as 1946. The campaign managers knew that in mosquitoes, regular exposure to DDT tended to produce widespread resistance in four to seven years. Since it took three years to clear malaria from a human population, that didn't leave a lot of leeway for the eradication effort. As it turned out, the logistics simply couldn't be made to work in large, heavily infested areas with high human populations, poor housing and roads, and generally minimal infrastructure. In 1969, the campaign was abandoned. Today, DDT resistance is widespread in *Anopheles,* as is resistance to many more recent pesticides.

Undoubtedly, the campaign saved millions of lives, and it did clear malaria from some areas. But its broadest legacy has been of much more dubious value. It engendered the idea of DDT as a first resort against mosquitoes and it established the unstable dynamic of DDT resistance in *Anopheles* populations. In mosquitoes, the genetic mechanism that confers resistance to DDT does not usually come at any great competitive "cost"—that is, when no DDT is being sprayed, the resistant mosquitoes may do just about as well as nonresistant mosquitoes. So once a population acquires resistance, the trait is not likely to disappear even if DDT isn't used for years. If DDT is reapplied to such a population, widespread resistance will reappear very rapidly. The rule of thumb among entomologists is that you may get seven years of resistance-free use the first time around, but you only get about seven months the second time. Even that limited respite, however, is enough to make the chemical an attractive option as an emergency measure— or to keep it in the arsenals of bureaucracies committed to its use.

Malaria Taxes

In December 2000, the POPs Treaty negotiators convened in Johannesburg, South Africa, even though, by an unfortunate coincidence, South Africa had suffered a potentially embarrassing setback earlier that year in its own POPs policies. In 1996, South Africa had switched its mosquito control programs from DDT to a less persistent group of pesticides known as pyrethroids. The move seemed solid and supportable at the time, since years of DDT use had greatly reduced *Anopheles* populations and largely eliminated one of the most troublesome local vectors, the appropriately named *A. funestus* ("funestus" means deadly). South Africa seemed to have beaten the DDT habit: the chemical had been used to achieve a worthwhile objective; it had then been discarded. And the plan worked—until a year before the POPs summit, when malaria infections rose to 61,000 cases, a level not seen in decades. *A. funestus* reappeared as well, in KwaZulu-Natal, and in a form resistant to pyrethroids. In early 2000, DDT was reintroduced, in an indoor spraying program. (This is now a standard way of using DDT for mosquito control; the pesticide is usually applied only to walls, where mosquitoes alight to rest.) By the middle of the year, the number of infections had dropped by half.

Initially, the spraying program was criticized, but what reasonable alternative was there? This is said to be the African predicament, and yet the South African situation is hardly representative of sub-Saharan Africa as a whole.

Malaria is considered endemic in 105 countries throughout the tropics and warm temperate zones, but by far the worst region for the disease is sub-Saharan Africa. The deadliest of the four parasite species, *Plasmodium falciparum,* is widespread throughout this region, as is one of the world's most effective malaria vectors, *Anopheles gambiae.* Nearly half the population of sub-Saharan Africa is at risk of infection, and in much of eastern and central Africa, and pockets of west Africa, it would be difficult to find anyone who has not been exposed to the parasites. Some 90 percent of the world's malaria infections and deaths occur in sub-Saharan Africa, and the disease now accounts for 30 percent of African childhood mortality. It is true that malaria is a grave problem in many parts of the world, but the African experience is misery on a very different order of magnitude. The average Tanzanian suffers more infective bites each *night* than the average Thai or Vietnamese does in a year.

As a broad social burden, malaria is thought to cost Africa between $3 billion and $12 billion annually. According to one economic analysis, if the disease had been eradicated in 1965, Africa's GDP would now be 35 percent higher than it currently is. Africa was also the gaping hole in the global eradication program: the WHO planners thought there was little they could do on the continent and limited efforts to Ethiopia, Zimbabwe, and South Africa, where eradication was thought to be feasible.

But even though the campaign largely passed Africa by, DDT has not. Many African countries have used DDT for mosquito control in indoor spraying programs, but the primary use of DDT on the continent has been as an agricultural insecticide. Consequently, in parts of west Africa especially, DDT resistance is now widespread in *A. gambiae.* But even if *A. gambiae* were not resistant, a full-bore campaign to suppress it would probably accomplish little, because this mosquito is so efficient at transmitting malaria. Unlike most *Anopheles* species, *A. gambiae* specializes in human blood, so even a small population would keep the disease in circulation. One way to get a sense for this problem is to consider the "transmission index"—the threshold number of mosquito bites necessary to perpetuate the disease. In Africa, the index overall is 1 bite per person per month. That's all that's necessary to keep malaria in circulation. In India, by comparison, the TI is 10 bites per person per month.

And yet Africa is not a lost cause—it's simply that the key to progress does not lie in the general suppression of mosquito populations. Instead of spraying, the most promising African programs rely primarily on "bednets"—mosquito netting that is treated with an insecticide, usually a pyrethroid, and that is suspended over a person's bed. Bednets can't eliminate malaria, but they can "deflect" much of the burden. Because *Anopheles* species generally feed in the evening and at night, a bednet can radically reduce the number of infective bites a person receives. Such a person would probably still be infected from time to time, but would usually be able to lead a normal life.

In effect, therefore, bednets can substantially reduce the disease. Trials in the use of bednets for children have shown a decline in malaria-induced mortality by 25 to 40 percent. Infection levels and the incidence of severe anemia also declined. In Kenya, a recent study has shown that pregnant women who use bednets tend to give birth to healthier babies. In parts of Chad, Mali, Burkina Faso, and Senegal, bednets are becoming standard household items. In the tiny west African nation of The Gambia, somewhere between 50 and 80 percent of the population has bednets.

Bednets are hardly a panacea. They have to be used properly and retreated with insecticide occasionally. And there is still the problem of insecticide resistance, although the nets themselves are hardly likely to be the main cause of it. (Pyrethroids are used extensively in agriculture as well.) Nevertheless, bednets can help transform malaria from a chronic disaster to a manageable public health problem—something a healthcare system can cope with.

So it's unfortunate that in much of central and southern Africa, the nets are a rarity. It's even more unfortunate that, in 28 African countries, they're taxed or subject to import tariffs. Most of the people in these countries would have trouble paying for a net even without the tax. This problem was addressed in the May 2000 "Abuja Declaration," a summit agreement on infectious diseases signed by 44 African countries. The Declaration included a pledge to do away with "malaria taxes." At last count, 13 countries have actually acted on the pledge, although in some cases only by reducing rather than eliminating the taxes. Since the Declaration was signed, an estimated 2 to 5 million Africans have died from malaria.

This failure to follow through with the Abuja Declaration casts the interest in DDT in a rather poor light. Of the 31 POPs treaty signatories that have reserved the right to use DDT, 21 are in Africa. Of those 21, 10 are apparently still taxing or imposing tariffs on bednets. (Among the African countries that have *not* signed the POPs treaty, some are almost certainly both using DDT and taxing bednets, but the exact number is difficult to ascertain because the status of DDT use is not always clear.) It is true that a case can be made for the use of DDT in situations like the one in South Africa in 1999—an infrequent flare-up in a context that lends itself to control. But the routine use of DDT against malaria is an exercise in toxic futility, especially when it's pursued at the expense of a superior and far more benign technology.

Learning to Live with the Mosquitoes

A group of French researchers recently announced some very encouraging results for a new anti-malarial drug known as G25. The drug was given to infected aotus monkeys, and it appears to have cleared the parasites from their systems. Although extensive testing will be necessary before it is known whether the drug can be safely given to people, these results have raised the hope of a cure for the disease.

Of course, it would be wonderful if G25, or some other new drug, lives up to that promise. But even in the absence of a cure, there are opportunities for progress that may one day make the current incidence of malaria look like some dark age horror. Many of these opportunities have been incorporated into an initiative that began in 1998, called the Roll Back Malaria (RBM) campaign, a collaborative effort between WHO, the World Bank, UNICEF, and the UNDP [United Nations Development Programme]. In contrast to the earlier WHO eradication program, RBM grew out of joint efforts between WHO and various African governments specifically to address African malaria. RBM focuses on household- and community-level intervention and it emphasizes apparently modest changes that could yield major progress. Below are four "operating principles" that are, in one way or another, implicit in RBM or likely to reinforce its progress.

1. Do away with all taxes and tariffs on bednets, on pesticides intended for treating bednets, and on antimalarial drugs. Failure to act on this front

certainly undercuts claims for the necessity of DDT; it may also undercut claims for antimalaria foreign aid.

2. Emphasize appropriate technologies. Where, for example, the need for mud to replaster walls is creating lots of pothole sized cavities near houses—cavities that fill with water and then with mosquito larvae—it makes more sense to help people improve their housing maintenance than it does to set up a program for squirting pesticide into every pothole. To be "appropriate," a technology has to be both affordable and culturally acceptable. Improving home maintenance should pass this test; so should bednets. And of course there are many other possibilities. In Kenya, for example, a research institution called the International Center for Insect Physiology and Ecology has identified at least a dozen native east African plants that repel *Anopheles gambiae* in lab tests. Some of these plants could be important additions to household gardens.

3. Use existing networks whenever possible, instead of building new ones. In Tanzania, for example, an established healthcare program (UNICEF's Integrated Management of Childhood Illness Program) now dispenses antimalarial drugs—and instruction on how to use them. The UNICEF program was already operating, so it was simple and cheap to add the malaria component. Reported instances of severe malaria and anemia in infants have declined, apparently as a result. In Zambia, the government is planning to use health and prenatal clinics as the network for a coupon system that subsidizes bednets for the poor. Qualifying patients would pick up coupons at the clinics and redeem them at stores for the nets.

4. Assume that sound policy will involve action on many fronts. Malaria is not just a health problem—it's a social problem, an economic problem, an environmental problem, an agricultural problem, an urban planning problem. Health officials alone cannot possibly just make it go away. When the disease flares up, there is a strong and understandable temptation to strap on the spray equipment and douse the mosquitoes. But if this approach actually worked, we wouldn't be in this situation today. Arguably the biggest opportunity for progress against the disease lies, not in our capacity for chemical innovation, but in our capacity for *organizational innovation*—in our ability to build an awareness of the threat across a broad range of policy activities. For example, when government officials are considering loans to irrigation projects, they should be asking: has the potential for malaria been addressed? When foreign donors are designing antipoverty programs, they should be asking: do people need bednets? Routine inquiries of this sort could go a vast distance to reducing the disease.

Where is the DDT in all of this? There isn't any, and that's the point. We now have half a century of evidence that routine use of DDT simply will not prevail against the mosquitoes. Most countries have already absorbed this lesson, and banned the chemical or relegated it to emergency only status. Now the RBM campaign and associated efforts are showing that the frequency and intensity of those emergencies can be reduced through systematic attention to the chronic aspects of the disease. There is less and less justification for DDT, and the futility of using it as a matter of routine is becoming increasingly apparent: in order to control a disease, why should we poison our soils, our waters, and ourselves?

Donald R. Roberts

NO

Statement before the U.S. Senate Committee on Environment & Public Works, Hearing on the Role of Science in Environmental Policy-Making

Thank you, Chairman Inhofe, and distinguished members of the Committee on Environment and Public Works, for the opportunity to present my views on the misuse of science in public policy. My testimony focuses on misrepresentations of science during decades of environmental campaigning against DDT.

Before discussing how and why DDT science has been misrepresented, you first must understand why this misrepresentation has not helped, but rather harmed, millions of people every year all over the world. Specifically you need to understand why the misrepresentation of DDT science has been and continues to be deadly. By way of explanation, I will tell you something of my experience.

I conducted malaria research in the Amazon Basin in the 1970s. My Brazilian colleague—who is now the Secretary of Health for Amazonas State—and I worked out of Manaus, the capitol of Amazonas State. From Manaus we traveled two days to a study site where we had sufficient numbers of cases for epidemiological studies. There were no cases in Manaus, or anywhere near Manaus. For years before my time there and for years thereafter, there were essentially no cases of malaria in Manaus. However, in the late 1980s, environmentalists and international guidelines forced Brazilians to reduce and then stop spraying small amounts of DDT inside houses for malaria control. As a result, in 2002 and 2003 there were over 100,000 malaria cases in Manaus alone.

Brazil does not stand as the single example of this phenomenon. A similar pattern of declining use of DDT and reemerging malaria occurs in other countries as well, Peru for example. Similar resurgences of malaria have occurred in rural communities, villages, towns, cities, and countries around the world. As illustrated by the return of malaria in Russia, South Korea, urban areas of the Amazon Basin, and increasing frequencies of outbreaks in the United States, our malaria problems are growing worse. Today there are 1 to 2 million malaria deaths each year and hundreds of millions of cases. The

U.S. Senate Committee on Environment & Public Works Hearing on the Tole of Science in Environmental Policy-Making, September 28, 2005.

poorest of the world's people are at greatest risk. Of these, children and pregnant women are the ones most likely to die.

We have long known about DDT's effectiveness in curbing insect-borne disease. Othmar Zeidler, a German chemistry student, first synthesized DDT in 1874. Over sixty years later in Switzerland, Paul Müller discovered the insecticidal property of DDT. Allied forces used DDT during WWII, and the new insecticide gained fame in 1943 by successfully stopping an epidemic of typhus in Naples, an unprecedented achievement. By the end of the war, British, Italian, and American scientists had also demonstrated the effectiveness of DDT in controlling malaria-carrying mosquitoes. DDT's proven efficacy against insect-borne diseases, diseases that had long reigned unchecked throughout the world, won Müller the Nobel Prize for Medicine in 1948. After WWII, the United States conducted a National Malaria Eradication Program, commencing operations on July 1, 1947. The spraying of DDT on internal walls of rural homes in malaria endemic counties was a key component of the program. By the end of 1949, the program had sprayed over 4,650,000 houses. This spraying broke the cycle of malaria transmission, and in 1949 the United States was declared free of malaria as a significant public health problem. Other countries had already adopted DDT to eradicate or control malaria, because wherever malaria control programs sprayed DDT on house walls, the malaria rates dropped precipitously. The effectiveness of DDT stimulated some countries to create, for the first time, a national malaria control program. Countries with pre-existing programs expanded them to accommodate the spraying of houses in rural areas with DDT. Those program expansions highlight what DDT offered then, and still offers now, to the malaria endemic countries. As a 1945 U.S. Public Health Service manual explained about the control of malaria: "Drainage and larviciding are the methods of choice in towns of 2,500 or more people. But malaria is a rural disease. Heretofore there has been no economically feasible method of carrying malaria control to the individual tenant farmer or sharecropper. Now, for the first time, a method is available—the application of DDT residual spray to walls and ceilings of homes." Health workers in the United States were not the only ones to recognize the particular value of DDT. The head of malaria control in Brazil characterized the changes that DDT offered in the following statement: "Until 1945–1946, preventive methods employed against malaria in Brazil, as in the rest of the world, were generally directed against the aquatic phases of the vectors (draining, larvicides, destruction of bromeliads, etc. . . .). These methods, however, were only applied in the principal cities of each state and the only measure available for rural populations exposed to malaria was free distribution of specific drugs."

DDT was a new, effective, and exciting weapon in the battle against malaria. It was cheap, easy to apply, long-lasting once sprayed on house walls, and safe for humans. Wherever and whenever malaria control programs sprayed it on house walls, they achieved rapid and large reductions in malaria rates. Just as there was a rush to quickly make use of DDT to control disease, there was also a rush to judge how DDT actually functioned to control malaria. That rush to judgment turned out to be a disaster. At the heart of the debate—to the extent there was a debate—was a broadly accepted model that established a mathematical framework for using DDT to kill mosquitoes and eradicate malaria. Instead of studying real data to see how DDT actually worked in controlling malaria, some scientists settled upon what they thought was a logical conclusion: DDT worked solely by killing mosquitoes. This conclusion was based on their belief in the model. Scientists who showed that DDT did not function by killing mosquitoes were ignored. Broad acceptance of the mathematical model led to strong convictions about DDT's toxic actions. Since they were convinced that DDT worked only by killing mosquitoes, malaria control specialists became very alarmed when a mosquito was reported to be resistant to DDT's toxic actions. As a result of concern about DDT resistance, officials decided to make rapid use of DDT before problems of resistance could eliminate their option to use DDT to eradicate malaria. This decision led to creation of the global malaria eradication program. The active years of the global malaria eradication program were from 1959 to 1969. Before, during, and after the many years of this program, malaria workers and researchers carried out their responsibilities to conduct studies and report their research. Through those studies, they commonly found that DDT was functioning in ways other than by killing mosquitoes. In essence, they found that DDT was functioning through mechanisms of repellency and irritancy. Eventually, as people forgot early observations of DDT's repellent actions, some erroneously interpreted new findings of repellent actions as the mosquitoes' adaptation to avoid DDT toxicity, even coining a term, "behavioral resistance," to explain what they saw. This new term accommodated their view that toxicity was DDT's primary mode of action and categorized behavioral responses of mosquitoes as mere adaptations to toxic affects. However this interpretation depended upon a highly selective use of scientific data. The truth is that toxicity is not DDT's primary mode of action when sprayed on house walls. Throughout the history of DDT use in malaria control programs there has always been clear and persuasive data that DDT functioned primarily as a spatial repellent. Today we know that there is no insecticide recommended for malaria control that rivals, much less equals, DDT's spatial repellent actions, or that is as long-acting, as cheap, as easy to apply, as safe for human exposure, or as efficacious in the control of malaria as DDT. . . . The 30 years of data from control programs of the Americas plotted . . . illustrate just how effective DDT is in malaria control. The period 1960s through 1979 displays a pattern of malaria controlled through house spraying. In 1979 the World Health Organization (WHO) changed its strategy for malaria control, switching emphasis from spraying houses to case detection and treatment. In other words, the WHO changed emphasis from malaria prevention to malaria treatment. Countries complied with WHO guidelines and started to dismantle their spray programs over the next several years. . . .

I find it amazing that many who oppose the use of DDT describe its earlier use as a failure. Our own citizens who suffered under the burden of malaria, especially in the rural south, would hardly describe it thus.

Malaria was a serious problem in the United States and for some localities, such as Dunklin County, Missouri, it was a very serious problem indeed. For four counties in Missouri, the average malaria mortality from 1910 to 1914 was 168.8 per 100,000 population. For Dunklin County, it was 296.7 per 100,000, a rate almost equal to malaria deaths in Venezuela and actually greater than the mortality rate for Freetown, Sierra Leone. Other localities in other states were equally as malarious. Growing wealth and improved living conditions were gradually reducing malaria rates, but cases resurged during WWII. The advent of DDT, however, quickly eradicated malaria from the United States.

DDT routed malaria from many other countries as well. The Europeans who were freed of malaria would hardly describe its use as a failure. After DDT was introduced to malaria control in Sri Lanka (then Ceylon), the number of malaria cases fell from 2.8 million in 1946 to just 110 in 1961. Similar spectacular decreases in malaria cases and deaths were seen in all the regions that began to use DDT. The newly formed Republic of China (Taiwan) adopted DDT use in malaria control shortly after World War II. In 1945 there were over 1 million cases of malaria on the island. By 1969 there were only 9 cases and shortly thereafter the disease was eradicated from the island and remains so to this day. Some countries were less fortunate. South Korea used DDT to eradicate malaria, but without house spray programs, malaria has returned across the demilitarized zone with North Korea. As DDT was eliminated and control programs reduced, malaria has returned to other countries such as Russia and Argentina. Small outbreaks of malaria are even beginning to appear more frequently in the United States.

These observations have been offered in testimony to document first that there were fundamental misunderstandings about how DDT functioned to exert control over malaria. Second, that regardless of systematic misunderstandings on the part of those who had influence over malaria control strategies and policies, there was an enduring understanding that DDT was the most cost-effective compound yet discovered for protecting poor rural populations from insect-borne diseases like malaria, dengue, yellow fever, and leishmaniasis. I want to emphasize that misunderstanding the mode of DDT action did not lead to the wholesale abandonment of DDT. It took an entirely new dimension in the misuse of science to bring us to the current humanitarian disaster represented by DDT elimination.

The misuse of science to which I refer has found fullest expression in the collection of movements within the environmental movement that seek to stop production and use of specific man-made chemicals. Operatives within these movements employ particular strategies to achieve their objectives. By characterizing and understanding the strategies these operatives use, we can identify their impact in the scientific literature or in the popular press.

The first strategy is to develop and then distribute as widely as possible a broad list of claims of chemical harm. This is a sound strategy because individual scientists can seldom rebut the scientific foundations of multiple and diverse claims. Scientists generally develop expertise in a single, narrow field and are disinclined to engage issues beyond their area of expertise. Even if an authoritative rebuttal of one claim occurs, the other claims still progress. A broad list of claims also allows operatives to tailor platforms for constituencies, advancing one set of claims with one constituency and a different combination for another. Clever though this technique is, a list of multiple claims of harm is hardly sufficient to achieve the objective of a ban. The second strategy then is to mount an argument that the chemical is not needed and propose that alternative chemicals or methods can be used instead. The third strategy is to predict that grave harm will occur if the chemical continues to be used.

The success of Rachel Carson's *Silent Spring* serves as a model for this tricky triad. In *Silent Spring*, Rachel Carson used all three strategies on her primary target, DDT. She described a very large list of potential adverse effects of insecticides, DDT in particular. She argued that insecticides were not really needed and that the use of insecticides produces insects that are insecticide resistant, which only exacerbates the insect control problems. She predicted scary scenarios of severe harm with continued use of DDT and other insecticides. Many have written rebuttals to Rachel Carson and others who have, without scientific justification, broadcast long lists of potential harms of insecticides. . . .

[T]ime and science have discredited most of Carson's claims. Rachel Carson's descriptions of inappropriate uses of insecticides that harmed wildlife are more plausible. However, harm from an inappropriate use does not meet the requirements of anti-pesticide activists. They can hardly lobby for eliminating a chemical because someone used it wrongly. No, success requires that even the proper use of an insecticide will cause a large and systematic adverse effect. However, the proper uses of DDT yield no large and systematic adverse effects. Absent such adverse actions, the activists must then rely on claims about insidious effects, particularly insidious effects that scientists will find difficult to prove one way or the other and that activists can use to predict a future catastrophe.

Rachel Carson relied heavily on possible insidious chemical actions to alarm and frighten the public. Many of those who joined her campaign to ban DDT and other insecticides made extensive use of claims of insidious effects. These claims were amplified by the popular press and became part of the public perception about modern uses of chemicals. For example, four well-publicized claims about DDT were:

1. DDT will cause the obliteration of higher trophic levels. If not obliterated, populations will undergo reproductive failure. Authors of this claim speculated that, even if the use of DDT were stopped, systematic and ongoing obliterations would still occur.
2. DDT causes the death of algae. This report led to speculations that use of DDT could result in global depletion of oxygen.
3. DDT pushed the Bermuda petrel to the verge of extinction and that full extinction might happen by 1978.
4. DDT was a cause of premature births in California sea lions.

Science magazine, the most prestigious science journal in the United States, published these and other phantasmagorical allegations and/or predictions of DDT harm. Nonetheless, history has shown that each and every one of these claims and predictions were false.

1. The obliteration of higher trophic levels did not occur; no species became extinct; and levels of DDT in all living organisms declined

precipitously after DDT was de-listed for use in agriculture. How could the prediction have been so wrong? Perhaps it was so wrong because the paper touting this view used a predictive model based on an assumption of no DDT degradation. This was a startling assertion even at the time as *Science* and other journals had previously published papers that showed DDT was ubiquitously degraded in the environment and in living creatures. It was even more startling that *Science* published a paper that flew so comprehensively in the face of previous data and analysis.

2. DDT's action against algae reportedly occurred at concentrations of 500 parts per billion. But DDT cannot reach concentrations in water higher than about 1.2 parts per billion, the saturation point of DDT in water.

3. Data on the Bermuda petrel did not show a cause-and-effect relationship between low numbers of birds and DDT concentrations. DDT had no affect on population numbers, for populations increased before DDT was de-listed for use in agriculture and after DDT was de-listed as well.

4. Data gathered in subsequent years showed that "despite relatively high concentrations [of DDT], no evidence that population growth or the health of individual California sea lions have been compromised. The population has increased throughout the century, including the period when DDT was being manufactured, used, and its wastes discharged off southern California."

If time and science have refuted all these catastrophic predictions, why do many scientists and the public not know these predictions were false? In part, we do not know the predictions were false because the refutations of such claims rarely appear in the literature.

When scientists hear the kinds of claims described above, they initiate research to confirm or refute the claims. After Charles Wurster published his claim that DDT kills algae and impacts photosynthesis, I initiated research on planktonic algae to quantify DDT's effects. From 1968–1969, I spent a year of honest and demanding research effort to discover that not enough DDT would even go into solution for a measurable adverse effect on planktonic algae. In essence, I conducted a confirmatory study that failed to confirm an expected result. I had negative data, and journals rarely accept negative data for publication. My year was practically wasted. Without a doubt, hundreds of other scientists around the world have conducted similar studies and obtained negative results, and they too were unable to publish their experimental findings. Much in the environmental science literature during the last 20–30 years indicates that an enormous research effort went into proving specific insidious effects of DDT and other insecticides. Sadly, the true magnitude of such efforts will never be known because while the positive results of research find their way into the scientific literature, the negative results rarely do. Research on insidious actions that produce negative results all too often ends up only in laboratory and field notebooks and is forgotten. For this reason, I place considerable weight on a published confirmatory study that fails to confirm an expected result.

The use of the tricky triad continues. A . . . recent paper . . . published in *The Lancet* illustrates the triad's modern application. Two scientists at the National Institute of Environmental Health Sciences, Walter Rogan and Aimin Chen, wrote this paper, entitled "Health risks and benefits of bis(4-chlorophenyl)-1,1, 1-trichloroethane (DDT)." It is interesting to see how this single paper spins all three strategies that gained prominence in Rachel Carson's *Silent Spring.*

The journal *Emerging Infectious Diseases* had already published a slim version of this paper, which international colleagues and I promptly rebutted. The authors then filled in some parts, added to the claims of harm, and republished the paper in the British journal, *The Lancet.* To get the paper accepted by editors, the authors described studies that support (positive results) as well as studies that do not support (negative results) each claim. Complying with strategy number 1 of the triad, Rogan and Chen produce a long list of possible harms, including the charge that DDT causes cancer in nonhuman primates. The literature reference for Rogan and Chen's claim that DDT causes cancer in nonhuman primates was a paper by Takayama et al. Takayama and coauthors actually concluded from their research on the carcinogenic effect of DDT in nonhuman primates that "the two cases involving malignant tumors of different types are inconclusive with respect to a carcinogenic effect of DDT in nonhuman primates." Clearly, the people who made the link of DDT with cancer were not the scientists who actually conducted the research.

The authors enacted strategy number two of the triad by conducting a superficial review of the role of DDT in malaria control with the goal of discrediting DDT's value in modern malaria control programs. The authors admitted that DDT had been very effective in the past, but then argued that malaria control programs no longer needed it and should use alternative methods of control. Their use of the second strategy reveals, in my opinion, the greatest danger of granting authority to anti-pesticide activists and their writings. As *The Lancet* paper reveals, the NIEHS scientists assert great authority over the topic of DDT, yet they assume no responsibility for the harm that might result from their erroneous conclusions. After many malaria control specialists have expressed the necessity for DDT in malaria control, it is possible for Rogan and Chen to conclude that DDT is not necessary in malaria control only if they have no sense of responsibility for levels of disease and death that will occur if DDT is not used.

Rogan and Chen also employ the third strategy of environmentalism. Their list of potential harms caused by DDT includes toxic effects, neurobehavior effects, cancers, decrements in various facets of reproductive health, decrements in infant and child development, and immunology and DNA damage. After providing balanced coverage of diverse claims of harm, the authors had no option but to conclude they could not prove that DDT caused harm. However, they then promptly negated this honest conclusion by asserting that if DDT is used for malaria control, then great harm might occur. So, in an amazing turn, they conclude they cannot prove DDT causes harm, but still predict severe harm if it is used.

Rogan and Chen end their paper with a call for more research. One could conclude that the intent of the whole paper is merely to lobby for research to better define DDT harm, and what's the harm in that? Surely increasing knowledge is a fine goal. However, if you look at the specific issue of the relative need for research, you will see that the harm of this technique is great. Millions of children and pregnant women die from malaria every year, and the disease sickens hundreds of millions more. This is an indisputable fact: impoverished people engage in real life and death struggles every day with malaria. This also is a fact: not one death or illness can be attributed to an environmental exposure to DDT. Yet, a National Library of Medicine literature search on DDT reveals over 1,300 published papers from the year 2000 to the present, almost all in the environmental literature and many on potential adverse effects of DDT. A search on malaria and DDT reveals only 159 papers. DDT is a spatial repellent and hardly an insecticide at all, but a search on DDT and repellents will reveal only 7 papers. Is this not an egregiously disproportionate research emphasis on non-sources of harm compared to the enormous harm of malaria? Does not this inequity contribute to the continued suffering of those who struggle with malaria? Is it possibly even more than an inequity? Is it not an active wrong?

Public health officials and scientists should not be silent about enormous investments into the research of theoretical risks while millions die of preventable diseases. We should seriously consider our motivations in apportioning research money as we do. For consider this: the U.S. used DDT to eradicate malaria. After malaria disappeared as an endemic disease in the United States, we became richer. We built better and more enclosed houses. We screened our windows and doors. We air conditioned our homes. We also developed an immense arsenal of mosquito control tools and chemicals. Today, when we have a risk of mosquito borne disease, we can bring this arsenal to bear and quickly eliminate risks. And, as illustrated by aerial spray missions in the aftermath of hurricane Katrina, we can afford to do so. Yet, our modern and very expensive chemicals are not what protect us from introductions of the old diseases. Our arsenal responds to the threat; it does not prevent the appearance of old diseases in our midst. What protects us is our enclosed, screened, air-conditioned housing, the physical representation of our wealth. Our wealth is the factor that stops dengue at the border with Mexico, not our arsenal of new chemicals. Stopping mosquitoes from entering and biting us inside our homes is critical in the prevention of malaria and many other insect-borne diseases. This is what DDT does for poor people in poor countries. It stops large proportions of mosquitoes from entering houses. It is, in fact, a form of chemical screening, and until these people can afford physical screening or it is provided for them, this is the only kind of screening they have.

DDT is a protective tool that has been taken away from countries around the world, mostly due to governments acceding to the whims of the anti-pesticide wing of environmentalism, but it is not only the anti-pesticide wing that lobbies against DDT. The activists have a sympathetic lobbying ally in the pesticide industry. As evidence of insecticide industry working to stop countries from using DDT, I am attaching an email message dated 23rd September and authored by a Bayer official. . . . The Bayer official states

"[I speak] Not only as the responsible manager for the vector control business in Bayer, being the market leader in vector control and pointing out by that we know what we are talking about and have decades of experiences in the evolution of this very particular market. [but] Also as one of the private sector representatives in the RBM Partnership Board and being confronted with that discussion about DDT in the various WHO, RBM et al circles. So you can take it as a view from the field, from the operational commercial level—but our companies [sic] point of view. I know that all of my colleagues from other primary manufacturers and internationally operating companies are sharing my view."

The official goes on to say that

"DDT use is for us a commercial threat (which is clear, but it is not that dramatical because of limited use), it is mainly a public image threat."

However the most damning part of this message was the statement that

"we fully support EU to ban imports of agricultural products coming from countries using DDT"

[There is] . . . clear evidence of international and developed country pressures to stop poor countries from using DDT to control malaria. This message also shows the complicity of the insecticide industry in those internationally orchestrated efforts.

Pressures to eliminate spray programs, and DDT in particular, are wrong. I say this not based on some projection of what might theoretically happen in the future according to some model, or some projection of theoretical harms, I say this based firmly on what has already occurred. The track record of the anti-pesticide lobby is well documented, the pressures on developing countries to abandon their spray programs are well documented, and the struggles of developing countries to maintain their programs or restart their uses of DDT for malaria control are well documented. The tragic results of pressures against the use of DDT, in terms of increasing disease and death, are quantified and well documented. How long will scientists, public health officials, the voting public, and the politicians who lead us continue policies, regulations and funding that have led us to the current state of a global humanitarian disaster? How long will support continue for policies and programs that favor phantoms over facts?

POSTSCRIPT

Should DDT Be Banned Worldwide?

Professor Roberts is not alone in his disapproval of the efforts to halt the use of DDT. Alexander Gourevitch, "Better Living Through Chemistry," *Washington Monthly* (March 2003), comes close to accusing environmentalists of condemning DDT on the basis of politics or ideology rather than of science. Angela Logomasini comes even closer in "Chemical Warfare: Ideological Environmentalism's Quixotic Campaign Against Synthetic Chemicals," in Ronald Bailey, ed., *Global Warming and Other Eco-Myths: How the Environmental Movement Uses False Science to Scare Us to Death* (Prima, 2002). Her admission that public health demands have softened some environmentalists' resistance to the use of DDT points to a basic truth about environmental debates: over and over again, they come down to what we should do first: Should we meet human needs regardless of whether or not species die and air and water are contaminated? Or should we protect species, air, water, and other aspects of the environment even if some human needs must go unmet? What if this means endangering the lives of children? In the debate over DDT, the human needs are clear, for insect-borne diseases have killed and continue to kill a great many people. Yet the environmental needs are also clear. The question is one of choosing priorities and balancing risks. See John Danley, "Balancing Risks: Mosquitoes, Malaria, Morality, and DDT," *Business and Society Review* (Spring 2002). It is worth noting that John Beard, "DDT and Human Health," *Science of the Total Environment* (February 2006), finds the evidence for the ill effects of DDT more convincing and says that it is still too early to say it does not contribute to human disease.

The serious nature of the malaria threat is well covered in Michael Finkel, "Bedlam in the Blood: Malaria," *National Geographic* (July 2007). Mosquitoes can be controlled in various ways: Swamps can be drained (which carries its own environmental price), and other breeding opportunities can be eliminated. Fish can be introduced to eat mosquito larvae. And mosquito nets can be used to keep the insects away from people. But these (and other) alternatives do not mean that there does not remain a place for chemical pesticides. In "Pesticides and Public Health: Integrated Methods of Mosquito Management," *Emerging Infectious Diseases* (January–February 2001), Robert I. Rose, an arthropod biotechnologist with the Animal and Plant Health Inspection Service of the U.S. Department of Agriculture, says, "Pesticides have a role in public health as part of sustainable integrated mosquito management. Other components of such management include surveillance, source reduction or prevention, biological control, repellents, traps, and pesticide-resistance management." "The most effective programs today rely on a range of tools," says McGinn in "Combating Malaria," *State of the World 2003* (W. W. Norton, 2003). Still, some countries see

DDT as essential. Today some countries see DDT as essential. See Tina Rosenberg, "What the World Needs Now Is DDT," *New York Times Magazine* (April 11, 2004). However when the World Health Organization (WHO) endorsed the use of DDT to fight malaria, there was immediate outcry; see Allan Schapira, "DDT: A Polluted Debate in Malaria Control," *Lancet* (December 16, 2006). So far, however, the outcry seems fruitless; see Apoorva Mandavilli, "DDT Is Back," *Discover* (January 2007).

It has proven difficult to find effective, affordable drugs against malaria; see Ann M. Thayer, "Fighting Malaria," *Chemical and Engineering News* (October 24, 2005), and Claire Panosian Dunavan, "Tackling Malaria," *Scientific American* (December 2005). A great deal of effort has gone into developing vaccines against malaria, but the parasite has demonstrated a persistent talent for evading all attempts to arm the immune system against it. See Z. H. Reed, M. Friede, and M. P. Kieny, "Malaria Vaccine Development: Progress and Challenges," *Current Molecular Medicine* (March 2006). A newer approach is to develop genetically engineered (transgenic) mosquitoes that either cannot support the malaria parasite or cannot infect humans with it; see George K. Christophides, "Transgenic Mosquitoes and Malarial Transmission," *Cellular Microbiology* (March 2005), and M. J. Friedrich, "Malaria Researchers Target Mosquitoes," *JAMA: Journal of the American Medical Association* (May 23, 2007). On the other hand, Nicholas J. White, "Malaria—Time to Act," *New England Journal of Medicine* (November 9, 2006), argues that rather than wait for a perfect solution, we should recognize that present tools—including DDT—are effective enough now that there is no excuse to avoid using them.

It is worth stressing that malaria is only one of several mosquito-borne diseases that pose threats to public health. Two others are yellow fever and dengue. A recent arrival to the United States is West Nile virus, which mosquitoes can transfer from birds to humans. However, West Nile virus is far less fatal than malaria, yellow fever, or dengue, and a vaccine is in development. See Dwight G. Smith, "A New Disease in the New World," *The World & I* (February 2002) and Michelle Mueller, "The Buzz on West Nile Virus," *Current Health 2* (April/May 2002). But "West Nile Virus Still a Threat," says *Clinical Infectious Diseases* (April 15, 2006).

It is also worth stressing that global warming means climate changes that may increase the geographic range of disease-carrying mosquitoes. Many climate researchers are concerned that malaria, yellow fever, and other now mostly tropical and subtropical diseases may return to temperate-zone nations and even spread into areas where they have never been known. See Atul A. Khasnis and Mary D. Nettleman, "Global Warming and Infectious Disease," *Archives of Medical Research* (November 2005). Sarah DeWeerdt, "Climate Change, Coming Home," *World Watch* (May/June 2007), notes that "by 2020 human-triggered climate change could kill 300,000 people worldwide every year," in large part because of the spread of diseases such as malaria.

ISSUE 17

Do Environmental Hormone Mimics Pose a Potentially Serious Health Threat?

YES: Michele L. Trankina, from "The Hazards of Environmental Estrogens," *The World & I* (October 2001)

NO: Michael Gough, from "Endocrine Disrupters, Politics, Pesticides, the Cost of Food and Health," *Daily Commentary* (December 15, 1997)

ISSUE SUMMARY

YES: Professor of biological sciences Michele L. Trankina argues that a great many synthetic chemicals behave like estrogen, alter the reproductive functioning of wildlife, and may have serious health effects—including cancer—on humans.

NO: Michael Gough, a biologist and expert on risk assessment and environmental policy, argues that only "junk science" supports the hazards of environmental estrogens.

Following World War II there was an exponential growth in the industrial use and marketing of synthetic chemicals. These chemicals, known as "xeno-biotics," were used in numerous products, including solvents, pesticides, refrigerants, coolants, and raw materials for plastics. This resulted in increasing environmental contamination. Many of these chemicals, such as DDT, PCBs, and dioxins, proved to be highly resistant to degradation in the environment; they accumulated in wildlife and were serious contaminants of lakes and estuaries. Carried by winds and ocean currents, these chemicals were soon detected in samples taken from the most remote regions of the planet, far from their points of introduction into the ecosphere.

Until very recently most efforts to assess the potential toxicity of synthetic chemicals to bio-organisms, including human beings, focused almost exclusively on their possible role as carcinogens. This was because of legitimate public concern about rising cancer rates and the belief that cancer causation was the most likely outcome of exposure to low levels of synthetic chemicals.

Some environmental scientists urged public health officials to give serious consideration to other possible health effects of xenobiotics. They were generally ignored because of limited funding and the common belief that toxic effects other than cancer required larger exposures than usually resulted from environmental contamination.

In the late 1980s Theo Colborn, a research scientist for the World Wildlife Fund who was then working on a study of pollution in the Great Lakes, began linking together the results of a growing series of isolated studies. Researchers in the Great Lakes region, as well as in Florida, the West Coast, and Northern Europe, had observed widespread evidence of serious and frequently lethal physiological problems involving abnormal reproductive development, unusual sexual behavior, and neurological problems exhibited by a diverse group of animal species, including fish, reptiles, amphibians, birds, and marine mammals. Through Colborn's insights, communications among these researchers, and further studies, a hypothesis was developed that all of these wildlife problems were manifestations of abnormal estrogenic activity. The causative agents were identified as more than 50 synthetic chemical compounds that have been shown in laboratory studies to either mimic the action or disrupt the normal function of the powerful estrogenic hormones responsible for female sexual development and many other biological functions.

Concern that human exposure to these ubiquitous environmental contaminants may have serious health repercussions was heightened by a widely publicized European research study, which concluded that male sperm counts had decreased by 50 percent over the past several decades (a result that is disputed by other researchers) and that testicular cancer rates have tripled. Some scientists have also proposed a link between breast cancer and estrogen disrupters.

In response to the mounting scientific evidence that environmental estrogens may be a serious health threat, the U.S. Congress passed legislation requiring that all pesticides be screened for estrogenic activity and that the Environmental Protection Agency (EPA) develop procedures for detecting environmental estrogenic contaminants in drinking water supplies; see the EPA's Endocrine Disruptor Screening Program Web site at http://www. epa.gov/scipoly/oscpendo/index.htm. Government-sponsored studies of synthetic endocrine disrupters and other hormone mimics are also under way in the United Kingdom and in Germany.

In the following selections, Michele L. Trankina argues that a great many synthetic chemicals behave like estrogen, alter the reproductive functioning of wildlife, and may have serious health effects—including cancer—on humans. She insists that regulatory agencies must minimize public exposure. Michael Gough argues that only "junk science" supports the hazards of environmental estrogens. Expensive testing and regulatory programs can only drive up the cost of food, he says, which will make it harder for the poor to afford fresh fruits and vegetables. Furthermore, health protection will not be increased.

YES

Michele L. Trankina

The Hazards of Environmental Estrogens

What do Barbie dolls, food wrap, and spermicides have in common? And what do they have to do with low sperm counts, precocious puberty, and breast cancer? "Everything," say those who support the notion that hormone mimics are disrupting everything from fish gender to human fertility. "Nothing," counter others who regard the connection as trumped up, alarmist chemophobia. The controversy swirls around the significance of a number of substances that behave like estrogens and appear to be practically everywhere—from plastic toys to topical sunscreens.

Estrogens are a group of hormones produced in both the female ovaries and male testes, with larger amounts made in females than in males. They are particularly influential during puberty, menstruation, and pregnancy, but they also help regulate the growth of bones, skin, and other organs and tissues.

Over the past 10 years, many synthetic compounds and plant products present in the environment have been found to affect hormonal functions in various ways. Those that have estrogenic activity have been labeled as environmental estrogens, ecoestrogens, estrogen mimics, or xenoestrogens (*xenos* means foreign). Some arise as artifacts during the manufacture of plastics and other synthetic materials. Others are metabolites (breakdown products) generated from pesticides or steroid hormones used to stimulate growth in livestock. Ecoestrogens that are produced naturally by plants are called phytoestrogens (*phyton* means plant).

Many of these estrogen mimics bind to estrogen receptors (within specialized cells) with roughly the same affinity as estrogen itself, setting up the potential to wreak havoc on reproductive anatomy and physiology. They have therefore been labeled as disruptors of endocrine function.

Bizarre Changes in Reproductive Systems

Heightened concern about estrogen-mimicking substances arose when several nonmammalian vertebrates began to exhibit bizarre changes in reproductive anatomy and fertility. Evidence that something was amiss came serendipitously in 1994, from observations by reproductive physiologist Louis Guillette of the University of Florida. In the process of studying the decline of alligator populations at Lake Apopka, Florida, Guillette and coworkers noticed that many male alligators had smaller penises than normal. In addition, females superovulated, with multiple nuclei in some of the surplus ova. Closer scrutiny linked these findings to a massive spill of DDT (dichloro-diphenyl-trichloroethane) into Lake Apopka in 1980. Guillette concluded that the declining alligator population was related to the effects of DDT exposure on the animals' reproductive systems.

Although DDT was banned for use in the United States in the early 1970s, it continues to be manufactured in this country and marketed abroad, where it is sprayed on produce that is then sold in U.S. stores. The principal metabolite derived from DDT is called DDE, a xenoestrogen that lingers in fat deposits in the human body for decades. Historically, there have been reports of estrogen-mimicking effects in various fish species, especially in the Great Lakes, where residual concentrations of DDT and PCBs (polychlorinated biphenyls) are high. These effects include feminization and hermaphroditism in males. In fact, fish serve as barometers of the effects of xenoestrogen contamination in bodies of water. An index of exposure is the presence of vitellogenin, a protein specific to egg yolk, in the blood of male fish. Normally, only females produce vitellogenin in their livers, upon stimulation by estrogen from the ovaries.

Ecoestrogens are further concentrated in animals higher up in the food chain. In the Great Lakes region, birds including male herring gulls, terns, and bald eagles exhibit hermaphroditic changes after feeding on contaminated fish. Increased embryo mortality among these avians has also been observed. In addition, evidence from Florida links sterility in male and female panthers to their predation on animals exposed to pesticides with estrogenic activity.

The harmful effects of DDT have also been observed in rodents. Female rodents treated with high concentrations of DDT become predisposed to mammary tumors, while males tend to develop testicular cancer. These observations raise the question of whether pharmacological (that is, low) doses of substances with estrogenic activity translate into physiological effects. Usually they do not, but chronic exposure may be enough to trigger such effects.

Dangers to Humans

If xenoestrogens cause such dramatic reproductive effects in vertebrates, including mammals, what might be the consequences for humans? Nearly a decade ago, Frederick vom Saal, a developmental biologist at the University of Missouri Columbia, cautioned that mammalian reproductive mechanisms are similar enough to warrant concern about the effects of hormone disruptors on humans.

A 1993 article in *Lancet* looked at decreasing sperm counts in men in the United States and 20 other countries and correlated these decreases with the growing concentration of environmental estrogens. The authors—Niels Skakkenbaek, a Danish reproductive endocrinologist, and Richard Sharpe, of the British Medical Research Council Reproductive Biology Unit in Scotland—performed a meta-analysis of 61 sperm-count studies published between 1938 and 1990 to make their connection.

"Nonbelievers" state that this interpretation is contrived. Others suggest alternative explanations. For instance, the negative effects on sperm counts could have resulted from simultaneous increases in the incidence of venereal diseases. Besides, there are known differences in steroid hormone metabolism between lower vertebrates (including nonprimate mammals) and primates, so one cannot always extrapolate from the former group to the latter.

Even so, the incidence of testicular cancer, which typically affects young men in their 20s and 30s, has increased worldwide. Between 1979 and 1991, over 1,100 new cases were reported in England and Wales—a 55 percent increase over previous rates. In Denmark, the rate of testicular cancer increased by 300 percent from 1945 to 1990. Intrauterine exposure to xenoestrogens during testicular development is thought to be the cause.

Supporting evidence comes from Michigan, where accidental contamination of cattle feed with PCBs in 1973 resulted in high concentrations in the breast milk of women who consumed tainted beef. Their sons exhibited defective genitalia. Furthermore, observers in England have noted increased incidences of cryptorchidism (undescended testes), which results in permanent sterility if left untreated, and hypospadias (urethral orifice on the underside of the penis instead of at the tip). The one area of agreement between those who attribute these effects to ecoestrogens and those who deny such a connection is that more research is needed.

Perhaps one of the most disturbing current trends is the alarming increase in breast cancer incidence. Fifty years ago, the risk rate was 1 woman in 20; today it is 1 in 8. Numerous studies have implicated xenoestrogens as the responsible agents. For instance, high concentrations of pesticides, especially DDT, have been found in the breast tissue of breast cancer patients on Long Island. In addition, it has long been known that under certain conditions, estrogen from any source can be tumor promoting, and that most breast cancer cell types have estrogen receptors.

Precocious Puberty

If that were not enough, the growing number of estrogen mimics in the environment has been linked to early puberty in girls. The normal, average age of onset is between 12 and 13. A recent study of 17,000 girls in the United States indicated that 7 percent of white and 27 percent of black girls exhibited physical signs of puberty by age seven. For 10-year-old girls, the percentages increased to 68 and 95, respectively. Studies from the United Kingdom, Canada, and New Zealand have shown similar changes in the age of puberty onset.

It is difficult, however, to elucidate the exact mechanisms that underlie these trends toward precocious puberty. One explanation, which applies especially to cases in the United States, points to the increasing number of children who are overweight or obese as a result of high-calorie diets and lack of regular exercise. Physiologically, an enhanced amount of body fat implies reproductive readiness and signals the onset of puberty in both boys and girls.

For girls, more body fat ensures that there is enough stored energy to support pregnancy and lactation. Young women with low percentages of body fat caused by heavy exercise, sports, or eating disorders usually do not experience the onset of menses until their body composition reflects adequate fat mass.

Many who study the phenomenon of premature puberty attribute it to environmental estrogens in plastics and secondhand exposure through the meat and milk of animals treated with steroid hormones. An alarming increase in the numbers of girls experiencing precocious puberty occurred in the 1970s and '80s in Puerto Rico. Among other effects, breast development occurred in girls as young as one year. Premature puberty was traced to consumption of beef, pork, and dairy products containing high concentrations of estrogen.

Another study from Puerto Rico revealed higher concentrations of phthalate— a xenoestrogen present in certain plastics—in girls who showed signs of early puberty, compared with controls.

It may be that excess body fat and exposure to estrogenic substances operate in concert to hasten puberty. Body fat is one site of endogenous estrogen synthesis. Exposure to environmental estrogens may add just enough exogenous hormone to exert the synergistic effect necessary to bring on puberty, much like the last drop of water that causes the bucket to overflow.

Although most xenoestrogens produce detrimental effects, at least one subgroup—the phytoestrogens—includes substances that can have beneficial effects. Phytoestrogens are generally weaker than natural estrogens. They are found in various foods, such as flax seeds, soybeans and other legumes, some herbs, and many fruits and vegetables. Some studies suggest that soy products may offer protection against certain cancers, including breast, prostate, uterus, and colon cancers. On the other hand, high doses of certain phytoestrogens— such as coumestrol (in sunflower seeds and alfalfa sprouts)—have been found to adversely affect the fertility and reproductive cycles of animals.

Unlike artificially produced xenoestrogens, phytoestrogens are generally not stored in the body but are readily metabolized and excreted. Their health effects should be evaluated on a case-by-case basis, considering such factors as the individual's age, medical and family history, and potential interactions with medications or supplements.

Plastics, Plastics Everywhere

It has been estimated that perhaps 100,000 synthetic chemicals are registered for commercial use in the world today and 1,000 new ones are formulated every year. While many are toxic and carcinogenic, little is known about the chronic effects of the majority of them. And there is growing concern about their potential hormone-disrupting effects. The problem of exposure is complicated by numerous carrier routes, including air, food, water, and consumer products.

Consider certain synthetics that turn up in familiar places, including food and consumer goods. Such seemingly inert products as plastic soda and water bottles, baby bottles, food wrap, Styrofoam, many toys, cosmetics, sunscreens, and even spermicides either contain or break down to yield xenoestrogens. In addition, environmental estrogens are among the byproducts created by such processes as the incineration of biological materials or industrial waste and chlorine bleaching of paper products.

In April 1999, Consumers Union confirmed information previously reported by the Food and Drug Administration regarding 95 percent of baby bottles sold in the United States. The bottles, made of a hard plastic known as polycarbonate, leach out the synthetic estrogen named bisphenol-A, especially when heated or scratched. Studies verifying the estrogenic activity of bisphenol-A were published in *Nature* in the 1930s, but it did not arouse much concern then. A 1993 report published in *Endocrinology* showed that bisphenol-A produced estrogenic effects in a culture of human breast cancer cells.

Additional studies published by vom Saal in 1997 and '98 have shown that bisphenol-A stimulates precocious puberty and obesity in mice. Others have detected leaching of bisphenol-A from polycarbonate products such as plastic tableware, water cooler jugs, and the inside coatings of certain cans (used for some canned foods) and bottle tops. Autoclaving in the canning process causes bisphenol-A to migrate into the liquid in cans.

Spokespersons from polycarbonate manufacturers have stated that they cannot replicate vom Saal's results, but he counters that industry researchers have not done the experiments correctly.

DEHA (di-[2-ethylhexyl] adipate) is a liquid plasticizer added to some plastic food wraps made of polyvinyl chloride (PVC). Scientific studies have shown that the fat-soluble DEHA can migrate into foods, especially luncheon meats, cheese, and other products with a high fat content. For a 45-pound child eating cheese wrapped in such plastic, the limit of safe intake is 1.5 ounces by European standards or 2.5 ounces by Environmental Protection Agency criteria. Studies conducted by Consumers Union indicate that DEHA leaching from commercial plastic wraps is eight times higher than European directives allow. Fortunately, alternative wraps—such as Handi-Wrap™ and Glad Cling Wrap—are made of polyethylene; chemicals in them do not appear to leach into foods.

Barbie dolls manufactured in the 1950s and '60s are made of PVC containing a stabilizer that degrades to a sticky, estrogen-mimicking residue and accumulates on the dolls' bodies. This phenomenon was noted by Danish museum officials in August 2000. Yvonne Shashoua, an expert in materials preservation at the National Museum of Denmark, warns that young children who play with older Barbie and Ken dolls expose themselves to this estrogenic chemical, and she suggests enclosing the dolls with xenoestrogen-free plastic wrap. Storing the dolls in a cool, dark place also helps prevent the harmful stabilizer from oozing out. Who would have thought that these models of glamour would become health hazards?

Another consumer nightmare is a group of chemical plasticizers known as phthalates. They have been associated with various problems, including testicular depletion of zinc, a necessary nutrient for spermatogenesis. Zinc deficiency results in sperm death and consequent infertility. Many products that previously contained phthalates have been reformulated to eliminate them, but phthalates continue to be present in vinyl flooring, medical tubing and bags, adhesives, infants' toys, and ink used to print on food wrap made of plastic and cardboard. They have been detected in fat-soluble foods such as infant formula, cheese, margarine, and chips.

Some environmental estrogens can be found in unusual places, such as contraceptive products containing the spermicide nonoxynol-9. This chemical degrades to nonylphenol, a xenoestrogen shown to stimulate breast cancer cells. Nonylphenol and other alkylphenols have been detected in human umbilical cords, plastic test tubes, and industrial detergents.

Various substances added to lotions (including sunscreens) and cosmetics serve as preservatives. Some of them, members of a chemical family called parabens, were shown to be estrogen mimics by a study at Brunel University in the United Kingdom. The researchers warned that "the safety of these chemicals should be reassessed with particular attention being paid to . . . levels of systemic exposure to humans." Officials of the European Cosmetic Toiletry and Perfumery Association dismissed the Brunel study as "irrelevant," on the grounds that parabens do not enter the systemic circulation. But they ignored the possibility of transdermal introduction.

Additional questions have been raised about the safety—in terms of xenoestrogen content—of recycled materials, especially plastics and paper. Because it is unlikely that a moratorium on chemical synthesis will occur anytime soon, such questions will continue to surface until the public is satisfied that regulatory agencies are doing all they can to minimize exposure. Fortunately, organizations such as the National Institutes of Health, the National Academy of Sciences, the Environmental Protection Agency, the Centers for Disease Control, many universities, and other institutions are involved in efforts to monitor and minimize the effects of environmental estrogens on wildlife and humans.

Michael Gough

 **NO**

Endocrine Disrupters, Politics, Pesticides, the Cost of Food and Health

Environmentalists and politicians and federal regulators have added environmental estrogens or endocrine disrupters to the "concerns" or scares that dictate "environmental health policy." That policy, from its beginning, has been based on ideology, not on science. To provide some veneer to the ideology, its proponents have spawned bad science and junk science that claims chemicals in the environment are a major cause of human illness. There is no substance to the claims, but the current policies threaten to cost billions of dollars in wasted estrogen testing programs and to drive some substantial proportion of pesticides from the market.

Rachel Carson, conjuring up a cancer-free, pre-industrial Garden of Eden launched the biggest environmental scare of all in the 1960s. She charged that modern industrial chemicals in the environment caused human cancers. It mattered not at all to her or to her readers that cancers are found in every society, pre-industrial and modern. What mattered were opinions of people such as Umberto Safiotti of the National Cancer Institute, who wrote:

> I consider cancer as a social disease, largely caused by external agents which are derived from our technology, conditioned by our societal life-style and whose control is dependent on societal actions and policies.

When Saffioti said "societal actions and policies," he meant government regulations.

By 1968, environmental groups and individuals—including some scientists—appeared on TV and on the floors of the House and Senate to say, over and over again, "The environment causes 90 percent of cancers." They didn't have to say "environment" meant pollution from modern industry and chemicals—especially pesticides—everyone already knew that. Saffioti and others had told them.

In the 1970s, the National Cancer Institute [NCI] released reports that blamed elevated rates for all kinds of cancers on chemicals in the workplace or in the environment. The institute did not have evidence to link those exposures to cancer. It didn't exist then, and it doesn't, except for a limited number of high exposures in the workplace, exist now. So what? The reports were gobbled up by the press, politicians, and the public.

In our ignorance of what causes most cancers, the "90 percent" misstatement provided great hope. If the carcinogenic agents in the environment could be identified and eliminated, cancer rates should drop. NCI scientists said so, and they said success was just around the corner if animal tests were used to identify carcinogens. Congress responded. It created the Environmental Protection Agency [EPA] and the Occupational Safety and Health Administration [OSHA]. Both agencies have lots of tasks, but both place an emphasis on controlling exposures to carcinogens. Congress passed and amended law after law. The Clean Air Act, the Clean Water Act, the Safe Drinking Water Act, amendments to the Fungicide, Insecticide, and Rodenticide Act—the euphonious FIFRA—the Toxic Substances Control Act, and the Resource Recovery and Control Act poured forth from Capitol Hill.

And in return, EPA and OSHA, to justify their existence, generated scare after scare. They are aided by all kinds of people eager for explanations about their health problems or for government grants and contracts for research or other work or for money to compensate for health effects or other problems that could be blamed on chemicals.

In 1978, we had the occupational exposure scare. Astoundingly, according to a government report, six workplace substances caused 38 percent of all the cancers in the United States. It was nonsense, of course, and many scientists ridiculed the report, but the government never retracted it. The government scientists who contributed to it never repudiated it.

At about the same time, wastes disposed in Love Canal near Niagara Fall, NY, spewed liquids and gases into a residential community. The chemicals were blamed as the cause of cancers, birth defects, miscarriages, skin diseases, you name it. None of it was true, but waste sites around the country were routinely identified as "another Love Canal" or a "Love Canal in the making," and Congress gave the nation the Superfund Law. Since its passage, Superfund has enriched lawyers and provided secure employment to thousands who wear moon suits and dig up, burn, and rebury wastes, and done nothing for the nation's health. For those who doubt the importance of politics in the environmental health saga, it's worth recalling that every state had two waste sites on the first list of sites slated for priority cleanup under Superfund.

By the 1980s, EPA was chucking out the scares. We had the 2,4,5-T scare, the dioxin scare, the 2,4-D scare, the asbestos in schools scare, the radon in homes scare, the Alar scare, the EMF scare. I've left some out, but the common thread that linked the scares together was cancer. Each scare prompted investigations by affected industries and non-government scientists. Each scare fell apart, revealed as a house of cards jerry-rigged from bad science, worse interpretations of the science, and terrible policy.

In fact, by the late 1970s, there was ample evidence that the much-talked about "cancer epidemic" and the 90 percent statement were simply wrong. Cancer rates were not increasing and rates for some cancers were higher in industrial countries and rates for others were higher in non-industrial countries. The U.S. fell in the middle of countries when ranked by cancer rates. Sure, there are some carcinogenic substances in the nation's workplaces, but the best estimates are that they cause four percent or less of all cancers, and the percentage is decreasing because the biggest occupational threat, asbestos, is gone. Environmental exposures might cause two or three percent of cancer—on the outside—and they might cause much less.

The research into causes of cancer—not stories designed to bolster the chemicals cause cancer myth—did reveal that there are preventable causes of cancer. Not smoking is a good idea, as is eating lots of fruits and vegetables, not gaining too much weight, restricting the number of sexual partners, and, for those who are fair-skinned, being careful about sun exposures. It's not a lot different from what your mother or grandmother told you.

The government can take a nanny role in urging us to behave, but that's not where the big bucks are. The big bucks are in regulation, and regulation doesn't seem to have much to do with cancer.

In any case, cancer death rates began to fall in 1990, they've fallen since, and the fall is growing steeper. Maybe that information is blunting the cancer scare. I somehow doubt it. I think that the public has become numb to the cancer scare or that it fatalistically accepts the notion that "everything causes cancer." In any case, the environmentalists and the regulators needed a new scare.

The collapse of the cancer scare wasn't good news to everyone. Government bureaucrats and scientists in the anti-carcinogen offices and programs at EPA and elsewhere have secure jobs. Congress easily finds the will to write laws establishing environmental protection activities, but it lacks the will or patience to examine those activities to see if they've accomplished anything. And, let's face it, Congress doesn't eliminate established programs. But the growth of programs slows, and money can become scarce, and that can squeeze researchers who depend on EPA grants and contracts to fund their often senseless surveys and testing programs. Moreover, the fading of scares doesn't benefit environmental organizations that utter shrill cries about scares and coming calamities in their campaigns for contributions.

Here's an example of just how disappointed some people can be that cancer isn't on the rise. Dr. Theo Colborn, a wildlife biologist working for the Conservation Foundation in the late 1980s, was convinced that the chemicals in the Great Lakes were causing human cancer. She set out to prove it by reviewing the available literature about cancer rates in that region. She couldn't. In fact, she found that the rates for some cancers in the Great Lakes region were lower than the rates for the same cancers in other parts of the United States and Canada. . . .

Failing to find cancer slowed her down but didn't stop her. She knew that those chemicals were causing something. All she had to do was find it.

And find it, she did. She collected every paper that described any abnormality in wildlife that live on or around the Great Lakes, and concluded that synthetic chemicals were mimicking the effects of hormones. They were causing every problem in the literature, whether it was homosexual behavior among gulls, crossed bills in other birds, cancer in fish, or increases or decreases in any wildlife population.

The chemicals that have those activities were called "environmental estrogens" or "endocrine disrupters." There was no more evidence to link them to every abnormality in wildlife than there had been in the 1960s to link every human cancer to chemicals. The absence of evidence wasn't much of a problem. Colborn and her colleagues believed that chemicals were the culprit, and the press and much of the public, nutured on the idea that chemicals were bad, didn't require evidence.

Even so, Colborn had a problem that EPA faced in its early days. Soon after EPA was established, the agency leaders realized that protecting wildlife and the environment might be a good thing, but that Congress might not decide to lavish funds on such activities. They were sure, however, that Congress would throw money at programs that were going to protect human health from environmental risks. Whether Colborn knew that history or not, she apparently realized that any real splash for endocrine disrupters depended on tying them to human health effects.

Using the same techniques she'd used to catalogue the adverse effects of endocrine disrupters on wildlife, she reviewed the literature about human health effects that some way or another might be related to disruption of hormone activity. The list was long, including cancers, birth defects, and learning disabilities, but the big hitter on the list was decreased sperm counts. According to Colborn and others' analyses of sperm counts made in different parts of the world under different conditions of nutrition and stress and at different time periods, sperm counts had decreased by 50 percent in the post–World War II period.

If there's anything that catches the attention of Congress, it's risks to males. Congress banned leaded gasoline after EPA released a report that said atmospheric lead was a cause of heart attacks in middle-aged men. The reported decrease in sperm counts leaped up for attention, and attention it got. Congressional hearings were held, magazine articles were written, experts opined about endocrine disrupters and sexual dysfunctions.

And then it fell apart. Scientists found large geographical variations in sperm counts that have not changed over time. Those geographical variations and poor study designs accounted for the reported decrease. That scare went away, but endocrine disrupters were here to stay.

Well-organized and affluent women's groups are convinced that breast cancer is unusually common on Long Island. . . . We know that obesity, estrogen replacement therapy, and late child-bearing or no child-bearing, all of which are more common in affluent women, are associated with breast cancer. Nevertheless, from the very beginning, environmental chemicals have been singled out as the cause of the breast cancer excess. The insecticide aldicarb, which is very resistant to degradation was blamed, but subsequent studies failed to confirm the link. A well-publicized study found a link between DDT and breast cancer, but larger, follow-up studies failed to confirm it. But there're lots of environmental chemicals, and no evidence is required to justify a suggestion of a link between the chemicals and cancer.

Senator Al D'Amato is from Long Island, and he shares his constituents' concerns. During a hearing about the Clean Water Act, Senator D'Amato heard testimony by Dr. Anna Soto from Tufts University about her "E-Screen." According to Dr. Soto, her quick laboratory test could identify chemicals that behave as environmental estrogens or endocrine disrupters for $500 a chemical. Since environmental estrogens seem to some people to be a likely cause of breast cancer, Soto's test appeared to be a real bargain.

Senator D'Amato pushed for an amendment to require E-Screen testing of chemicals that are regulated under the Clean Water Act, but he was unsuccessful. Later in 1995, a senior Senate staffer, Jimmy Powell, took the E-Screen amendment to a very junior Senate staffer and told her to incorporate it into the Safe Drinking Water Act as an "administrative amendment." She did, it passed the Senate, and, for the first time, there was a legislative requirement for endocrine testing.

In the spring of 1996, House Committees were considering legislation to amend the Safe Drinking Water Act and new legislation related to pesticides in food. Aware of the Senate's action, some members of the House committees were eager to include endocrine testing in their legislation, but there was resistance as well. Chemical companies viewed the imposition of yet another test as certain to be an expense, unlikely to cost as little as $500 a chemical, and bound to raise new concerns about chemicals that would require far more extensive tests and research to understand or discount.

Furthermore, so far as food was concerned, there was a general conviction that all the safety factors built into the testing of pesticides and other chemicals that might end up in food provided adequate margins of protection. That conviction was shattered by rumors that reached the House in May 1996. According to the rumors, Dr. John McLachlan and his colleagues at Tulane University had shown that mixtures of pesticides and other environmental chemicals such as PCBs were far more potent in activating estrogen receptors, the first step in estrogen modulation of biochemical pathways than were single chemicals. In the most extreme case, two chemicals at concentrations considered safe by all conventional toxicity tests were 1600 times as potent as estrogen receptor activators as either chemical by itself. The powerful synergy raised new alarms.

In May, everyone concerned about pesticides knew that EPA had a draft of the Tulane paper, and EPA staff were drifting around House offices, but they refused to answer questions about the Tulane results. The silence signaled the expected significance of the paper. A month later, in June, the paper appeared in *Science*. It was a big deal. *Science* ran a news article about the research with a picture of the Tulane researchers. It also ran an editorial by a scientist from the National Institutes of Health who offered some theoretical explanations for how combinations of pesticides at very low levels could affect cells and activate the estrogen receptors. *The New York Times, The Washington Post*, other major newspapers, and newsmagazines and TV reported the news. If there was ever any doubt that FQPA [Food Quality Protection Act] would require tests for endocrine activity, the flurry of news about the Tulane results erased them.

While the House was drafting the FQPA, Dr. Lynn Goldman, an assistant administrator at EPA, established a committee called the "Endocrine Disrupter Screening and Testing Advisory Committee" (EDSTAC) [which is now] considering tests for all of the 70,000 chemicals that it estimates are present in commerce, and it's not limiting its recommendations to tests for estrogenic activity. It's adding tests for testosterone and thyroid hormone activity as well as for anti-estrogenic, anti-testosterone, and anti-thyroid activity. The relatively simple E-Screen, which is run on cultured cells, is to be supplemented by some whole animal tests. Tests on single compounds will have to be complemented by tests of mixtures of compounds. The FQPA requires that "valid" tests be used. None of the tests being considered by EDSTAC has been validated; many of them have never been done.

EDSTAC's estimate of 70,000 chemicals in commerce is on the high side—some of those chemicals are used in such small amounts and under such controlled conditions that there's no exposure to them. Dr. Dan Byrd has estimated that 50,000 is a more realistic number. He's also looked at the price lists from commercial testing laboratories to see how much they would charge for a battery of tests something like EDSTAC is considering. Some of the tests haven't been developed, but assuming they can be, Dr. Byrd estimates testing each chemical will cost between $100,000 and $200,000. The total cost would be between $5 and $10 billion. . . .

The Tulane results played some major role in the passage of FQPA, the focus on endocrine disrupters, and Dr. Goldman's establishment of EDSTAC. The Tulane results are wrong. Several groups of scientists tried to replicate the Tulane results. None could. At first, Dr. McLachlan insisted his results were correct. He said that the experiments he reported required expertise and finesse and suggested that the scientists who couldn't repeat his findings were at fault, essentially incompetent. That changed. In July 1997, just 13 months after he published his report, he threw in the towel, acknowledging that neither his laboratory nor anyone else had been able to produce the results that had created such a stir.

Whether the initial results were caused by a series of mistakes or a willful desire to show, once and for all, that environmental chemicals, especially pesticides are bad, bad, bad, we don't know. We do know that the results were wrong.

No matter, EPA now assumes as a matter of policy that synergy occurs. Good science, repeatable science that showed the reported synergy didn't occur has been brushed aside. In its place, we have bad science or junk science. If the Tulane results were the products of honest mistakes, they're bad science; if they flowed from ideology, they're junk science. The effect is the same, but the reasons are different.

The estrogenic disrupter testing under FQPA is going to cost a lot of money and cause a lot of mischief. But the effects of that testing are off somewhere in the future. More immediately, a combination of ideology-driven science and congressional misreading of that science threatens to drive between 50 and 80 percent of all pesticides from the market.

In 1993, a committee of the National Research Council spun together the facts that childhood developmental takes place at specific times as an infant matures into a toddler and then a child, that infants, toddlers, and children eat, proportionally, far larger amounts of foods such as apple juice and apple sauce and orange juice than do adults, and that pesticides can be present in those processed foods. From those three observations, the committee concluded that an additional safety factor should be included in setting acceptable levels for pesticides in those foods. Left out from the analysis was any evidence that current exposures cause any harm to any infants, toddlers, or children. No matter.

Most people who worry about pesticides expected EPA and the Food and Drug Administration to react to the NRC recommendation by reducing the allowable levels of pesticides on foods that are destined for consumption by children. Maybe they would have. We'll never know. In the FQPA, Congress directed that a new ten-fold safety factor be incorporated into the evaluation of the risks from pesticides.

Safety factors are a fundamental part in the evaluation of pesticide risks. Pesticides are tested in laboratory animals to determine what concentrations to cause effects on the nervous, digestive, endocrine, and other systems. At some, sufficiently low dose that varies from pesticide to pesticide, the chemical does not cause those adverse effects. That dose, called the "No Observed Adverse Effect Level" (or NOAEL), is then divided by 100 to set the acceptable daily limit for human ingestion of the chemical. The FQPA requires division by another factor of 10, so the acceptable daily limit will be the NOAEL divided by 1000 instead of 100. Acceptable limits will be ten-fold less.

Dr. Byrd has estimated that up to 80 percent of all currently permitted uses of pesticides would be eliminated by an across the board application of the 1000-fold safety factor. He cites another toxicologist who estimates that 50 percent of all pesticides would be eliminated from the market. The extent to which these draconian reductions will be forced remains to be seen, but pesticide manufacturers and users can look forward to a period of even-greater limbo as EPA sorts through it new responsibilities and decides how to implement FQPA.

There's no convincing evidence that pesticides in food contribute to cancer causation and none that they cause other adverse health effects. Restrictions on pesticides in food will not have a demonstrable effect on human health. On the other hand, the estrogen testing program and the new safety factor will drive pesticide costs up and pesticide availability down.

Some manufacturers may lose profitable product lines; some may even lose their businesses. Farmers will pay more. They will pass those costs onto middlemen and processors, who, in turn, will pass them onto consumers. Increases in the costs of fruits and vegetables won't change the food purchasing habits of the middle class, but they may and probably will affect the purchases of the poor. The poor are already at greater risks because of poor diets, and the increased costs can be expected to further decrease their consumption of fresh fruits and vegetables.

POSTSCRIPT

Do Environmental Hormone Mimics Pose a Potentially Serious Health Threat?

Stephen H. Safe's "Environmental and Dietary Estrogens and Human Health: Is There a Problem?" *Environmental Health Perspectives* (April 1995) is often cited to support the contention that there is no causative link between environmental estrogens and human health problems. He draws a cautious conclusion, calling the link "implausible" and "unproven." Gough's belief that the battle against environmental estrogens is motivated by environmentalist ideology rather than facts is repeated by Angela Logomasini in "Chemical Warfare: Ideological Environmentalism's Quixotic Campaign Against Synthetic Chemicals," in Ronald Bailey, ed., *Global Warming and Other Eco-Myths: How the Environmental Movement Uses False Science to Scare Us to Death* (Prima, 2002). Some caution is certainly warranted, for the complex and variable manner by which different compounds with estrogenic properties may affect organisms makes projections from animal effects to human effects risky.

Sheldon Krimsky, in "Hormone Disruptors: A Clue to Understanding the Environmental Cause of Disease," *Environment* (June 2001), summarizes the evidence that many chemicals released to the environment affect—both singly and in combination or synergistically—the endocrine systems of animals and humans and may threaten human health with cancers, reproductive anomalies, and neurological effects. He cautions that the regulatory machinery is likely to move very slowly, adding that we cannot wait for scientific certainty about the hazards before we act. See also M. Gochfeld, "Why Epidemiology of Endocrine Disruptors Warrants the Precautionary Principle," *Pure & Applied Chemistry* (December 1, 2003).

Theo Colborn, a senior scientist with the World Wildlife Fund, first drew public attention to the potential problems of environmental estrogens with the book *Our Stolen Future* (Dutton, 1996), coauthored by Dianne Dumanoski and John Peterson Myers. Colborn clearly believes that the problem is real; she finds the evidence that extensive damage is being done to wildlife by synthetic estrogenic chemicals convincing and thinks it likely that humans are experiencing similar health problems. Recent data reinforce her and Krimsky's points; see Rebecca Renner, "Human Estrogens Linked to Endocrine Disruption," *Environmental Science and Technology* (January 1, 1998) and Ted Schettler, et al., *Generations at Risk: Reproductive Health and the Environment* (MIT Press, 1999). When researchers added synthetic estrogen to a Canadian lake, they found that the reproduction of fathead minnows was so severely affected that the population nearly died out; see Karen A. Kidd, et al.,

"Collapse of a Fish Population After Exposure to a Synthetic Estrogen," *Proceedings of the National Academy of Sciences of the United States of America* (May 22, 2007). In 1999 the National Research Council published *Hormonally Active Agents in the Environment* (National Academy Press), in which the council's Committee on Hormonally Active Agents in the Environment reports on its evaluation of the scientific evidence pertaining to endocrine disruptors. The National Environmental Health Association has called for more research and product testing; see Ginger L. Gist, "National Environmental Health Association Position on Endocrine Disrupters," *Journal of Environmental Health* (January–February 1998).

Elisabete Silva, Nissanka Rajapakse, and Andreas Kortenkamp, in "Something From 'Nothing'—Eight Weak Estrogenic Chemicals Combined at Concentrations Below NOECs Produce Significant Mixture Effects," *Environmental Science and Technology* (April 2002), find synergistic effects of exactly the kind dismissed by Gough. Also, after reviewing the evidence, the U.S. National Toxicology Program found that low-dose effects had been demonstrated in animals (see Ronald Melnick, et al., "Summary of the National Toxicology Program's Report of the Endocrine Disruptors Low-Dose Peer Review," *Environmental Health Perspectives* [April 2002]). Some researchers say there is reason to think endocrine disruptors may be linked to the epidemic of obesity in the United States (Retha R. Newbold, et al., "Developmental Exposure to Endocrine Disruptors and the Obesity Epidemic," *Reproductive Toxicology*, April 2007) and to a decline in the proportion of male births in the United States and Japan (Julia R. Barrett, "Shift in the Sexes," *Environmental Health Perspectives*, June 2007). The evidence seems to favor the view that environmental hormone mimics have potentially serious effects.

ISSUE 18

Is the Superfund Program Successfully Protecting Human Health from Hazardous Materials?

YES: **Robert H. Harris, Jay Vandeven, and Mike Tilchin,** from "Superfund Matures Gracefully," *Issues in Science & Technology,* Summer 2003, Vol. 19, Issue 4

NO: **Randall Patterson,** from "Not in Their Back Yard," *Mother Jones* (May/June 2007)

ISSUE SUMMARY

YES: Environmental consultants Robert H. Harris, Jay Vandeven, and Mike Tilchin argue that although the Superfund program still has room for improvement, it has made great progress in risk assessment and treatment technologies.

NO: Journalist Randall Patterson argues that the Superfund Program is not applied to some appropriate situations, largely because people resist its application.

The potentially disastrous consequences of ignoring hazardous wastes and disposing of them improperly burst upon the consciousness of the American public in the late 1970s. The problem was dramatized by the evacuation of dozens of residents of Niagara Falls, New York, whose health was being threatened by chemicals leaking from the abandoned Love Canal, which was used for many years as an industrial waste dump. Awakened to the dangers posed by chemical dumping, numerous communities bordering on industrial manufacturing areas across the country began to discover and report local sites where chemicals had been disposed of in open lagoons or were leaking from disintegrating steel drums. Such esoteric chemical names as dioxins and PCBs have become part of the common lexicon, and numerous local citizens' groups have been mobilized to prevent human exposure to these and other toxins.

The expansion of the industrial use of synthetic chemicals following World War II resulted in the need to dispose of vast quantities of wastes laden with organic and inorganic chemical toxins. For the most part, industry adopted a casual attitude toward this problem and, in the absence of regulatory restraint, chose the least expensive means of disposal available. Little attention was paid to the ultimate fate of chemicals that could seep into surface water or groundwater. Scientists have estimated that less than 10 percent of the waste was disposed of in an environmentally sound manner.

The magnitude of the problem is truly mind-boggling: Over 275 million tons of hazardous waste is produced in the United States each year; as many as 10,000 dump sites may pose a serious threat to public health, according to the federal Office of Technology Assessment; and other government estimates indicate that more than 350,000 waste sites may ultimately require corrective action at a cost that could easily exceed $500 billion.

Congressional response to the hazardous waste threat is embodied in two complex legislative initiatives. The Resource Conservation and Recovery Act (RCRA) of 1976 mandated action by the Environmental Protection Agency (EPA) to create "cradle to grave" oversight of newly generated waste, and the Comprehensive Environmental Response, Compensation, and Liability Act of 1980 (CERCLA), commonly called "Superfund," gave the EPA broad authority to clean up existing hazardous waste sites. The implementation of this legislation has been severely criticized by environmental organizations, citizens' groups, and members of Congress who have accused the EPA of foot-dragging and a variety of politically motivated improprieties. Less than 20 percent of the original $1.6 billion Superfund allocation was actually spent on waste cleanup.

Amendments designed to close RCRA loopholes were enacted in 1984, and the Superfund Amendments and Reauthorization Act (SARA) added $8.6 billion to a strengthened cleanup effort in 1986 and an additional $5.1 billion in 1990. While acknowledging some improvement, both environmental and industrial policy analysts remain very critical about the way that both RCRA and Superfund/SARA are being implemented. Efforts to reauthorize and modify both of these hazardous waste laws have been stalled in Congress since the early 1990s. But the work went on. The Superfund program continued to identify hazardous waste sites that warranted cleanup and to clean up sites; see http://www.epa.gov/superfund/ for the latest news. In 2004, it even declared that the infamous Love Canal site was finally safe.

In the following selections, Robert H. Harris, Jay Vandeven, and Mike Tilchin argue that the Superfund program has had to struggle with unexpectedly large cleanup tasks such as mines, harbors, and the ruins of the World Trade Center in New York; is subject to greater resource demands than ever before; and still has room for improvement. But, they say, the program has made great progress in risk assessment and treatment technologies. Journalist Randall Patterson makes it clear that not all places made dangerous to human health by chemicals are industrial sites. They can be suburbs such as El Dorado Hills, CA, where ordinary construction activities produce hazardous waste of another sort, in this case dust containing asbestos. He argues that El Dorado Hills has not been labeled a Superfund site largely because people deny there is a problem and resist the EPA's intervention.

YES

Robert H. Harris, Jay Vandeven, and Mike Tilchin

Superfund Matures Gracefully

Superfund, one of the main programs used by the Environmental Protection Agency (EPA) to clean up serious, often abandoned, hazardous waste sites, has been improved considerably in recent years. Notably, progress has been made in two important areas: the development of risk assessments that are scientifically valid yet flexible, and the development and implementation of better treatment technologies.

The 1986 Superfund Amendments and Reauthorization Act (SARA) provided a broad refocus to the program. The act included an explicit preference for the selection of remediation technologies that "permanently and significantly reduce the volume, toxicity, or mobility of hazardous substances." SARA also required the revision of the National Contingency Plan (NCP) that sets out EPA's rules and guidance for site characterization, risk assessment, and remedy selection.

The NCP specifies the levels of risk to human health that are allowable at Superfund sites. However, "potentially responsible parties"—companies or other entities that may be forced to help pay for the cleanup—have often challenged the risk assessment methods used as scientifically flawed, resulting in remedies that are unnecessary and too costly. Since SARA was enacted, fundamental changes have evolved in the policies and science that EPA embraces in evaluating health risks at Superfund sites, and these changes have in turn affected which remedies are most often selected. Among the changes are three that collectively can have a profound impact on the selected remedy and attendant costs: EPA's development of land use guidance, its development of guidance on "principal threats," and the NCP requirement for the evaluation of "short-term effectiveness."

Before EPA's issuance in 1995 of land use guidance for evaluating the potential future public health risks at Superfund sites, its risk assessments usually would assume a future residential use scenario at a site, however unrealistic that assumption might be. This scenario would often result in the need for costly soil and waste removal remedies necessary to protect against hypothetical risks, such as those to children playing in contaminated soil or drinking contaminated ground water, even at sites where future residential use was highly improbable. The revised land use guidance provided a basis for selecting more realistic future use scenarios, with projected exposure patterns that may allow for less costly remedies.

Potentially responsible parties also complained that there was little room to tailor remedies to the magnitude of cancer risk at a site, and that the same costly remedies would be chosen for sites where the cancer risks may differ by several orders of magnitude. However, EPA's guidance on principal threats essentially established a risk-based hierarchy for remedy selection. For example, if cancer risks at a site exceed 1 in 1,000, then treatment or waste removal or both might be required. Sites that posed a lower lifetime cancer risk could be managed in other ways, such as by prohibiting the installation of drinking water wells, which likely would be far less expensive than intrusive remedies.

Revisions to the NCP in 1990 not only codified provisions required by the 1986 Superfund amendments, but also refined EPA's evolving remedy-selection criteria. For example, these revisions require an explicit consideration of the short-term effectiveness of a remedy, including the health and safety risks to the public and to workers associated with remedy implementation. EPA had learned by bitter experience that to ignore implementation risks, such as those associated with vapor and dust emissions during the excavation of wastes, could lead to the selection of remedies that proved costly and created unacceptable risks.

Although these changes in risk assessment procedures have brought greater rationality to the evaluation of Superfund sites, EPA still usually insists on the use of hypothetical exposure factors (for example, the length of time that someone may come in contact with the site) that may overstate risks. The agency has been slow in embracing other methodologies, such as probabilistic exposure analysis, that might offer more accurate assessments. Thus, some remedies are still fashioned on risk analyses that overstate risk.

Technological Evolution

Cleanup efforts in Superfund's early years were dominated by containment and excavation-and-disposal remedies. But over the years, cooperative work by government, industry, and academia have led to the development and implementation of improved treatment technologies.

The period from the mid-1980s to the early 1990s was marked by a dramatic increase in the use of source control treatment, reflecting the preference expressed in SARA for "permanent solutions and alternative treatment technologies or resource recovery technologies to the maximum extent practicable." Two types of source control technologies that have been widely used are incineration and soil vapor extraction. Although the use of incineration decreased during the 1990s because of cost and other factors, soil vapor extraction remains a proven technology at Superfund sites.

Just as early source control remedies relied on containment or excavation and disposal offsite, the presumptive remedy for groundwater contamination has historically been "pump and treat." It became widely recognized in the early 1990s that conventional pump-and-treat technologies had significant limitations, including relatively high costs. What emerged to fill the gap was an approach called "monitored natural attenuation" (MNA), which makes use of a variety of technologies, such as biodegradation, dispersion, dilution, absorption, and volatilization. As the name suggests, monitoring the effectiveness of the process is a key element of this technology. And although cleanup times still may be on the order of years, there is evidence that MNA can achieve comparable results in comparable periods and at significantly lower costs than conventional pump-and-treat systems. EPA has taken an active role in promoting this technology, and its use has increased dramatically in recent years.

As suggested by the MNA example, what may prove an even more formidable challenge than selecting specific remedies is the post-remedy implementation phase—that is, the monitoring and evaluation that will be required during coming decades to ensure that the remedy chosen is continuing to protect human health and the environment. Far too few resources have been devoted to this task, which will require not only monitoring and maintaining the physical integrity of the technology used and ensuring the continued viability of institutional controls, but also evaluating and responding to the developing science regarding chemical detection and toxicity.

Coming Challenges

In recent years, the rate at which waste sites are being added to the National Priorities List (NPL) has been decreasing dramatically as compared with earlier years. In fiscal years 1983 to 1991, EPA placed an average of 135 sites on the NPL annually. The rate dropped to an average of 27 sites per year between 1992 and 2001. Although many factors have contributed to this trend, three stand out:

- There was a finite group of truly troublesome sites before Superfund's passage, and after a few years most of those were identified.
- The program's enforcement authority has had a profound impact on how wastes are managed, significantly reducing, although not eliminating, the types of waste management practices that result in the creation of Superfund sites.
- A range of alternative cleanup programs, such as voluntary cleanup programs and those for brownfields, have evolved at both the federal and state levels. No longer is Superfund the only path for cleaning up sites.

But such programmatic changes are about more than just site numbers. In 1988, most NPL sites were in the investigation stage, and the program was widely criticized as being too much about studies and not enough about cleanup. Superfund is now a program predominantly focused on the design and construction of cleanup remedies.

This shift reflects the natural progress of sites through the Superfund pipeline, the changes in NPL listing activity, and a deliberate emphasis on achieving "construction completion," which is the primary measure of achievement for the program as established under the Government Performance and Results Act. It is a truism in regulatory matters that what gets done is what gets measured, and Superfund is no exception.

In the late 1990s, many observers believed that the demands on Superfund were declining and that it would be completed sometime in the middle of the first decade of the new century. But this is not proving to be true. Although expenditures have not been changing dramatically over time, the resource demands on the program are greater today than ever before.

Few people would have predicted, for example, that among the biggest technical and resource challenges facing Superfund at this date would be the cleanup of hard-rock mining sites and of large volumes of sediments from contaminated waterways and ports. These sites tend to be very costly to clean up, with the driver behind these great costs weighted more toward the protection of natural resources than of human health. In mapping the future course of the program, Congress and EPA must address the question of whether Superfund is the most appropriate program for cleaning up these types of sites.

There are other uncertainties, as well. The substantial role that Superfund has played in emergency response in the aftermath of 9/11, the response to the anthrax attacks of October 2001, and the program's role in the recovery of debris from the crash of the space shuttle Columbia were all totally unforeseeable. Although many valuable lessons have been learned over the past 20 years of the program, there remain substantial opportunities for improvement as well as considerable uncertainty about the kinds of environmental problems Superfund will tackle in the coming decade.

Randall Patterson  **NO**

Not in Their Back Yard

When the EPA discovered asbestos in their Little League fields, the residents of idyllic El Dorado Hills rushed to protect themselves—from reality.

The land in question is a high, rolling hillside, where California's Central Valley slopes into the Sierra foothills. Many have been drawn here, and after the first stampede, for gold in 1849, most left disappointed. Then the land lay virtually empty for many years until, as the state capital grew, its emptiness became valuable. In the late 1970s, as the local Chamber of Commerce tells it, a developer was traveling along U.S. Highway 50 when, 25 miles east of Sacramento, he looked up and had a vision.

Grand homes started to appear, followed by fine schools, lush golf courses, and a central shopping district modeled after a Tuscan village. The entire hillside was transformed into a patchwork of gated communities, and over the past 15 years, El Dorado Hills, as developers dubbed the spot, has been one of the fastest-growing areas in California. "High per capita income along with low crime rates have attracted many new families to this vibrant and desirable destination of the future," says the Chamber of Commerce's website. "As the name 'El Dorado' translates, this is a land of golden opportunity!"

Some residents of the bedroom community like to think of it as the next Silicon Valley, but the president of the Chamber of Commerce admits there are few jobs, and "the one thing very different about El Dorado Hills" is that no one has to be here. Almost no one leaves either, which is surprising when you consider that what lies beneath these hills isn't gold but something potentially deadly.

Housing developers and real estate agents don't like talking about the dark side of paradise, so I had to rely on Terry Trent to give me an unofficial tour. Trent, a 50-ish retired construction consultant, no longer lives in El Dorado County but admits that his former home has become something of an obsession. "There are some really dangerous things in the ground," he said as we drove down the main street, El Dorado Hills Boulevard. Talking nonstop, Trent pointed to the hillside that forms the spine of the town, which, according to his map from the state's Division of Mines and Geology, contains a vast deposit of naturally occurring asbestos.

There are similar deposits all over California; serpentine, a primary source of asbestos, is the state rock. It's not dangerous—unless it becomes airborne dust and gets in your lungs. The trouble is, in El Dorado Hills, dust isn't just a fact of life but a sign of progress. "Here in the mountains, we don't let no rocks stand in our way," Trent said. Over the past 10 years, as the town's population has doubled to around 35,000, hundreds of bulldozers have scraped over this ground, clearing land for thousands of new homes. In that earth is a form of asbestos—amphibole—that's significantly more toxic than the type commonly used commercially.

Trent's map of the amphibole deposits was made 50 years ago. Still, most residents were unaware of the asbestos until 11 years ago, when Trent uncovered a huge, glistening vein of it in his yard. He went to the local papers with the news, and there was a brief outcry. But the construction went on, and most of El Dorado Hills has been built in the years since.

Turning onto Harvard Way, Trent stopped before a grassy embankment that he claims is shot through with asbestos. "You can see how they carved through it to build this road," he said. He recalled watching as construction crews gouged out a spot for a new middle school nearby, crushing the excavated rock and transporting the gravel to become the foundation for the new community center. A half-mile stretch of Harvard Way was covered in dust, he said, for months.

Nine years ago, tests commissioned by the *Sacramento Bee* found high concentrations of amphibole fibers inside Trent's home. He moved soon afterward. To this day, he still can't understand why few of his former neighbors seem concerned about the threat beneath them, or why the town is still fully occupied, its homes still worth hundreds of thousands of dollars. "I don't like to be mean," he said, "but I'm a little cranky with these people. It's like they're in shock. They don't want this stuff to exist; therefore, they put their heads in the sand. They say, 'Well, it's not here, and it's not bad, and here, I'll prove that it's not here and it's not bad.'"

In the fall of 2004, agents from the Environmental Protection Agency descended on El Dorado Hills in respirators and protective suits and headed for a town park. There, they began playing as children would—tossing baseballs, kicking soccer balls, biking, and running. All the while, they took air samples, and all the while they were quietly watched by the citizens of El Dorado Hills.

The civic leaders of El Dorado Hills had spent many months trying to stave off these tests, scrambling to protect the community not from potentially toxic substances, but from the EPA's potentially toxic information. Taking the lead was Vicki Barber, the superintendent of schools. A stout woman with compressed lips and an unwavering gaze, she recently won an award for being "a person who does not accept the word 'no' when it comes to what is good for students." After asbestos was found during the construction of a high school soccer field in 2002, Barber questioned a costly EPA-mandated cleanup. When a citizen formally asked the EPA to test the town's public areas for asbestos in 2003, Barber quickly emerged as the agency's most determined local foe. Before the study was even under way, she began writing to the EPA as well as to senators and congressmen, questioning whether the agency had the "legal and scientific authority" to conduct what she called a "science experiment" with "limited benefit to the residents." At least four state legislators and one congressman responded by putting pressure on the EPA, which in turn agreed not to declare El Dorado Hills a Superfund site, regardless of what it might find there.

In May 2005, the EPA announced its findings: Almost every one of more than 400 air samples collected at the park, as well as along a hiking trail and in three school yards, contained asbestos fibers. Most of the samples contained amphibole fibers known as tremolite—long, thin strands that, when inhaled, tend to wedge in the lungs for much longer and at higher concentrations than other forms of asbestos, even at very low exposure levels.

The correlation between amphibole asbestos and lung disease has been demonstrated around the world, from Canada to Cyprus, and perhaps most convincingly in the mining town of Libby, Montana, where hundreds of residents have been diagnosed with asbestos-related diseases and many have died. After reviewing the EPA's findings in El Dorado Hills, a Canadian epidemiologist told the *Fresno Bee* that the area's exposure levels were comparable to those in towns where mining had gone on for a century. "You can certainly say people are going to die, and there are going to be increased cases of cancer," he said. "I wouldn't live there. I wouldn't want my family to live there."

The EPA, however, avoided making such dire statements. Its report neither specified how toxic the samples were nor quantified the health risk to the residents of El Dorado Hills. Rather, it simply noted that the exposure levels were "of concern." Jere Johnson, who oversaw the study, explained that the agency was trying not to sound alarmist. "We wanted to inform the community without scaring them," she said.

But local officials received the EPA's findings almost as a declaration of war on El Dorado Hills and its way of life. Jon Morgan, the director of the county's environmental management department, thundered to a local television station that the tremolite announcement "may unnecessarily scare the living daylights out of every man, woman, and child in El Dorado County and could possibly devastate the county for years to come."

When I visited El Dorado Hills, more than a year after the EPA had run its tests, I was told over and over that the biggest threat facing the town was big-government intrusion. "The issue is not asbestos," declared James Sweeney, chairman of the county Board of Supervisors. "The issue is the EPA." Superintendent Barber said that El Dorado Hills is like any other community in California, "and yet why is it that the EPA has decided to focus on only here?" One county supervisor claimed that an EPA official had told her he was going to make El Dorado Hills "a poster child for asbestos." "What I heard," town general manager Wayne Lowery said, "is that the Libby, Montana, project is being wrapped up and they have all these employees with training, and they're looking for a place to keep them employed."

What were officials supposed to do with the EPA's information, wondered Debbie Manning, the president of the Chamber of Commerce. "Are you just going to make it a ghost town?" There was no need to rush to judgment until all the facts had been determined. "I'm sure you know that asbestos is a complicated issue," she told me, explaining that there had been a locally led effort to get the information out about asbestos. Fortunately, El Dorado Hills is a highly educated community, she said, and you can trust that "with good information, people can make proper decisions."

One of those struggling to do just that was Vicki Summers, a 47-year-old mom who said that her chief duty was to protect her two children from "any type of health issues." Surrounded by stuffed animals, she sat on a couch in her spacious living room, her eyes darting as she spoke of the dangers they faced. "Every day in the newspaper, there's something new," she said—chemicals in the water, mad cow disease, bird flu. She often felt overwhelmed, especially when experts said different things. Summers had guzzled green tea for its anti-cancer benefits until she heard it caused colon cancer. She had gorged on fish during her first pregnancy, thinking it was "brain food," only to be told, while expecting her second child, that mercury could cause brain damage.

The same sort of thing had happened to her dad. When she was a girl, he had made a point of buying margarine because he wanted to protect his family from cholesterol. "And now he's had three strokes," she explained, "and they say it's the plaque buildup from the hydrogenated vegetable oil." So much conflicting information was almost paralyzing. "It would be better to be ignorant," she said, "because then you wouldn't have to be stressed out by all this."

Summers called her home her "dream house," and said she had long considered El Dorado Hills "heaven on Earth," but when she heard about the asbestos, her first thought was, "If this is Love Canal, and there's going to be a mass exodus, I don't want to be the last one out."

But no exodus from El Dorado Hills ever began. Many residents seemed oddly comforted by the apparent uncertainty of the situation. Just after the EPA report came out, Summers joined a thousand other citizens in the town gym, looking for answers. There, school superintendent Barber reassured the crowd that the town was "deeply committed to maintaining public health and safety." Then she ripped into the EPA report for offering no solid "risk information." She pointed out there had been no abundance of "pulmonary cases" in the area. In other words, with nobody getting sick or dying, there was no real evidence of any hazard, so why worry? "Risk is a part of all of our lives," Barber said. "But we also need to keep it in perspective." She received enthusiastic applause.

Scientists, however, are more certain about the dangers of asbestos. There is no known safe exposure level to asbestos, whether it is in a commercial or a natural form; even low doses can cause malignant mesothelioma, a rare lung cancer. And the evidence keeps mounting. In the summer of 2005, researchers from the University of California-Davis Department of Public Health Sciences announced the findings of a study comparing the addresses of 2,900 Californians suffering from malignant mesothelioma against a geological map of the state. They concluded that the risk of developing lung cancer was directly related to how close the patients had lived to areas of rock associated with naturally occurring tremolite asbestos. (Likewise, the odds of getting mesothelioma drop 1 percent for every mile one moves away from an asbestos source.) As one of the authors explained, "We showed that breathing asbestos in your community is not magically different from breathing asbestos in an industrial setting."

All of this passed over El Dorado Hills like clouds in the blue sky. County supervisor Helen Baumann's reaction to the UC Davis report was that "there are studies counter to that." Local government would "err on the side of public health and safety," she said, "until some better science comes forward."

So began the effort to live in El Dorado Hills without coming into contact with the earth. In the park and the school playing fields where the EPA had collected samples, workers trucked in clean soil to replace two feet of topsoil, as the agency had recommended. The county banned using leaf blowers on town property, "except for emergency situations." Residents were encouraged to take precautions such as removing their shoes before entering their homes, driving slowly with the windows up on unpaved roads, and "limiting time spent on dirt." Home-builders agreed to spray down building sites and rinse off their trucks' tires after work. Any sightings of "fugitive dust" were to be reported to a new "dust enforcement" team.

Life in El Dorado Hills under the new regime was "manageable," said Baumann. That seemed to be the point of the show: If you took a few precautions, there was no need to panic. She proudly pointed out that the dust hot line rarely received a call. Gerri Silva, the interim director of the county department of environmental management, told me the measures had worked so well that in all of El Dorado Hills, "there is no dust."

In the town's main coffee shop, Bella Bru, languor pervaded as well. An engineer named Matt Parisek said that asbestos was "pretty muck a fake issue." "It's pretty obvious [the EPA] selected this county because of our conservative Republican reputation," he said. And, wondered retiree Carole Gilmore, "If asbestos is all over California, I don't know why they zero in on El Dorado Hills."

Real estate broker Charles Hite, who is the current resident of Terry Trent's old house, pointed out that none of his neighbors had died. "If it is such a health hazard," he asked, "why are people still buying and building? Why are real estate prices going up? Why doesn't government shut the city down?"

There was a circular logic at work. The local government had told everyone to stay calm, and so residents weren't afraid. And if no one was scared, then perhaps there was nothing to fear. "The vast, vast majority" trusted that they were safe, said Baumann. But, she added, there remained "a very, very small centralized group that keeps pushing issues and pushing issues to the extreme." She specifically mentioned Terry Trent, who to this day sends regular, dire emails to residents, reporters, and scientists. The group of extremists who "do not represent our community" also included, in Baumann's view, some members of a community group that meets monthly to discuss asbestos.

Vicki Summers had joined the Asbestos Community Advisory Group as part of her effort to educate herself. When someone would mention the low body count as evidence that nothing was wrong, she could now cite the UC Davis study, or note that the latency period for asbestos-related diseases is about 30 years. Who lived in these hills back then, she asked, "a dozen ranchers?" Other parents assumed that if they couldn't see asbestos in the air, it wasn't there; Summers could picture invisible fibers burrowing into her kids' lungs. "It never disappears," she said, wide-eyed. "That's the scary thing. And it's a known carcinogen. It's a known carcinogen. And if we know that, we know it's not good for our kids."

Summers said she was thinking only of her family when, at a community meeting, she passed a note to an EPA official, asking, "Should I move?" After that was reported by a newspaper, she received an answer in an anonymous, late-night phone call: "Vicki Summers, I think you should move!"

Short of evacuating, there are only two ways to solve El Dorado Hills' asbestos problem: Either pave over every asbestos deposit in town—an impossible task—or make the asbestos magically disappear. Soon after the EPA report came out, Superintendent Barber got in touch with the National Stone, Sand and Gravel Association, an industry trade group, which commissioned the R.J. Lee Group to scrutinize the EPA's data. In 2000, it was reported that R.J. Lee's president, Richard Lee, had been paid about $7 million for testifying more than 250 times on behalf of the asbestos industry. When the *Seattle Post-Intelligencer* found asbestos in Crayola crayons, it was Lee who did a study that discovered none. When people began getting sick and dying in Libby, Montana, R.J. Lee found that local asbestos levels had been overstated by the EPA.

As in Libby, R.J. Lee declared that what the EPA had been calling asbestos in El Dorado Hills was, in fact, not asbestos at all. Some of the fibers were not the proper size; others contained a bit too much aluminum. The EPA had, in other words, completely goofed.

Supervisor Baumann hailed "a study that has profound information in it." Barber bustled off to Washington, D.C., at taxpayer expense, carrying news of "this startling scientific development." And Wayne Lowery, the town manager, admitted it was hard to get excited about controlling asbestos "when you've got two different interpretations of the same data."

At a meeting of the Asbestos Community Advisory Group in the fire station last winter, Summers and a handful of other residents met with the EPA's Jere Johnson to try to understand how asbestos could vanish before their eyes. Two builders were also at the table. Baumann seemed to have been thinking of them when, at an earlier county supervisors' meeting, she had issued a "public thank you" to "the very, very smart people attending those meetings, making sure another thought process is heard."

One of the builders introduced R.J. Lee's report and started to expound on "cleavage fragments," mineral composition, microns, and the like. "I don't want to get bogged down with these semantics," a fireman interrupted. "We know we have it here." The group's chairman shushed him, though, saying he didn't want to get bogged down in semantics either, "but I think we almost have to." The builder persisted with his mantra of scientific uncertainty. "We all wish science to be this nice black-and-white thing," he said, "and sometimes it isn't." The debate seemed essentially over by the time the EPA's Johnson clarified the topic: It was not geology but public health. "The body can't tell the difference between a fragment and a fiber," Johnson said. Whatever you called it, it would still get stuck in your lungs.

For Summers, the starkness of the issue had again faded to gray. When she started looking for a safer place to live, she was alarmed at first to realize there were cancer clusters and crime everywhere, and then she had taken comfort in danger. If nowhere on Earth was completely safe, she reasoned, why should she leave El Dorado Hills? Perhaps it was as Vicki Barber had said: Risk is a part of all of our lives. Her family could stay, she decided, as long as she took precautions, such as wiping down her home with damp rags, and kept pushing to get all the facts. And if new studies revealed the asbestos to be "just fragments and dust," well, she admitted, she'd be "thrilled to death."

But she knew otherwise, didn't she? She knew that asbestos was here, and she knew that it was bad? The question silenced her. "I don't want to know it, though," she finally answered. "I don't want to know it. That's when I can't sleep. I mean, I love it here. I love it here. That's why it's hard. Part of me wants to be in denial."

After reviewing R.J. Lee's report on El Dorado Hills, the EPA declared last spring that the company had violated "generally accepted scientific principles." R.J. Lee, in turn, discovered "a number of important differences of opinion as well as factual misstatements" in the EPA's response. The EPA at last asked the U.S. Geological Survey to step in and do its own testing. In December, the USGS announced that its analysis confirmed that "material that can be classified as tremolite asbestos is" in El Dorado Hills. But the geologists were uncomfortable assessing the health risks, and so the controversy continues.

Supervisor Baumann thought the EPA should just leave El Dorado Hills be. Her constituents had "really calmed down, been educated." Why, she wondered, would anyone want to excite them again? The publichealth debate seemed to have come full circle when Sweeney, the chairman of the Board of Supervisors, griped about the "extreme cost" of dust controls, saying, "We're putting our public at risk by telling them to do things that are absolutely unnecessary."

Up on the ridge, behind the high school, Terry Trent stood beside a driveway with a piece of tremolite in his hand. As he tossed the rock back to the ground, there was the sound of a small engine starting. "Watch this," Trent said, and a smile broke over his face as a gardener passed by with a leaf blower, dust billowing all around.

POSTSCRIPT

Is the Superfund Program Successfully Protecting Human Health from Hazardous Materials?

Superfund cleanups, when those responsible for contaminated sites could not pay or could not be found, were to have been funded by taxes on industry (e.g., the Crude Oil Tax, the Chemical Feedstock Tax, the Toxic Chemicals Importation Tax, and the Corporate Environmental Income Tax). These taxes expired in 1995, and Congress has so far refused to reauthorize them; the Senate voted down the latest attempt in March 2004. New legislation was proposed in 2007; see "Hinchey Renews Push for Corporate Superfund Tax," *Chemical Week* (May 2, 2007). The program exhausted its funds in September 2003 and is now running on government revenues. See "Superfund Program: Current Status and Future Fiscal Challenges," Report to the Chairman, Subcommittee on Oversight of Government Management, the Federal Workforce, and the District of Columbia, Committee on Governmental Affairs, U.S. Senate (GAO, July 2003). Activist groups, such as the Public Interest Research Group (PIRG), have issued calls for the Bush administration to reinstate funding without delay. But, says PIRG, "The Bush administration opposes reauthorization of the polluter pays taxes, supports a steep increase in the amount paid by taxpayers, and has dramatically slowed down the pace of cleanups at the nation's worst toxic waste sites." The Competitive Enterprise Institute objects that such taxes are an assault on consumer pocketbooks, as is the Comprehensive Environmental Response, Compensation, and Liability Act's (CERCLA's) "joint and several liability" clause, which can make minor contributors to toxics sites liable for large cleanup costs even when they acted according to all laws and regulations in force at the time.

Meanwhile, the hazardous waste problem takes new forms. Even in the 1990s, an increasingly popular method of disposing of hazardous wastes was to ship them from the United States to "dumping grounds" in developing countries. Iwonna Rummel-Bulska's "The Basel Convention: A Global Approach for the Management of Hazardous Wastes," *Environmental Policy and Law* (vol. 24, no. 1, 1994) describes an international treaty designed to prevent or at least limit such waste dumping. But eight years later, in February 2002, the Basel Action Network and the Silicon Valley Toxics Coalition (http://svtc.org) published *Exporting Harm: The High-Tech Trashing of Asia*. This lengthy report documents the shipping of electronics wastes, including defunct personal computers, monitors, and televisions, as well as circuit boards and other products rich in lead, beryllium, cadmium, mercury, and other toxic materials. Some 50 to 80 percent of the "e-waste" collected for recycling in the western United States is shipped to destinations such as China, India, and Pakistan, where recycling and disposal methods lead to widespread human and environmental contamination. An updated version of this report, *Poison PCs and Toxic TVs: E-Waste Tsunami to Roll Across the US: Are We Prepared?*, was released in February 2004. In August 2005, Greenpeace released K. Brigden, et al., *Recycling of Electronic Wastes in China and India: Workplace and Environmental Contamination* (http://www.e-takeback.org/press_open/greenpeace.pdf). Currently the problem of electronic waste continues to grow; see Elisabeth Jeffries, "E-Wasted," *World Watch* (July/August 2006).

Among the solutions that have been urged to address the hazardous waste problem are "take-back" and "remanufacturing" practices. Gary A. Davis and Catherine A. Wilt, in "Extended Product Responsibility," *Environment* (September 1997), urge such solutions as crucial to the minimization of waste and describe how they are becoming more common in Europe. Brad Stone, "Tech Trash, E-Waste: By Any Name, It's an Issue," *Newsweek* (December 12, 2005), describes their appearance in the United States, where some states are making take-back programs mandatory for computers, televisions, and other electronic devices. After *Exporting Harm* was published and drew considerable attention from the press, some industry representatives hastened to emphasize such practices as Hewlett Packard's recycling of printer ink cartridges. See Doug Bartholomew's "Beyond the Grave," *Industry Week* (March 1, 2002), which also stressed the need to minimize waste by intelligent design. The Institute of Industrial Engineers published in its journal *IIE Solutions* Brian K. Thorn's and Philip Rogerson's "Take It Back" (April 2002), which stressed the importance of designing for reuse or remanufacturing. Anthony Brabazon and Samuel Idowu, in "Costing the Earth" *Financial Management (CIMA)* (May 2001), note that "take-back schemes may [both] provide opportunities to build goodwill and [help] companies to use resources more efficiently."

ISSUE 19

Should the United States Reprocess Spent Nuclear Fuel?

YES: Phillip J. Finck, from Statement before the House Committee on Science, Energy Subcommittee, Hearing on Nuclear Fuel Reprocessing (June 16, 2005)

NO: Matthew Bunn, from "The Case against a Near-Term Decision to Reprocess Spent Nuclear Fuel in the United States," Testimony for the House Committee on Science, Energy Subcommittee, Hearing on Nuclear Fuel Reprocessing (June 16, 2005)

ISSUE SUMMARY

YES: Phillip J. Finck argues that by reprocessing spent nuclear fuel, the United States can enable nuclear power to expand its contribution to the nation's energy needs while reducing carbon emissions, nuclear waste, and the need for waste repositories such as Yucca Mountain.

NO: Matthew Bunn argues that there is no near-term need to embrace nuclear spent fuel reprocessing, costs are highly uncertain, and there is a worrisome risk that the increased availability of bomb-grade nuclear materials will increase the risk of nuclear war and terrorism.

Nuclear waste is generated when uranium and plutonium atoms are split to make energy in nuclear power plants, when uranium and plutonium are purified to make nuclear weapons, and when radioactive isotopes useful in medical diagnosis and treatment are made and used. These wastes are radioactive, meaning that as they break down they emit radiation of several kinds. Those that break down fastest are most radioactive.

According to the U.S. Department of Energy, high-level waste includes spent reactor fuel (52,000 tons) and waste from weapons production (91 million gallons). Low- and mixed-level waste includes waste from hospitals and research labs, remnants of decommissioned nuclear plants, and air filters (472 million cubic feet). The high-level waste is the most hazardous and poses the most severe disposal problems and experts say such materials must be kept away from people and other living things.

The Nuclear Age began in the 1940s. As nuclear waste accumulated, there also developed a sense of urgency about finding a place to put it where it would not threaten humans or ecosystems for a quarter million years or more. Among the potential answers was reprocessing, which separates (and recycles) unused fuel from spent fuel and thereby reduces the quantity of waste while also extending the supply of fuel. After the Nuclear Nonproliferation Treaty went into force in 1970, it became United States policy not to reprocess spent nuclear fuel and thereby to limit the availability of bomb-grade material. As a consequence, spent fuel was not recycled, and the waste continued to accumulate.

In 1982, the Nuclear Waste Policy Act called for locating candidate disposal sites for high-level wastes and choosing one by 1998. Since no state chosen as a candidate site was happy, and many sites were for various reasons less than ideal, the schedule proved impossible to meet. In 1987, Congress attempted to settle the matter by designating Yucca Mountain, Nevada, as the one site to be intensively studied and developed. It would be opened for use in 2010. However, problems have plagued the project, as summarized by Chuck McCutcheon, "High-Level Acrimony in Nuclear Storage Standoff," *Congressional Quarterly Weekly Report* (September 25, 1999), and Sean Paige, "The Fight at the End of the Tunnel," *Insight on the News* (November 15, 1999).

In February 2002, U.S. Secretary of Energy Spencer Abraham recommended to the President that the nation go ahead with development of the Yucca Mountain site. His report made the points that a disposal site is necessary, that Yucca Mountain has been thoroughly studied, and that moving ahead with the site best serves "our energy future, our national security, our economy, our environment, and safety." Nevadans and activists of several kinds have objected strenuously; early in 2005, reports that researchers had lied about the data that showed Yucca Mountain to be a safe, long-term repository gave them hope that approval of the site would be delayed, perhaps indefinitely.

But the need to dispose of nuclear waste is not about to go away, especially if the United States expands its reliance on nuclear power (see Issue 12). In the following selections, Phillip J. Finck, Deputy Associate Laboratory Director, Applied Science and Technology and National Security, Argonne National Laboratory, argues that by reprocessing spent nuclear fuel the United States can enable nuclear power to expand its contribution to the nation's energy needs while reducing carbon emissions, nuclear waste, and the need for waste repositories such as Yucca Mountain. Matthew Bunn, a Senior Research Associate at Harvard University's John F. Kennedy School of Government, argues that there is no near-term need to embrace nuclear spent fuel reprocessing, costs are highly uncertain, and there is a worrisome risk that the increased availability of bomb-grade nuclear materials will increase the risk of nuclear war and terrorism.

YES

Phillip J. Finck

Statement before the House Committee on Science, Energy Subcommittee, Hearing on Nuclear Fuel Reprocessing

Summary

Management of spent nuclear fuel from commercial nuclear reactors can be addressed in a comprehensive, integrated manner to enable safe, emissions-free, nuclear electricity to make a sustained and growing contribution to the nation's energy needs. Legislation limits the capacity of the Yucca Mountain repository to 70,000 metric tons from commercial spent fuel and DOE defense-related waste. It is estimated that this amount will be accumulated by approximately 2010 at current generation rates for spent nuclear fuel. To preserve nuclear energy as a significant part of our future energy generating capability, new technologies can be implemented that allow greater use of the repository space at Yucca Mountain. By processing spent nuclear fuel and recycling the hazardous radioactive materials, we can reduce the waste disposal requirements enough to delay the need for a second repository until the next century, even in a nuclear energy growth scenario. Recent studies indicate that such a closed fuel cycle may require only minimal increases in nuclear electricity costs, and are not a major factor in the economic competitiveness of nuclear power (The University of Chicago study, "The Economic Future of Nuclear Power," August 2004). However, the benefits of a closed fuel cycle can not be measured by economics alone; resource optimization and waste minimization are also important benefits. Moving forward in 2007 with an engineering-scale demonstration of an integrated system of proliferation-resistant, advanced separations and transmutation technologies would be an excellent first step in demonstrating all of the necessary technologies for a sustainable future for nuclear energy.

Nuclear Waste and Sustainability

World energy demand is increasing at a rapid pace. In order to satisfy the demand and protect the environment for future generations, energy sources

House Committee on Science, Energy Subcommittee, Hearing on Nuclear Fuel Reprocessing, June 16, 2005.

must evolve from the current dominance of fossil fuels to a more balanced, sustainable approach. This new approach must be based on abundant, clean, and economical energy sources. Furthermore, because of the growing world-wide demand and competition for energy, the United States vitally needs to establish energy sources that allow for energy independence.

Nuclear energy is a carbon-free, secure, and reliable energy source for today and for the future. In addition to electricity production, nuclear energy has the promise to become a critical resource for process heat in the production of transportation fuels, such as hydrogen and synthetic fuels, and desalinated water. New nuclear plants are imperative to meet these vital needs.

To ensure a sustainable future for nuclear energy, several requirements must be met. These include safety and efficiency, proliferation resistance, sound nuclear materials management, and minimal environmental impacts. While some of these requirements are already being satisfied, the United States needs to adopt a more comprehensive approach to nuclear waste management. The environmental benefits of resource optimization and waste minimization for nuclear power must be pursued with targeted research and development to develop a successful integrated system with minimal economic impact. Alternative nuclear fuel cycle options that employ separations, transmutation, and refined disposal (e.g., conservation of geologic repository space) must be contrasted with the current planned approach of direct disposal, taking into account the complete set of potential benefits and penalties. In many ways, this is not unlike the premium homeowners pay to recycle municipal waste.

The spent nuclear fuel situation in the United States can be put in perspective with a few numbers. Currently, the country's 103 commercial nuclear reactors produce more than 2000 metric tons of spent nuclear fuel per year (masses are measured in heavy metal content of the fuel, including uranium and heavier elements). The Yucca Mountain repository has a legislative capacity of 70,000 metric tons, including spent nuclear fuel and DOE defense-related wastes. By approximately 2010 the accumulated spent nuclear fuel generated by these reactors and the defense-related waste will meet this capacity, even before the repository starts accepting any spent nuclear fuel. The ultimate technical capacity of Yucca Mountain is expected to be around 120,000 metric tons, using the current understanding of the Yucca Mountain site geologic and hydrologic characteristics. This limit will be reached by including the spent fuel from current reactors operating over their lifetime. Assuming nuclear growth at a rate of 1.8% per year after 2010, the 120,000 metric ton capacity will be reached around 2030. At that projected nuclear growth rate, the U.S. will need up to nine Yucca Mountain-type repositories by the end of this century. Until Yucca Mountain starts accepting waste, spent nuclear fuel must be stored in temporary facilities, either storage pools or above ground storage casks.

Today, many consider repository space a scarce resource that should be managed as such. While disposal costs in a geologic repository are currently quite affordable for U.S. electric utilities, accounting for only a few percent of the total cost of electricity, the availability of U.S. repository space will likely remain limited.

Only three options are available for the disposal of accumulating spent nuclear fuel:

- Build more ultimate disposal sites like Yucca Mountain.
- Use interim storage technologies as a temporary solution.
- Develop and implement advanced fuel cycles, consisting of separation technologies that separate the constituents of spent nuclear fuel into elemental streams, and transmutation technologies that destroy selected elements and greatly reduce repository needs.

A responsible approach to using nuclear power must always consider its whole life cycle, including final disposal. We consider that temporary solutions, while useful as a stockpile management tool, can never be considered as ultimate solutions. It seems prudent that the U.S. always have at least one set of technologies available to avoid expanding geologic disposal sites.

Spent Nuclear Fuel

The composition of spent nuclear fuel poses specific problems that make its ultimate disposal challenging. Fresh nuclear fuel is composed of uranium dioxide (about 96% U238, and 4% U235). During irradiation, most of the U235 is fissioned, and a small fraction of the U238 is transmuted into heavier elements (known as "transuranics"). The spent nuclear fuel contains about 93% uranium (mostly U238), about 1% plutonium, less than 1% minor actinides (neptunium, americium, and curium), and 5% fission products. Uranium, if separated from the other elements, is relatively benign, and could be disposed of as low-level waste or stored for later use. Some of the other elements raise significant concerns:

- The fissile isotopes of plutonium, americium, and neptunium are potentially usable in weapons and, therefore, raise proliferation concerns. Because spent nuclear fuel is protected from theft for about one hundred years by its intense radioactivity, it is difficult to separate these isotopes without remote handling facilities.
- Three isotopes, which are linked through a decay process (Pu241, Am241, and Np237), are the major contributors to the estimated dose for releases from the repository, typically occurring between 100,000 and 1 million years, and also to the long-term heat generation that limits the amount of waste that can be placed in the repository.
- Certain fission products (cesium, strontium) are major contributors to the repository's shortterm heat load, but their effects can be mitigated by providing better ventilation to the repository or by providing a cooling-off period before placing them in the repository.
- Other fission products (Tc99 and I129) also contribute to the estimated dose.

The time scales required to mitigate these concerns are daunting: several of the isotopes of concern will not decay to safe levels for hundreds of thousands of years. Thus, the solutions to long-term disposal of spent nuclear fuel are limited to three options: the search for a geologic environment that will remain stable for that period; the search for waste forms that can contain these elements for that period; or the destruction of these isotopes. These three options underlie the major fuel cycle strategies that are currently being developed and deployed in the U.S. and other countries.

Options for Disposing of Spent Nuclear Fuel

Three options are being considered for disposing of spent nuclear fuel: the once-through cycle is the U.S. reference; limited recycle has been implemented in France and elsewhere and is being deployed in Japan; and full recycle (also known as the closed fuel cycle) is being researched in the U.S., France, Japan, and elsewhere.

1. Once-through Fuel Cycle

This is the U.S. reference option where spent nuclear fuel is sent to the geologic repository that must contain the constituents of the spent nuclear fuel for hundreds of thousands of years. Several countries have programs to develop these repositories, with the U.S. having the most advanced program. This approach is considered safe, provided suitable repository locations and space can be found. It should be noted that other ultimate disposal options have been researched (e.g., deep sea disposal; boreholes and disposal in the sun) and abandoned. The challenges of long-term geologic disposal of spent nuclear fuel are well recognized, and are related to the uncertainty about both the long-term behavior of spent nuclear fuel and the geologic media in which it is placed.

2. Limited Recycle

Limited recycle options are commercially available in France, Japan, and the United Kingdom. They use the PUREX process, which separates uranium and plutonium, and directs the remaining transuranics to vitrified waste, along with all the fission products. The uranium is stored for eventual reuse. The plutonium is used to fabricate mixed-oxide fuel that can be used in conventional reactors. Spent mixed-oxide fuel is currently not reprocessed, though the feasibility of mixed-oxide reprocessing has been demonstrated. It is typically stored or eventually sent to a geologic repository for disposal. Note that a reactor partially loaded with mixed-oxide fuel can destroy as much plutonium as it creates. Nevertheless, this approach always results in increased production of americium, a key contributor to the heat generation in a repository. This approach has two significant advantages:

- It can help manage the accumulation of plutonium.
- It can help significantly reduce the volume of spent nuclear fuel (the French examples indicate that volume decreases by a factor of 4).

Several disadvantages have been noted:

- It results in a small economic penalty by increasing the net cost of electricity a few percent.
- The separation of pure plutonium in the PUREX process is considered by some to be a proliferation risk; when mixed-oxide use is insufficient, this material is stored for future use as fuel.
- This process does not significantly improve the use of the repository space (the improvement is around 10%, as compared to a factor of 100 for closed fuel cycles).
- This process does not significantly improve the use of natural uranium (the improvement is around 15%, as compared to a factor of 100 for closed fuel cycles).

3. Full Recycle (the Closed Fuel Cycle)

Full recycle approaches are being researched in France, Japan, and the United States. This approach typically comprises three successive steps: an advanced separations step based on the UREX+ technology that mitigates the perceived disadvantages of PUREX, partial recycle in conventional reactors, and closure of the fuel cycle in fast reactors.

The first step, UREX+ technology, allows for the separations and subsequent management of highly pure product streams. These streams are:

- Uranium, which can be stored for future use or disposed of as low-level waste.
- A mixture of plutonium and neptunium, which is intended for partial recycle in conventional reactors followed by recycle in fast reactors.
- Separated fission products intended for short-term storage, possibly for transmutation, and for long-term storage in specialized waste forms.
- The minor actinides (americium and curium) for transmutation in fast reactors.

The UREX+ approach has several advantages:

- It produces minimal liquid waste forms, and eliminates the issue of the "waste tank farms."
- Through advanced monitoring, simulation and modeling, it provides significant opportunities to detect misuse and diversion of weapons-usable materials.
- It provides the opportunity for significant cost reduction.
- Finally and most importantly, it provides the critical first step in managing all hazardous elements present in the spent nuclear fuel.

The second step—partial recycle in conventional reactors—can expand the opportunities offered by the conventional mixed-oxide approach. In particular, it is expected that with significant R&D effort, new fuel forms can be developed that burn up to 50% of the plutonium and neptunium present in spent nuclear fuel. (Note that some studies also suggest that it might be possible to recycle fuel in these reactors many times—i.e., reprocess and recycle the irradiated advanced fuel—and further destroy plutonium and neptunium; other studies also suggest possibilities for transmuting americium in these reactors. Nevertheless, the practicality of these schemes is not yet established and requires additional scientific and engineering research.) The advantage of the second step is that it reduces the overall cost of the closed fuel cycle by burning plutonium in conventional reactors, thereby reducing the number of fast reactors needed to complete the transmutation mission of minimizing hazardous waste. This step can be entirely bypassed, and all transmutation performed in advanced fast reactors, if recycle in conventional reactors is judged to be undesirable.

The third step, closure of the fuel cycle using fast reactors to transmute the fuel constituents into much less hazardous elements, and pyroprocessing technologies to recycle the fast reactor fuel, constitutes the ultimate step in reaching sustainable nuclear energy. This process will effectively destroy the transuranic elements, resulting in waste forms that contain only a very small fraction of the transuranics (less than 1%) and all fission products. These technologies are being developed at Argonne National Laboratory and Idaho National Laboratory, with parallel development in Japan, France, and Russia.

The full recycle approach has significant benefits:

- It can effectively increase use of repository space by a factor of more than 100.
- It can effectively increase the use of natural uranium by a factor of 100.
- It eliminates the uncontrolled buildup of all isotopes that are a proliferation risk.
- The fast reactors and the processing plant can be deployed in small co-located facilities that minimize the risk of material diversion during transportation.
- The fast reactor does not require the use of very pure weapons usable materials, thus increasing their proliferation resistance.
- It finally can usher the way towards full sustainability to prepare for a time when uranium supplies will become increasingly difficult to ensure.
- These processes would have limited economic impact; the increase in the cost of electricity would be less than 10% (ref: OECD).
- Assuming that demonstrations of these processes are started by 2007, commercial operations are possible starting in 2025; this will require adequate funding for demonstrating the separations, recycle, and reactor technologies.
- The systems can be designed and implemented to ensure that the mass of accumulated spent nuclear fuel in the U.S. would always remain below 100,000 metric tons—less than the technical capacity of Yucca Mountain—thus delaying, or even avoiding, the need for a second repository in the U.S.

Conclusion

A well engineered recycling program for spent nuclear fuel will provide the United States with a long-term, affordable, carbon-free energy source with low environmental impact. This new paradigm for nuclear power will allow us to manage nuclear waste and reduce proliferation risks while creating a sustainable energy supply. It is possible that the cost of recycling will be slightly higher than direct disposal of spent nuclear fuel, but the nation will only need one geologic repository for the ultimate disposal of the residual waste.

Matthew Bunn

 **NO**

The Case against a Near-Term Decision to Reprocess Spent Nuclear Fuel in the United States

Madam chairwoman and members of the committee: It is an honor to be here today to discuss a subject that is very important to the future of nuclear energy and efforts to stem the spread of nuclear weapons—reprocessing of spent nuclear fuel.

I believe that, while research and development (R&D) on advanced concepts that may offer promise for the future should continue, a near-term decision to reprocess U.S. commercial spent nuclear fuel would be a serious mistake, with costs and risks far outweighing its potential benefits. Let me make seven points to support that view.

First, reprocessing by itself does not make any of the nuclear waste go away. Whatever course we choose, we will still need a nuclear waste repository such as Yucca Mountain. Reprocessing is simply a chemical process that separates the radioactive materials in spent fuel into different components. In the traditional process, known as PUREX, reprocessing produces separated plutonium (which is weapons-usable), recovered uranium, and high-level waste (containing all the other transuranic elements and fission products). In the process, intermediate and low-level wastes are also generated. More advanced processes now being examined, such as UREX+ and pyroprocessing, attempt to address some of the problems of the PUREX process, but whether they will do so successfully remains to be seen. Once the spent fuel has been reprocessed, the plutonium and uranium separated from the spent fuel can in principle be recycled into new fuel; in the more advanced processes, some other long-lived species would also be irradiated in reactors (or accelerator-driven assemblies) to transmute them into shorterlived species.

More Expensive

Second, reprocessing and recycling using current or near-term technologies would substantially increase the cost of nuclear waste management, even if the cost of both uranium and geologic repositories increase significantly. In a recent Harvard study, we concluded, even making a number of assumptions

House Committee on Science, Energy Subcommittee, Hearing on Nuclear Fuel Reprocessing, June 16, 2005.

that were quite favorable to reprocessing, that shifting to reprocessing and recycling would increase the costs of spent fuel management by more than 80% (after taking account of appropriate credits or charges for recovered plutonium and uranium from reprocessing). Reprocessing (at an optimistic reprocessing price) would not become economic until uranium reached a price of over $360 per kilogram—a price not likely to be seen for many decades, if then. Government studies even in countries such as France and Japan have reached similar conclusions. The UREX+ technology now being pursued adds a number of complex separation steps to the traditional PUREX process, in order to separate important radioactive isotopes for storage or transmutation, and there is little doubt that reprocessing and transmutation using this process would be even more expensive. Other processes might someday reduce the costs, but this remains to be demonstrated, and a number of recent official studies have estimated costs for reprocessing and transmutation that are far higher than the costs of traditional reprocessing and recycling, not lower.

To follow this course, either the current 1 mill/kilowatt-hour nuclear waste fee would have to be substantially increased, or billions of dollars in tax money would have to be used to subsidize the effort. Since facilities required for reprocessing and transmutation would not be economically attractive for private industry to build, the U.S. government would either have to build and operate these facilities itself, give private industry large subsidies to do so, or impose onerous regulations requiring private industry to do so with its own funds. All of these options would represent dramatic government intrusions into the nuclear fuel industry, and the implications of such intrusions have not been appropriately examined. I am pleased that the subcommittee plans a later hearing with representatives from the nuclear industry to discuss these economic and institutional issues.

Unnecessary Proliferation Risks

Third, traditional approaches to reprocessing and recycling pose significant and unnecessary proliferation risks, and even proposed new approaches are not as proliferation-resistant as they should be. It is crucial to understand that any state or group that could make a bomb from weapon-grade plutonium could make a bomb from the reactor-grade plutonium separated by reprocessing. Despite the remarkable progress of safeguards and security technology over the last few decades, processing, fabricating, and transporting tons of weapons-usable separated plutonium every year—when even a few kilograms is enough for a bomb—inevitably raises greater risks than not doing so. The dangers posed by these operations can be reduced with sufficient investment in security and safeguards, but they cannot be reduced to zero, and these additional risks are unnecessary.

Indeed, contrary to the assertion in the Energy and Water appropriations subcommittee report that plutonium reprocessing in other countries poses little risk because the plutonium is immediately recycled as fresh fuel—a conclusion that would not be correct even if the underlying assertion were true—the fact is that reprocessing is far outpacing the use of the resulting plutonium as fuel, with the result that over 240 tons of separated, weapons-usable civilian plutonium now exists in the world, a figure that will soon surpass the amount of plutonium in all the world's nuclear weapons arsenals combined. The British Royal Society, in a 1998 report, warned that even in an advanced industrial state like the United Kingdom, the possibility that plutonium stocks might be "accessed for illicit weapons production is of extreme concern."

Moreover, a near-term U.S. return to reprocessing could significantly undermine broader U.S. nuclear nonproliferation policies. President Bush has announced an effort to convince countries around the world to forego reprocessing and enrichment capabilities of their own; has continued the efforts of past administrations to convince other states to avoid the further accumulation of separated plutonium, because of the proliferation hazards it poses; and has continued to press states in regions of proliferation concern not to reprocess (including not only states such as North Korea and Iran, but also U.S. allies such South Korea and Taiwan, both of which had secret nuclear weapons programs closely associated with reprocessing efforts in the past). A U.S. decision to move toward reprocessing itself would make it more difficult to convince other states not to do the same.

Advocates argue that the more advanced approaches now being pursued would be more proliferation-resistant. Technologies such as pyroprocessing are undoubtedly better than PUREX in this respect. But the plutonium-bearing materials that would be separated in either the UREX+ process or by pyroprocessing would not be radioactive enough to meet international standards for being "self-protecting" against possible theft. Moreover, if these technologies were deployed widely in the developing world, where most of the future growth in electricity demand will be, this would contribute to potential proliferating states building up expertise, realworld experience, and facilities that could be readily turned to support a weapons program.

Proponents of reprocessing and recycling often argue that this approach will provide a nonproliferation benefit, by consuming the plutonium in spent fuel, which would otherwise turn geologic repositories into potential plutonium mines in the long term. But the proliferation risk posed by spent fuel buried in a safeguarded repository is already modest; if the world could be brought to a state in which such repositories were the most significant remaining proliferation risk, that would be cause for great celebration. Moreover, this risk will be occurring a century or more from now, and if there is one thing we know about the nuclear world a century hence, it is that its shape and contours are highly uncertain. We should not increase significant proliferation risks in the near term in order to reduce already small and highly uncertain proliferation risks in the distant future.

As-Yet-Unexamined Safety and Terrorism Risks

Fourth, reprocessing and recycling using technologies available in the near term would be likely to raise additional safety and terrorism risks. Until Chernobyl, the world's worst nuclear accident had been the explosion at the reprocessing plant at Khystym in 1957, and significant accidents at both Russian and Japanese reprocessing plants occurred as recently as the 1990s. No complete life-cycle study of the safety and terrorism risks of reprocessing and recycling compared to those of direct disposal has yet been done by disinterested parties. But it seems clear that extensive processing of intensely radioactive spent fuel using volatile chemicals presents more opportunities for release of radionuclides than does leaving spent fuel untouched in thick metal or concrete casks.

Limited Waste Management Benefits

Fifth, the waste management benefits that might be derived from reprocessing and transmutation are quite limited. Two such benefits are usually claimed: decreasing the repository volume needed per kilowatt-hour of electricity generated (potentially eliminating the need for a second repository after Yucca Mountain); and greatly reducing the radioactive dangers of the material to be disposed.

It is important to recognize that reprocessing and recycling as currently practiced (with only one round of recycling the plutonium as uranium-plutonium mixed oxide (MOX) fuel) does not have either of these benefits. The size of a repository needed for a given amount of waste is determined not by the volume of the waste but by its heat output. Because of the build-up of heat-emitting higher actinides when plutonium is recycled, the total heat output of the waste per kilowatt-hour generated is actually higher—and therefore the needed repositories larger and more expensive—with one round of reprocessing and recycling than it is for direct disposal. And the estimated long-term doses to humans and the environment from the repository are not noticeably reduced.

Newer approaches that might provide a substantial reduction in radiotoxic hazards and in repository volume are complex, likely to be expensive, and still in an early stage of development. Most important, even if they achieved their goals, the benefits would not be large. The projected long-term radioactive doses from a geologic repository are already low. No credible study has yet been done comparing the risk of increased doses in the near term from the extensive processing and operations required for reprocessing and transmutation to the reduction in doses thousands to hundreds of thousands of years in the future that might be achieved by this method.

With respect to reducing repository volume, while the Department of Energy (DOE) has not yet performed any detailed study of the maximum amount of spent fuel that could be emplaced at Yucca Mountain, there is little doubt that even without reprocessing, the mountain could hold far more than the current legislative limit. There are a variety of approaches to providing additional capacity at Yucca Mountain or elsewhere without recycling. Indeed, as a recent American Physical Society report noted, it is possible that even if all existing reactors receive license extensions allowing them to operate for 60 years, Yucca Mountain will be able to hold all the spent fuel they will generate in their lifetimes, without reprocessing. While proponents of reprocessing and transmutation point to the likely difficulty of licensing a second repository in the United States after Yucca Mountain's capacity is filled, it is likely to be at least as difficult to gain public acceptance and licenses for the facilities needed for reprocessing and transmutation—particularly as such facilities will likely pose more genuine hazards to their neighbors than would a nuclear waste repository.

Limited Energy Benefits

Sixth, the energy benefits of reprocessing and recycling would also be limited. Additional energy can indeed be generated from the plutonium and uranium in spent fuel. But in today's market, spent fuel is like oil shale: getting the energy out of it costs far more than the energy is worth. In the only approach to recycling that is commercially practiced today—which involves a single round of recycling as MOX fuel in existing light-water reactors—the amount of energy generated from each ton of uranium mined is increased by less than 20%. In principle, if, in the future, fast-neutron breeder reactors become economic, so that the 99.3% of natural uranium that is U-238 could be turned to plutonium and burned, the amount of energy that could be derived from each ton of uranium mined might be increased 50-fold.

But there is no near-term need for this extension of the uranium resource. World resources of uranium likely to be economically recoverable in future decades at prices far below the price at which reprocessing would be economic are sufficient to fuel a growing global nuclear enterprise for many decades, relying on direct disposal without recycling.

Nor does reprocessing serve the goal of energy security, even for countries such as Japan, which have very limited domestic energy resources. If energy security means anything, it means that a country's energy supplies will not be disrupted by events beyond that country's control. Yet events completely out of the control of any individual country—such as a theft of poorly guarded plutonium on the other side of the world—could transform the politics of plutonium overnight and make major planned programs virtually impossible to carry out. Japan's experience following the scandal over BNFL's falsification of safety data on MOX fuel, and following the accidents at Monju and Tokai, all of which have delayed Japan's plutonium programs by many years, makes this point clear. If anything, plutonium recycling is much *more* vulnerable to external events than reliance on once-through use of uranium, whose supplies are diverse, plentiful, and difficult to cut off.

Premature to Decide—and No Need to Rush

Seventh, there is no need to rush to make this decision in 2007, or in fact any time in the next few decades. Dry storage casks offer the option of storing spent fuel cheaply, safely, and securely for decades. During that time, technology will develop; interest will accumulate on fuel management funds set aside today, reducing the cost of whatever we choose to do in the long run; political and economic circumstances may change in ways that point clearly in one direction or the other; and the radioactivity of the spent fuel will decay, making it cheaper to process in the future, if need be. Our generation has an obligation to set aside sufficient funds so that we are not passing unfunded obligations on to our children and grandchildren, but it is not our responsibility to make and implement decisions prematurely, thereby depriving future generations of what might turn out to be better options developed later. Indeed, because the repository will remain open for 50–100 years, with the spent fuel readily retrievable, moving forward with direct disposal will still leave all options open for decades to come.

Similarly, there is no need to rush to set up new interim storage sites on DOE or military sites, and no possibility of performing the needed reviews and getting the needed licenses to do so by 2006, as the Energy and Water appropriations subcommittee proposed. There is a legitimate debate as to whether such interim spent fuel storage prior to emplacement in a geologic repository should be centralized at one or two sites, or whether in most cases the fuel should continue to be stored at existing reactor sites. In any case, the government should fulfill its obligations to the utilities by taking title to the fuel and paying the cost of storage. At the same time, we should continue to move toward opening a permanent geologic repository as quickly as we responsibly can—in part because public acceptance of interim spent fuel storage facilities is only likely to be forthcoming if the public is convinced that they will not become permanent waste dumps.

Nor is there any need to rush on deciding whether a second nuclear waste repository will be needed. While existing nuclear power plants will have discharged enough fuel to fill the current legislated capacity limit within a few years, the reality is that it will be decades before sufficient fuel to fill Yucca Mountain has in fact been emplaced. We can and should defer this decision, and take the time to consider the options in detail. Congress should consider amending current law and giving the Secretary of Energy another decade or more before reporting on the need for a second repository.

Proponents of deciding quickly on reprocessing sometimes argue that such decisions are necessary because no new nuclear reactors will be purchased unless sufficient geologic repository capacity for all the spent fuel they will generate throughout their lifetimes has already been provided. I do not believe this is correct. I believe that if the government is fulfilling its obligation to take title to spent fuel and pay the costs of managing it, and clear progress is being made toward opening and operating a nuclear waste repository, investors will have sufficient confidence that they will not be saddled with unexpected spent fuel obligations to move forward. By contrast, if the government were seriously considering drastic changes in spent fuel management approaches which might major increases in the nuclear waste fee, investors might well wish to wait to see the outcome of those decisions before investing in new nuclear plants.

It is a good thing there is no need to rush, as we simply do not have the information that would be needed to make a decision on reprocessing in 2007. The advanced reprocessing technologies now being pursued are in a very early stage of development. As of a year ago, UREX+ had been demonstrated on a total of one pin of real spent fuel, in a small facility—and had not met all of its processing goals in that test. Frankly, in my judgment there is little prospect that further development of complex multi-stage aqueous separations processes such as UREX+ will result in processes that will provide low costs, proliferation resistance, and waste management benefits sufficient to make them worth implementing in competition with direct disposal. Pyroprocessing has been tried on a somewhat larger scale over the years, but the process is designed for processing metals, and significant development is still needed to be confident in industrial-scale application to the oxide spent fuel from current reactors. Other, longer-term processes might offer more promise, but too little is known about them to know for sure.

So far, we do not have a credible life-cycle analysis of the cost of a reprocessing and transmutation system compared to that of direct disposal; DOE has yet to do any detailed estimate of how much spent fuel can be placed in Yucca Mountain, and of non-reprocessing approaches to extending that capacity; we do not have a realistic evaluation of the impact of a reprocessing and transmutation on the existing nuclear fuel industry; we do not have a serious evaluation of the licensing and public acceptance issues facing development and deployment of such a system; we do not have any serious assessment of the safety and terrorism risks of a reprocessing and transmutation system, compared to those of direct disposal; and we do not yet have assessments of the proliferation implications of the proposed systems that are detailed enough to support responsible decision-making. In short, now is the time for continued research and development, and additional systems analysis, not the time for committing to processing using any particular technology.

Recommendations

For the reasons just outlined, I recommend that we follow the advice of the bipartisan National Commission on Energy Policy, which reflected a broad spectrum of opinion on energy matters generally and on nuclear energy in particular, and recommended that the United States should:

1. "continue indefinitely the U.S. moratoria on commercial reprocessing of spent nuclear fuel and construction of commercial breeder reactors";
2. establish expanded interim spent fuel storage capacities "as a complement and interim backup" to Yucca Mountain;
3. proceed "with all deliberate speed" toward licensing and operating a permanent geologic waste repository; and
4. continue research and development on advanced fuel cycle approaches that might improve nuclear waste management and uranium utilization, without the huge disadvantages of traditional approaches to reprocessing.

At the same time, the U.S. government should redouble its efforts to: (a) limit the spread of reprocessing and enrichment technologies, as a critical element of a strengthened nonproliferation effort; (b) ensure that every nuclear warhead and every kilogram of separated plutonium and highly enriched uranium (HEU) worldwide are secure and accounted for, as the most critical step to prevent nuclear terrorism; and (c) convince other countries to end the accumulation of plutonium stockpiles, and work to reduce stockpiles of both plutonium and HEU around the world. The Bush administration should, in particular, resume the effort to negotiate a 20-year U.S.-Russian moratorium on separation of plutonium that was almost completed at the end of the Clinton administration.

Similar recommendations have been made in the MIT study on the future of nuclear energy, and in the American Physical Society study of nuclear energy and nuclear weapons proliferation.

It remains possible that someday approaches to reprocessing and recycling will be developed that make security, economic, political, and environmental sense. Research and development should explore such possibilities. Continued investment in R&D on advanced fuel cycle technologies is justified, in part to ensure that the United States will have the technological expertise and credibility to play a leading role in limiting the proliferation risks of the fuel cycle around the world. But the leverage of these technologies in meeting the most serious energy challenges of the 21st century is likely to be somewhat limited in comparison to the promise of other potential future energy technologies, and the emphasis that nuclear fuel cycle R&D should receive in the overall energy R&D portfolio should reflect that.

The global nuclear energy system would have to grow substantially if nuclear energy was to make a substantial contribution to meeting the world's 21st century needs for carbon-free energy. Building the support from governments, utilities, and publics needed to achieve that kind of growth will require making nuclear energy as cheap, as simple, as safe, as proliferation-resistant, and as terrorism-proof as possible. Reprocessing using any of the technologies likely to be available in the near term points in the wrong direction on every count. Those who hope for a bright future for nuclear energy, therefore, should oppose near-term reprocessing of spent nuclear fuel.

POSTSCRIPT

Should the United States Reprocess Spent Nuclear Fuel?

The nuclear waste disposal problem is real and it must be dealt with. If it is not, we may face the same kinds of problems created by the former Soviet Union, which disposed of some nuclear waste simply by dumping it at sea. For a recent summary of the nuclear waste problem and the disposal controversy, see Michael E. Long, "Half Life: The Lethal Legacy of America's Nuclear Waste," *National Geographic* (July 2002). The need for care in nuclear waste disposal is underlined by Tom Carpenter and Clare Gilbert, "Don't Breathe the Air," *Bulletin of the Atomic Scientists* (May/June 2004); they describe the Hanford Site in Hanford, Washington, where wastes from nuclear weapons production were stored in underground tanks. Leaks from the tanks have contaminated groundwater, and an extensive cleanup program is under way. But cleanup workers are being exposed to both radioactive materials and toxic chemicals, and they are falling ill. And in June 2004, the U.S. Senate voted to ease cleanup requirements. Per F. Peterson, William E. Kastenberg, and Michael Corradini, "Nuclear Waste and the Distant Future," *Issues in Science and Technology* (Summer 2006), argue that the risks of waste disposal have been sensibly addressed and we should be focusing more attention on other risks (such as those of global warming).

In November 2005, President Bush signed the budget for the Department of Energy, which contained $50 million to start work toward a reprocessing plant; see Eli Kintisch, "Congress Tells DOE to Take Fresh Look at Recycling Spent Reactor Fuel," Science (December 2, 2005). Reprocessing spent nuclear fuel will be expensive, but the costs do not seem to be great enough to make nuclear power unacceptable; see "The Economic Future of Nuclear Power," University of Chicago (August 2004) (http://www.anl.gov/Special_Reports/NuclEconSumAug04.pdf). Matthew L. Wald, "A New Vision for Nuclear Waste," *Technology Review* (December 2004), says that in the absence of Yucca Mountain, reprocessing, or other solutions to the problem of what to do with spent fuel, utilities have been storing the fuel on site. The latest techniques involve massive casks sitting on concrete pads and exposed to the air. The casks contain the waste, even against the threats of earthquakes and bombs, while the air carries away the heat generated by radioactive decay. Over time, the waste in the casks becomes less hazardous. Other potential solutions exist as well. Steven Ashley, in "Divide and Vitrify," *Scientific American* (June 2002), describes work on potential methods of separating the most hazardous components of nuclear waste. One such approach is to expose nuclear waste to neutrons from particle accelerators or special nuclear reactors and thereby greatly hasten the process of radioactive decay. William H. Hannum,

Gerald E. Marsh, and George S. Stanford, "Smarter Use of Nuclear Waste," *Scientific American* (December 2005), discuss the use of fast-neutron reactors to accomplish this.

It is an unfortunate truth that the reprocessing of nuclear spent fuel does indeed increase the risks of nuclear proliferation. On February 28, 2004, *The Economist* ("The World Wide Web of Nuclear Danger") wrote that the risk that someone, somewhere, might detonate a bomb in anger is arguably greater than at any time since the 1962 Cuban missile crisis brought the cold-war world soberingly close to the brink. Both nations and terrorists itch to possess nuclear weapons, whose destructive potential makes present members of the "nuclear club" tremble. Can the risk be controlled? John Deutch, Arnold Kanter, Ernest Moniz, and Daniel Poneman, in "Making the World Safe for Nuclear Energy," *Survival* (Winter2004/2005), argue that present nuclear nations could supply fuel and reprocess spent fuel for other nations; nations that refuse to participate would be seen as suspect and subject to international action. In February 2006, the United States Department of Energy announced the Global Nuclear Energy Partnership, to be operated by the United States, Russia, Great Britain, and France. It would lease nuclear fuel to other nations, reprocess spent fuel without generating material that could be diverted to making nuclear bombs, reduce the amount of waste that must be disposed of, and help meet future energy needs. See Stephanie Cooke, "Just Within Reach?" *Bulletin of the Atomic Scientists* (July/August 2006), and Jeff Johnson, "Reprocessing Key to Nuclear Plan," *Chemical & Engineering News* (June 18, 2007). Critics such as Karen Charman, "Brave Nuclear World, Parts I and II," *World Watch* (May/June and July/August 2006), insist that nuclear power is far too expensive and carries too serious risks of breakdown and exposure to wastes to rely upon, especially when cleaner, cheaper, and less dangerous alternatives exist.

Contributors to This Volume

EDITOR

THOMAS A. EASTON is a professor of science at Thomas College in Waterville, Maine, where he has been teaching environmental science, science, technology, and society, emerging technologies, and computer science since 1983. He received a B.A. in biology from Colby College in 1966 and a Ph.D. in theoretical biology from the University of Chicago in 1971. He writes and speaks frequently on scientific and futuristic issues. His books include *Focus on Human Biology*, 2nd ed., coauthored with Carl E. Rischer (Harper-Collins, 1995), *Careers in Science*, 4th ed. (VGM Career Horizons, 2004), *Taking Sides: Clashing Views on Controversial Issues in Science, Technology and Society* (McGraw-Hill, 8th ed., 2008), and *Classic Editions Sources: Environmental Studies* (McGraw-Hill, 3rd ed., 2008). Dr. Easton is also a well-known writer and critic of science fiction.

AUTHORS

JAMES ALLEN is a research analyst at Climate Change Capital, a London investment banking group that aims "to make the world's environment cleaner while delivering attractive financial returns."

MARK ANSLOW is a reporter for *The Ecologist*.

RONALD BAILEY is a science correspondent for *Reason* magazine. A member of the Society of Environmental Journalists, his articles have appeared in many popular publications, including the *Wall Street Journal, The Public Interest,* and *National Review.* He has produced several series and documentaries for PBS television and *ABC News,* and he was the Warren T. Brookes Fellow in Environmental Journalism at the Competitive Enterprise Institute in 1993. He is the editor of *Earth Report 2000: Revisiting the True State of the Planet* (McGraw-Hill, 1999) and the author of *Global Warming and Other Eco-Myths: How the Environmental Movement Uses False Science to Scare Us to Death* (Prima, 2002).

JON BOONE is a retired university administrator and the producer and director, with David Beaudoin, of the documentary video, *Life Under a Windplant* (http://www.stopillwind.org/). A lifelong environmentalist, he helped found the North American Bluebird Society and is a consultant with the Roger Tory Peterson Institute in New York. He is currently writing a book on the Dutch artist, Johannes Vermeer. He has no financial involvement with wind power but does seek informed, effective public policy.

ROBERT D. BULLARD, one of the most influential leaders of the environmental justice movement, is Ware Professor of Sociology and director of the Environmental Justice Resource Center at Clark Atlanta University. He is the author of numerous articles, scholarly papers, and books (notably *Dumping in Dixie: Race, Class, and Environmental Quality* [Westview Press, 1990]) that address the inequitable treatment of African-Americans and other minorities in environmental planning and decision making.

MATTHEW BUNN is a senior research associate in the Project on Managing the Atom in the Belfer Center for Science and International Affairs at Harvard University's John F. Kennedy School of Government. His current research interests include nuclear theft and terrorism; disposition of excess plutonium; and nuclear waste storage, disposal, and reprocessing.

KAREN CHARMAN is a journalist specializing in environmental issues. She is also the managing editor of the journal *Capitalism Nature Socialism*.

GERALD D. COLEMAN is the former rector of St. Patrick's Seminary and University in Menlo Park, California. His books include several on Catholic views of sexuality.

CHARLI E. COON is Senior Policy Analyst, Thomas A. Roe Institute for Economic Policy Studies, The Heritage Foundation.

GIULIO A. De LEO is an associate professor of applied ecology and environmental impact assessment in the Dipartimento di Scienze Ambientali at the Universit degli Studi di Parma in Parma, Italy.

BOB DINNEEN is president and CEO of the Renewable Fuels Association, the national trade association representing the U.S. ethanol industry.

MYRON EBELL is director of energy and global warming policy at the Competitive Enterprise Institute (CEI), a Washington, D.C.-based public policy organization dedicated to advancing the principles of free enterprise and limited government.

PHILLIP J. FINCK is deputy associate laboratory director, applied science and technology and national security, Argonne National Laboratory.

MONITA FONTAINE is a member of the board of directors of the National Endangered Species Act Reform Coalition and the National Marine Manufacturers Association's vice president for government relations.

DAVID FRIEDMAN is Research Director at the Union of Concerned Scientists.

MARINO GATTO is a professor of applied ecology in the Dipartimento di Elettronica e Informazione at Politecnico di Milano in Milan, Italy. His main research interests include ecological models and the management of renewable resources. Gato is associate editor of *Theoretical Population Biology*.

BERNARD D. GOLDSTEIN is Professor of Environmental and Occupational Health at the University of Pittsburgh. From 2001 to 2005, he was dean of the Graduate School of Public Health. He has published a number of papers on the precautionary principle and risk assessment.

MICHAEL GOUGH, a biologist and expert on risk assessment and environmental policy, has participated in science policy issues at the Congressional Office of Technology Assessment, in Washington think tanks, and on various advisory panels. He most recently edited *Politicizing Science: The Alchemy of PolicyMaking* (Hoover Institution Press, 2003).

BRIAN HALWEIL is a senior researcher at the Worldwatch Institute. His latest book is *Eat Here: Reclaiming Homegrown Pleasures in a Global Supermarket* (W. W. Norton, 2004).

ROBERT H. HARRIS is a principal with ENVIRON International Corporation. He has over 25 years of experience in the area of environmental health and toxic chemicals, with particular emphasis on water and air pollution and hazardous waste issues. He is recognized nationally as an expert consultant on the treatment and disposal of municipal solid and hazardous waste, as well as on air, soil, and groundwater contamination.

CHARLES KOMANOFF is an internationally known energy-economist and transport-economist and an environmental activist in New York City.

JOHN KOSTYACK is senior counsel and director of Wildlife Conservation Campaigns with the National Wildlife Federation.

DWIGHT R. LEE is the Ramsey Professor of Economics and Private Enterprise in the Terry College of Business at the University of Georgia. He received his Ph.D. from the University of California at San Diego in 1972 and his research has covered a variety of areas, including personal finance, public finance, the economics of political decision making, and the economics of the environment and natural resources. Lee has published over 100 articles and commentaries in academic journals, magazines, and newspapers. He is coauthor, with Richard B. McKenzie, of *Getting Rich in America: Eight Simple Rules for Building a Fortune and a Satisfying Life* (HarperBusiness, 2000).

JOHN E. LOSEY is an associate professor in the Department of Entomology at Cornell University.

SEAN McDONAGH is a Columban priest and ex-missionary to the Philippines. His latest book is *The Death of Life: The Horror of Extinction* (Columba Press, Dublin, 2004).

ANNE PLATT McGINN is a senior researcher at the Worldwatch Institute and the author of "Why Poison Ourselves? A Precautionary Approach to Synthetic Chemicals," Worldwatch Paper 153 (November 2000).

MICHAEL MEYER, the European editor for *Newsweek International*, is a member of the New York Council on Foreign Relations and was an inaugural fellow at the American Academy in Berlin. He won the Overseas Press Club's Morton Frank Award for business/economic reporting from abroad in 1986 and 1988. He is the author of *The Alexander Complex* (Times Books, 1989), an examination of the psychology of American empire builders.

JOHN J. MILLER is a political reporter for the *National Review,* a contributing editor for *Reason* magazine, a former vice president of the Center for Equal Opportunity, and a Bradley Fellow at the Heritage Foundation. His most recent book is *A Gift of Freedom: How the John M. Olin Foundation Changed America* (Encounter, 2005).

IAIN MURRAY is a senior fellow at the Competitive Enterprise Institute (CEI), a Washington, D.C.-based public policy organization dedicated to advancing the principles of free enterprise and limited government.

NANCY MYERS is communications director for the Science and Environmental Health Network. She is the co-editor, with Carolyn Raffensperger, of *Precautionary Tools for Reshaping Environmental Policy* (MIT Press, 2005).

GRANTA NAKAYAMA is Assistant Administrator for Enforcement and Compliance Assurance (OECA) at the U.S. Environmental Protection Agency (EPA). His office is responsible for enforcing the nation's environmental laws. It also serves as the National Program Manager for environmental justice. Before his appointment, he was a partner in the Kirkland & Ellis LLP law firm. He has degrees in both engineering and law.

DAVID NICHOLSON-LORD is an environmental writer, formerly with *The Times, The Independent,* and *The Independent on Sunday,* where he was environment editor. He is the author of *The Greening of the Cities* (Routledge, 1987) and of *Green Cities—And Why We Need Them* (New Economics Foundation, 2003). He is a member of UNESCO's UK Man and the Biosphere Urban Forum, an executive of the Urban Wildlife Network, and a trustee of the National Wildflower Centre. He also teaches environment in the journalism faculty at City University, London.

IVAN OSORIO is editorial director at the Competitive Enterprise Institute (CEI), a Washington, D.C.-based public policy organization dedicated to advancing the principles of free enterprise and limited government.

RANDALL PATTERSON is a journalist who writes frequently for *Mother Jones* and other magazines.

JEREMY RIFKIN is the president of the Foundation on Economic Trends in Washington, D.C., and has written many books on the impact of scientific and technological changes on the economy, the workforce, society, and the environment. Among his latest books is *The European Dream: How Europe's Vision of the Future Is Quietly Eclipsing the American Dream* (Tarcher/Penguin, 2004).

DONALD R. ROBERTS is a professor in the Division of Tropical Public Health, Department of Preventive Medicine and Biometrics, Uniformed Services University of the Health Sciences.

MIKE TILCHIN is a vice president of CH2M HILL.

BRIAN TOKAR is an associate faculty member at Goddard College in Plainfield, Vermont. A regular correspondent for *Z* magazine, he has been an activist for over 20 years in the peace, antinuclear, environmental, and green politics movements. He is the author of *The Green Alternative: Creating an Ecological Future,* 2d. ed. (R & E Miles, 1987).

MICHELE L. TRANKINA is a professor of biological sciences at St. Maryís University and an adjunct associate professor of physiology at the University of Texas Health Science Center, both in San Antonio, Texas.

JAY VANDEVEN is a principal with ENVIRON International Corporation. He has 16 years of experience in the assessment and remediation of soil and groundwater contamination, contaminant fate and transport, environmental cost allocation, and environmental insurance claims.

MACE VAUGHAN is an entomologist and Conservation Director at the Xerces Society for Invertebrate Conservation (http://www.xerces.org/).

MICHAEL J. WALLACE is executive vice president of constellation energy, a leading supplier of electricity to large commercial and industrial customers.

ANTHONY WHITE is Managing Director of Market Development and Chairman of Advisory at Climate Change Capital, a London investment banking group that aims "to make the world's environment cleaner while delivering attractive financial returns." He was a founding member of the UK Government's Energy Advisory Panel and is a current member of the UK Government's Commission on Environmental Markets and Economic Performance. He also sits on the Advisory Boards of the UK Energy Research Centre and Sussex University's Energy Group.